T0211506

INTERNATIONAL CENTRE FOR MECHANICAL SCIENCES

COURSES AND LECTURES - No. 325

SHAPE AND LAYOUT OPTIMIZATION OF STRUCTURAL SYSTEMS AND OPTIMALITY CRITERIA METHODS

EDITED BY

G.I.N. ROZVANY
ESSEN UNIVERSITY

SPRINGER-VERLAG WIEN GMBH

Le spese di stampa di questo volume sono in parte coperte da
contributi del Consiglio Nazionale delle Ricerche.

This volume contains 261 illustrations.

In order to make this volume available as economically and as
rapidly as possible the authors' typescripts have been
reproduced in their original forms. This method unfortunately
has its typographical limitations but it is hoped that they in no
way distract the reader.

ISBN 978-3-211-82363-7   ISBN 978-3-7091-2788-9 (eBook)
DOI 10.1007/978-3-7091-2788-9

PREFACE

The original aim of this international course was to discuss layout and shape optimization, which constitute some of the most complex problems of structural optimization but, at the same time, are estremely important owing to the significant cost savings involved.

It was decided later to include a treatment of optimality criteria methods because they play a significant role in layout optimization.

The first five chapters by the Editor review continuum-based optimality criteria (COC), both at an analytical level and at a computational level using iterative procedures. The next five chapters by the Editor cover various aspects of layout optimization. In Chapters 11 and 12, important aspects of topology optimization are discussed by other authors.

Shape optimization is treated extensively in Chapters 13-15, 17 and 20-22 and a brief review of domain composition is given in Chapter 23. The problem of structural reanalysis and theorems of structural variation are discussed in Chapters 18 and 19. Finally, an interesting new development, applications of artificial neural nets in structural optimization, is presented in Chapter 16.

The Editor is grateful to all authors for contributing valuable chapters; to the participants of the course for showing a high level of interest and for stimulating discussions; to E. Becker for preparing the diagrams for ten chapters written by him; to A. Fischer for part of the text-processing; and, in particular, to his wife Susann for her untiring efforts in processing some of his chapters, editing the text of entire volume, and organizing the scientific program.

G.I.N. Rozvany

CONTENTS

Chapter 1

AIMS, PROBLEMS AND METHODS OF STRUCTURAL
OPTIMIZATION

G.I.N. Rozvany
Essen University, Essen, Germany

The *main aim* of this course is to discuss
- optimality criteria (OC) methods, and
- shape and layout optimization

in structural design.

A *structure* is a solid body that is subject to stresses and deformations.

The *aim of structural optimization* is
- the minimization (or maximization) of an *objective function* (e.g. cost of materials and labour, structural weight, storage capacity, etc.),
- subject to

  (i) *geometrical constraints* (e.g. restriction on height, prescribed variation of the cross-sections over given "segments"), and

  (ii) *behavioural constraints* (e.g. restrictions on stresses, displacements, buckling load, natural frequency, etc.).

Behavioural constraints can be
- *local constraints*, in which only stresses or stress resultants for a given cross-section are involved, or
- *global constraints*, which contain integrals of stresses or stress resultants for the entire structure. System instability (buckling) and natural frequency constraints are global ones, as are deflection constraints if expressed in terms of work equations.

In the design of complex, real structural systems, *discretization* and the use of numerical methods is unavoidable. However, in order to achieve a reasonable accuracy it is necessary to use a *very large number of elements*.

Whereas the *analysis capability* of modern finite element software is between *ten thousand* and *hundred thousand* degrees of freedom, the *optimization capability* for highly nonlinear and nonseparable problems is restricted to a *few hundred variables*, if conventional techniques (e.g. primal methods of mathematical programming) are used. This results in a *discrepancy between analysis capability and optimization capability*, which was pointed out repeatedly by Berke and Khot (e.g. 1987).[*]

Depending on the relative proportions of their dimensions, structures may be idealised as
○ *one-dimensional* continua [e.g. bars, beams, arches, rings, frames, trusses, beam-grids (grillages), shell grids, cable nets];
○ *two-dimensional* continua (plates, disks, structures subject to plane stress or plane strain, shells, folded plates, truss-like continua, grillage-like continua, shell-grid-like continua, etc.);
○ *three-dimensional* continua (stressed systems with dimensions of the same order of magnitude in the three directions).

Problems of structural optimization may be classified as
○ *sizing* or optimization of the *cross-sectional dimensions* of one- or two-dimensional structures, for which the cross-sectional geometry is partially prescribed, so that the cross-section can be fully described by a finite number of variables;
○ *shape optimization* (e.g. the shape of the centroidal axis of bars and the middle surface of shells, boundaries of continua or interfaces between different materials in composites);
○ *layout optimization* which consists of three simultaneous operations:
(a) topological optimization (spatial sequence or configuration of members and joints),
(b) geometrical optimization (location of joints and shape of member axes), and
(c) optimization of the cross-sections.

One of the difficulties in *shape optimization* is that the optimal shape may represent a *multiply connected* set with internal boundaries whose topology is not known and is difficult to determine, because new internal boundaries cannot be easily generated. Moreover, in many shape optimization problems the theoretical optimal shape contains an infinite number of internal boundaries. The determination of the topology of such internal boundaries becomes a *layout optimization* problem.

Similarly, unconstrained cross-section (thickness) optimization of plates and shells

---

[*] A list of References for all chapters by Rozvany (and co-authors) can be found at the end of Chapter 10.

often results in a, theoretically, infinite number of rib-like formations whose *layout* must be optimized.

This means that both cross-section and shape optimization may require, in effect, layout optimization.

Layout optimization is the most complex task in structural optimization because
○ one has a choice of an infinite number of possible topologies, and
○ for each point of the structural domain, there exists an infinite number of member directions. The cross-sections of non-vanishing (optimal) members must be optimized simultaneously.

Layout optimization is important because it enables much greater material savings than pure cross-section optimization.

Basic formulations of structural optimization problems are
○ *continuum formulation* using analytical methods for solving differential equations of continuum mechanics, or
○ *discretized formulation* using finite difference (FD), finite element (FE) or boundary element (BE) methods.

The *methods of structural optimization* fall into two major categories:
○ *Direct minimization techniques* (e.g. mathematical programming, MP).
○ *Indirect methods* (e.g. optimality criteria, OC, methods).

In *mathematical programming* (MP) methods each iteration consists of two basic steps:
(1) calculation of the value of the *objective function*, and *its gradients* with respect to all design variables, for a feasible solution; and
(2) calculation of a *locally optimal feasible change* of the design variables.

Steps (1) and (2) are repeated until a local minimum of the objective function is reached.

The *main advantage of MP methods* is their *robustness* which means that they are readily applicable to most problems within and outside the field of structural optimization. However, many MP methods use analytical sensitivities (gradients) whose efficient derivation can be highly problem-dependent.

MP methods can be divided into so-called *primal methods*, in which the original design variables are considered, and *dual methods*, in which a modified problem is solved.

The *main disadvantage of primal and dual MP methods* is their very limited optimization capability in terms of the number of variables and number of active constraints, respectively.

Optimality criteria are necessary (and sometimes sufficient) conditions for minimality of the objective function.

Applications of optimality criteria methods include

○ *idealized systems*, for which closed-form analytical solutions are obtained either
   (i) by hand, or
   (ii) by special analytically based computer programs
   with a view to determining
   (a) fundamental features of optimal solutions,
   (b) the range of validity and applicability of various numerical methods, and
   (c) the relative economy of realistic designs (basis of comparison); and
○ *large, real systems*, where OC methods can be used for
   (a) checking the validity of solutions determined by other methods, and
   (b) developing efficient iterative re-sizing strategies.

Optimality criteria (OC) methods fall into two distinct categories, i.e. discretized and continuum-based OC-methods.

Discretized optimality criteria (DOC) methods have been developed since the late sixties by aero-space engineers (e.g. Berke, Khot, Venkayya) who use the Kuhn-Tucker minimality condition in *finite dimensional* design space, expressed in terms of *nodal forces*. Earlier OC approaches included such *intuitive methods* as fully stressed design, stress ratio methods and constant mutual energy design.

Continuum-based optimality criteria (COC) methods employ Euler-Lagrange type minimality conditions in *infinite dimensional design spaces*, using calculus of variations, control-theory and functional analysis. The optimality conditions are differential equations in terms of *generalized stresses* (stresses or stress resultants), re-interpreted by introducing the concept of an *"adjoint structure"*, also termed after its originator "Pragerian field" in some earlier publications (e.g. Rozvany, 1989).

A more important difference between DOC and COC formulations, however, is the fact that in the former the variables are the *design parameters* whilst in the latter they also *include the real and virtual internal forces*. Moreover, in the COC method only static admissibility is required in the original problem formulation and kinematic admissibility becomes an optimality criterion (Rozvany and Zhou, 1992).

An *analytical solution* of the equilibrium, compatibility and strain-stress relations for the real and adjoint systems and the optimality criteria can only be obtained for *smaller, idealized problems*. For *large, practical systems*, therefore, an iterative solution becomes necessary.

In *iterative COC-methods*, the above equations are discretized and then each iteration consists of two basic steps:
(a) the analysis of the real and adjoint systems, and
(b) updating of the cross-sectional dimensions.

The analysis phase can be carried out by using a standard FE package. Since the update operation usually involves uncoupled and explicit equations, the storage and CPU time requirement of this step is a fraction of that required for the analysis phase (a). A large number of design constraints can be treated by using the *active set strategy.*

The main advantage of the COC method is that it can highly efficiently optimize an extremely large number of elements and variables.

A *disadvantage of the COC method* is that for each type of structure and for each type of design condition, a lengthy analytical derivation of the relevant optimality criteria is necessary. This disadvantage can be removed by using a *computerized method* that generates analytically the required optimality criteria.

Initial achievements of the COC-method, in special test examples, are:

○ *Cross-section optimization* of systems with up to *one million* elements, one million variables, and a similar number of local and global constraints.

○ *Additional refinements,* such as upper and lower limits on the cross-sectional dimensions, segmentation, allowance for selfweight and cost of supports, non-linear and non-separable objective and stiffness functions, combination of local and global (e.g. stress and deflection) constraints.

○ *Layout optimization* with an initial set ("structural universe") of many thousand members.

The main advantages of COC over dual MP and DOC methods, however, is the fact that calculation of the Lagrangians for stress constraints is carried out at an element level and hence *the computational effort only depends essentially on the number of active global constraints.*

It should be clarified at this stage that, in COC test examples, we are assessing the capability of an *optimizer* which is independent of the capability of the *analyser* used. If we use a *general-purpose FE package* (e.g. ANSYS) as analyser, then the analysis capability is rather restricted in test examples. In the upper range of the COC capability, therefore, special-purpose *FE simulators* were used. These give the same results as general-purpose FE programs within the capability of the latter, but can also handle, owing to their higher efficiency for the restricted boundary and loading conditions, certain structures (e.g. beams) with up to several million elements.

A comprehensive list of optimality criteria for various design requirements and constraints is given in a recent book (Rozvany, 1989).

The development of generally applicable and reliable OC methods would represent an unprecedented break-through in structural design.

Structural optimization has important *applications* in all fields of technology, including aero-space, structural, mechanical, naval and civil engineering, and covers such diverse applications as aeroplanes, motor cars, sporting equipment, building structures and artificial organs.

In Chapters 2-5 various aspects of the COC algorithm are outlined. In Chapters 6-10 applications of COC in layout optimization are discussed.

Chapter 2

CONTINUUM-BASED OPTIMALITY CRITERIA (COC)
METHODS - AN INTRODUCTION

G.I.N. Rozvany
Essen University, Essen, Germany

1 Preliminaries

The aim of this lecture is to discuss some fundamental aspects of continuum-based
optimality criteria (COC) methods. As mentioned in the opening lecture, the above
criteria are usually derived by *variational or control-theoretic methods*. However,
Prager usually based his proofs (e.g. Prager, 1974; Prager and Rozvany, 1977a) on
known *energy theorems* of structural mechanics. Some of these optimality criteria
are then reinterpreted as equilibrium, compatibility and generalized strain-stress relations for a fictitious system called *adjoint structure*. This mathematical device has
two advantages: first, engineers can visualize such quantities as bending moments
and deflections more easily than abstract mathematical entities; second, the analysis of the adjoint structure can be carried out after discretization by using existing
numerical algorithms and programs (e.g. FE software).

The optimality criteria under consideration are also called *static-kinematic optimality* criteria, since they are partly interpreted as static and kinematic continuity
conditions (i.e. equilibrium and compatibility requirements) for the adjoint structure.

Before discussing optimality criteria for elastic structures, we shall review briefly
some early developments in *optimal plastic design*, in which only a stress (strength)
condition is prescribed but elastic compatibility is not required. It will be explained
later that the results of optimal plastic design are often also valid for optimal elastic
design.

2 A Review of Early Optimality Criteria for Optimal Plastic Design

The first optimality criterion for a continuum was proposed by Michell (1904), who considered *minimum weight trusses*. Since his solutions usually consist of an infinite number of members, Prager (1974) termed Michell's frameworks "truss-like continua". Michell's optimality condition takes the form

$$\bar{\epsilon} = k \, \text{sgn} \, N \qquad (\text{for } |N| > 0) , \qquad (1)$$

$$|\bar{\epsilon}| \leq k \qquad (\text{for } N = 0) , \qquad (2)$$

where N is the member force, $\bar{\epsilon}$ is the longitudinal strain in the members and k is a constant. The "sgn" function has the usual meaning (sgn $N = 1$ for $N > 0$ and sgn $N = -1$ for $N < 0$). Expressed in words, Michell's criterion requires *the axial strain to be of a constant absolute value* (k) in directions of non-zero axial forces, and in the directions of zero axial forces the absolute value of the strain must not exceed the above value (k). The overbars in (1)-(5) signify that we are dealing with "adjoint" fields, but in (1) and (4) they can also represent deflections for the corresponding elastic structure.

The next optimality condition was introduced, half a century later, by Foulkes (1954) for the optimal plastic design of *least-weight prismatic beams*. His criterion can be stated as

$$\frac{\sum_i |\bar{\theta}|}{L_i} = k , \qquad (3)$$

where i denotes a segment of constant cross-section, $\bar{\theta}$ are the plastic hinge rotations, L_i the segment length and k a constant. This means that for segment-wise prismatic frames the *average absolute curvature* for each segment must be the same ($\bar{\kappa}_{av\,i} = k$). The displacements corresponding to (3) can be regarded as an "adjoint" field, but they also represent the velocity field for a plastic mechanism.

Foulkes' optimality criterion was extended to plastic *beams with a freely varying rectangular cross-section of given depth but varying width* by Heyman (1959), who obtained an optimality condition somewhat similar to that of Michell:

$$\bar{\kappa} = k \, \text{sgn} \, M \qquad (\text{for } |M| > 0) , \qquad (4)$$

$$|\bar{\kappa}| \leq k \qquad (\text{for } |M| = 0) , \qquad (5)$$

where $\bar{\kappa} = -\bar{u}''$ is the "curvature" of the beam deflection $\bar{u}$, M is the beam bending moment and k is a constant. The inequality condition in (5) was not stated explicitly by Heyman and was used only later by the author (e.g. Rozvany, 1976) and others. Whereas $\bar{u}$ is a fictitious or "adjoint" deflection, it also represents one possible velocity field at plastic collapse.

A general optimality criterion for the optimal plastic design of structures with freely varying cross-sectional dimensions was proposed by Prager and Shield (1967),

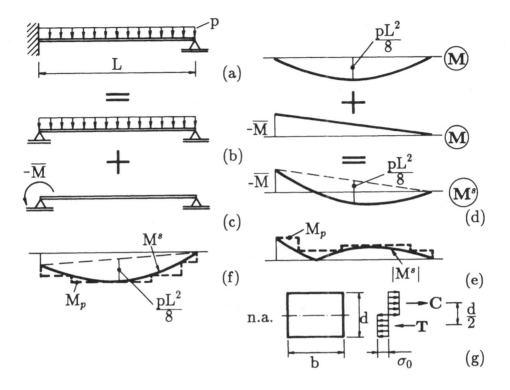

Fig. 1 An example illustrating the plastic design procedure based on the lower bound theorem of limit design.

who generalized an earlier optimality condition by Marcal and Prager (1964). Similar optimality criteria were derived by Mroz (1963), Prager and Taylor (1968), Masur (1970), and Save (1972). In order to explain the Prager-Shield condition in greater detail, we review here briefly the *lower bound theorem of plastic design*. According to this theorem, the calculated load is always smaller than or at most equal to the actual collapse load if the corresponding stress field Q (a) is statically admissible (satisfying the equilibrium equations and static boundary conditions), and (b) fulfils the yield inequality at all cross-sections.

On the basis of the above theorem, we can adopt the following *plastic design procedure* for redundant structures (Fig. 1a):

(i) Make the structure statically determinate by releasing all redundant connections (Fig. 1b, in which the clamping action at the left-hand end of the beam has been removed). The resulting structure is termed "primary structure".

(ii) Apply arbitrary actions (moments or forces) corresponding to the redundancies released in step (i), see Fig. 1c.

(iii) Determine the generalized stresses (e.g. moments) caused by the external load in the primary structure (Fig. 1d, top).

(iv) Calculate the generalized stresses caused by the arbitrarily chosen redundant actions (Fig. 1d, middle).

(v)   Superimpose the generalized stresses obtained in steps (iii) and (iv) (Fig. 1d, bottom). These generalized stresses are now statically admissible and are denoted by $\mathbf{Q}^S$ (in the case of a beam, by M^S).

(vi) Adopt cross-sectional dimensions such that the yield inequality is satisfied everywhere (Fig. 1e). For a plastic beam designed for flexural stresses only, the yield inequality is

$$|M^S| \leq M_p ,\tag{6}$$

where M_p is the plastic moment capacity of the beam. At the latter moment value, the entire cross-section is in yield, if a rigid-perfectly plastic behaviour is assumed (Fig. 1g). For the considered beam, an M^S diagram with only positive moments would also give an admissible (safe) design (Fig. 1f), but the latter would be clearly uneconomical.

In an optimal plastic design for freely varying cross-sectional dimensions, the above yield inequality is replaced by the equality

$$|M^S| = M_p .\tag{7}$$

Clearly, a design with $|M^S| < M_p$ at some cross-sections cannot be optimal if the cross-sectional dimensions are not subjected to geometrical constraints.

Using Prager's terminology (e.g. Prager, 1974), the *basic variables of structural mechanics* are generalized stresses $\mathbf{Q}(\mathbf{x}) = [Q_1(\mathbf{x}), \ldots, Q_n(\mathbf{x})]$, generalized strains $\mathbf{q}(\mathbf{x}) = [q_1(\mathbf{x}), \ldots, q_n(\mathbf{x})]$, loads $\mathbf{p}(\mathbf{x}) = [p_1(\mathbf{x}), \ldots, p_m(\mathbf{x})]$ and displacements $\mathbf{u}(\mathbf{x}) = [u_1(\mathbf{x}), \ldots, u_m(\mathbf{x})]$, where $\mathbf{x}$ are the spatial coordinates. A generalized stress may represent a local stress or stress resultant (e.g. a bending moment or shear force) and a generalized strain may refer to a local strain or to an entire cross-section (e.g. curvature or twist of a bar).

In the general formulation of Prager and Shield (1967), the *total cost* Φ to be minimized can be expressed as the integral of the *specific cost* ψ over the structural domain D. The specific cost (cost per unit length, area or volume, which may represent, for example, the cross-sectional area), in turn, can be expressed as a function of the statically admissible generalized stress vector $\mathbf{Q}^S$ at the considered cross-section $\psi = \psi(\mathbf{Q}^S)$. It follows that the optimal plastic design problem can be formulated as

$$\min \Phi = \int_D \psi(\mathbf{Q}^S) \, \mathrm{d}\mathbf{x} ,\tag{8}$$

where $\mathbf{x}$ are the spatial coordinates.

Before stating the Prager-Shield condition, we review briefly the *fundamental relations of structural mechanics*. Considering an *elastic* system, for example, these

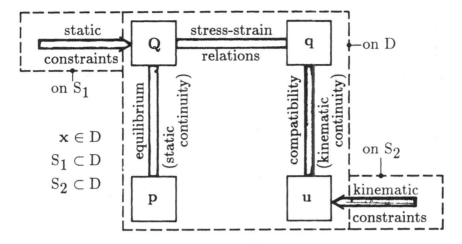

$Q(x)$: generalized stresses
$q(x)$: generalized strains
$p(x)$: generalized loads
$u(x)$: generalized displacements

Fig. 2   Fundamental relations of structural mechanics.

relations are shown in Fig. 2. On the structural domain D, we must satisfy the equilibrium (or static continuity) conditions (p, Q), the compatibility (or kinematic continuity) conditions (u, q) and the generalized strain-stress relations (Q, q). In addition, on some subset S_1 of the domain D static constraints (or static boundary conditions) and on some other subset $S_2 \subset D$ kinematic constraints (or kinematic boundary conditions) must be fulfilled.

The above relations are illustrated in the context of a *Bernoulli-type cantilever beam* in Fig. 3, for which the generalized stress is the bending moment $(Q \rightarrow M)$ and the generalized strain the curvature $(q \rightarrow \kappa = -u''$ where u is the beam deflection). Then the equilibrium, compatibility and generalized strain-stress relations are

$$M'' = -p ,   \tag{9}$$

$$u'' = -\kappa ,   \tag{10}$$

$$EI\kappa = M ,   \tag{11}$$

where p is the load, E is Young's modulus, I is the moment of inertia and other symbols have been defined earlier. At the free end of the beam (S_1), only static boundary conditions need to be observed

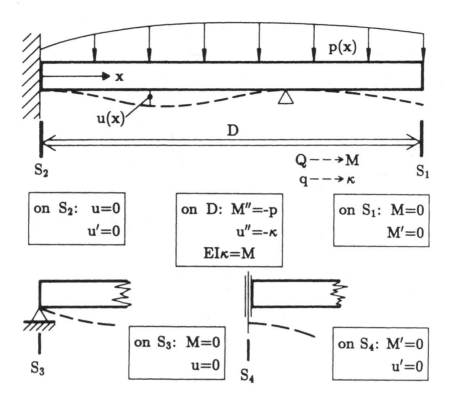

Fig. 3   An example illustrating the fundamental relations of structural mechanics.

$$M = M' = 0 \,, \tag{12}$$

and at the clamped end (S_2) the kinematic boundary conditions are

$$u = u' = 0 \,. \tag{13}$$

At the bottom of Fig. 3, supports representing some mixed boundary conditions are indicated.

The requirements for (lower bound) *plastic design* are shown in Fig. 4, in which the stress field must satisfy static boundary conditions (constraints), equilibrium (static continuity) conditions and the yield inequality $Y(Q) \leq Y_0$.

In *optimal plastic design* (Fig. 5), we are still required to satisfy static boundary conditions and equilibrium, but the yield inequality is replaced with the cost minimality condition in which the "specific cost function" $\psi(Q)$ is based on the assumption that yield equality is satisfied at all cross-sections.

Optimal plastic design via Prager-Shield *optimality criteria* is represented schematically in Fig. 6. The above optimality conditions were derived by Prager and

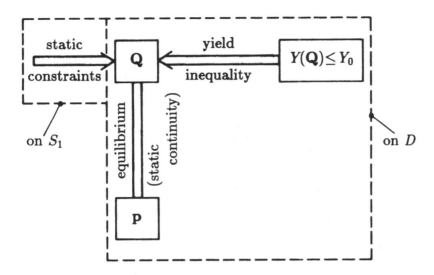

Fig. 4 Fundamental relations of plastic design.

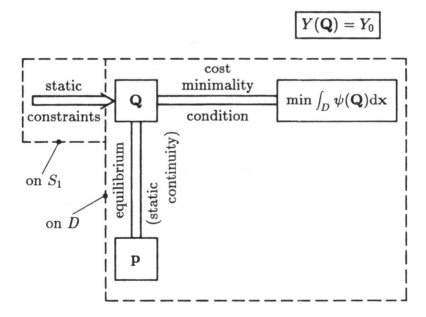

Fig. 5 Fundamental relations of optimal plastic design by direct cost minimization.

Shield (1967) from energy theorems and from variational principles by the author (e.g. Rozvany, 1976) who formulated it in its considerably extended form presented here.

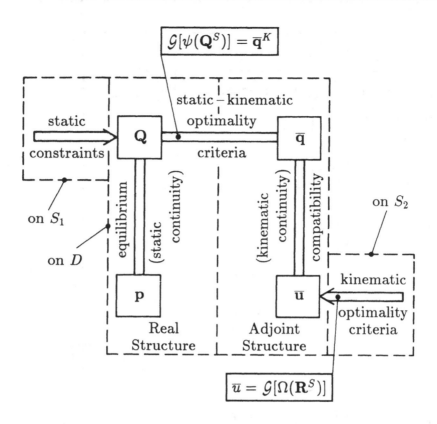

Fig. 6 Fundamental relations of optimal plastic design via optimality criteria.

In Fig. 6, we have the actual loads $\mathbf{p}$ and generalized stresses $\mathbf{Q}$ of the real structure and the displacements $\bar{\mathbf{u}}$ and generalized strains $\bar{\mathbf{q}}$ of the adjoint structure. The latter two must satisfy the usual compatibility conditions. For structures with rigid and costless supports, the kinematic boundary conditions for $\bar{\mathbf{u}}$ are the same as for a usual structure, but for non-zero support cost the adjoint kinematic boundary conditions are given in Fig. 6, in which $\Omega(\mathbf{R})$ is the cost of the reaction $\mathbf{R} = (R_1, \ldots, R_m)$. In the optimal strain-stress relation (Fig. 6), the adjoint strains $\bar{\mathbf{q}}$ must be kinematically admissible (denoted by $\bar{\mathbf{q}}^K$) and are given by the gradient of the specific cost function $\psi(\mathbf{Q})$, in which the stress vector must be statically admissible ($\mathbf{Q}^S$):

$$\bar{\mathbf{q}}^K = \mathcal{G}[\psi(\mathbf{Q}^S)] . \tag{14}$$

The optimality criterion in (14) represents a considerable extension of the original Prager-Shield (1967) condition. Both in the adjoint boundary conditions for $\bar{\mathbf{u}}$ and in the strain-stress relation $(\mathbf{Q}, \bar{\mathbf{q}})$ in Fig. 6, we make use of the *generalized gradient* or "G-gradient operator" $\mathcal{G}$ (e.g. Rozvany, 1976, 1981a). For differentiable functions

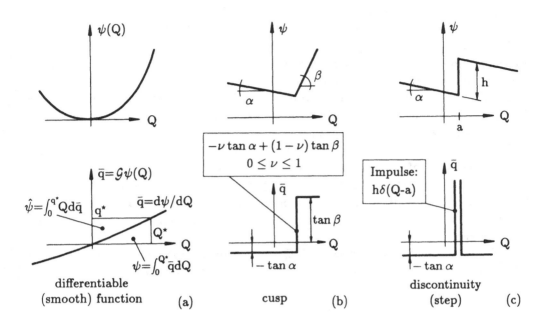

Fig. 7   G-gradient operator for functions and generalized functions with a single variable.

$\psi(\mathbf{Q})$ or $\Omega(\mathbf{R})$, the G-gradient has the meaning

$$\mathcal{G}[\psi(\mathbf{Q})] = \mathbf{grad\ Q} = \left(\frac{\partial \psi}{\partial Q_1} , \ldots, \frac{\partial \psi}{\partial Q_n} \right) , \tag{15}$$

$$\mathcal{G}[\Omega(\mathbf{R})] = \left(\frac{\partial \Omega}{\partial R_1} , \ldots, \frac{\partial \Omega}{\partial R_m} \right) , \tag{16}$$

but for non-differentiable and discontinuous functions the G-gradient has an extended meaning. This is shown, in the context of *functions with a single variable* (Q), in Fig. 7. It can be seen that for *differentiable* (smooth) *functions* the adjoint strain q (scalar quantity in this case) is given by

$$q = \frac{d\psi}{dQ} , \tag{17}$$

implying (Fig. 7a)

$$\psi = \int_0^{Q^*} q \, dQ , \tag{18}$$

where Q^* is some arbitrary stress value. In Fig. 7a, we also define the *"complementary cost"*

$$\hat{\psi} = \int_0^{q^*} Q \, dq , \tag{19}$$

where the strain q^* corresponds to the stress value Q^*. The concept of complementary cost will be used in duality principles [see the relation (20) later].

In the case of a *cusp* of a piece-wise differentiable function, the G-gradient, and the corresponding adjoint strain value, are non-unique and equal the convex combination of the adjacent slopes (gradients).

Finally, in the case of a *discontinuity* (step) in the specific cost function, the G-gradient, and the adjoint strain, contain an impulse (Dirac distribution) whose (Lebesque) integral equals the magnitude of the step.

The optimality criteria approach in Fig. 6 converts *an optimization problem* (Fig. 5) *into a problem of analysis* which has both conceptual and computational advantages. Whilst in plastic optimal design only the static aspects of the structure are "real" and only the kinematic aspects of the structure are "adjoint", we shall see later that in optimal elastic design we must consider a complete real and a complete adjoint structure.

For convex specific cost functions $\psi(\mathbf{Q})$, the minimum total cost can be calculated either from the primal formula in (9) or from the following *dual formula*

$$\Phi_{\min} = \int_D [\mathbf{p} \mathbf{u}^K - \hat{\psi}(\mathbf{q}^K)]dx , \qquad (20)$$

in which $\mathbf{u}$ and $\mathbf{q}$ satisfy the Prager-Shield condition (14). Moreover, (8) and (20) furnish *upper and lower bounds* on Φ for any statically admissible stress field $\mathbf{Q}^S$ and any kinematically admissible strain/displacement field $(\mathbf{q}^K, \mathbf{u}^K)$.

3 Illustrative Example: Heyman Beam

3.1 Solution by Optimality Criteria Method

To illustrate the analytical COC method, we consider a propped cantilever in Fig. 8a with the specific cost function

$$\psi = k|M| , \qquad (21)$$

where ψ is the cross-sectional area, M is the bending moment and k is a given constant. For rectangular cross-sections of given depth and variable width, we have

$$\psi = bd , \quad M_p = bd^2/4 , \qquad (22)$$

where M_p is the plastic moment capacity, b is the width and d is the depth (see Fig. 3g). It follows that for $|M| = M_p$, we have in (21)

$$k = 4/d . \qquad (23)$$

If we now apply the Prager-Shield condition in (14) to the specific cost function in (21), we obtain exactly Heyman's optimality criteria in (4) and (5). The specific

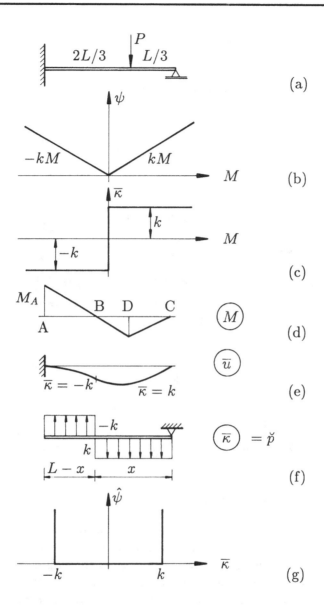

Fig. 8 Example: Heyman beam.

cost function and the above optimality criteria are represented graphically in Figs. 8b and c.

For the considered beam, a statically admissible moment diagram is given in Fig. 8d in which M_A can take on any arbitrary value. The corresponding adjoint deflection field is shown in Fig. 8e, since by (4) the adjoint structure has constant curvatures of $+k$ or $-k$ depending on the sign of the bending moment.

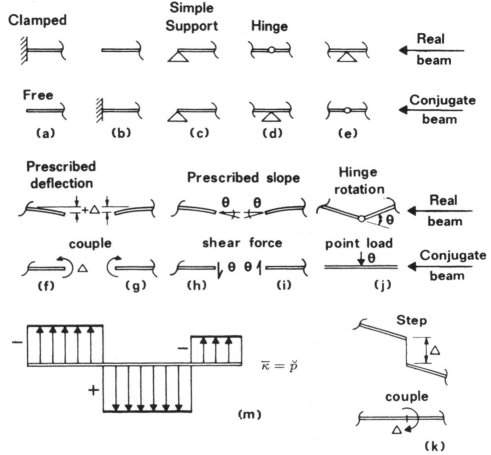

Fig. 9   The conjugate beam method.

Using the *method of conjugate beams*, we can regard the curvatures as fictitious loads $\check{p}$ whereas the support and end conditions are to be modified in accordance with Fig. 9. Then the conjugate shear force $\check{V}$ gives the slopes of the beams and the conjugate moments $\check{M}$ the beam deflections. It follows that the conjugate beam for our problem becomes the one in Fig. 8f. The equilibrium condition of this conjugate beam is identical to the compatibility condition for the adjoint field in Fig. 8e. This means that the conjugate moment $\check{M}_C$ for the point C must be zero, that is

$$\check{M}_C = k[-(L-x)(L+x)/2 + x^2/2] = 0,\qquad(24)$$

implying

$$x = L/\sqrt{2},\qquad(25)$$

where x is the distance of the zero moment point from the simple support and L is the span. Since from statical considerations (see Fig. 8d)

$$M_A - V_A(L - x) = 0, \tag{26}$$

where the shear force V_A at point A is

$$V_A = \left(-\frac{M_A}{L} + \frac{P}{3}\right), \tag{27}$$

we have by (25)-(27)

$$M_A + \left(-\frac{M_A}{L} + \frac{P}{3}\right)L\left(1 - \frac{1}{\sqrt{2}}\right) = 0, \tag{28}$$

or

$$M_A = -PL(\sqrt{2} - 1)/3 = -0.1381PL. \tag{29}$$

The value of the maximum positive moment M_D in Fig. 8d then becomes

$$M_D = M_A\frac{1/\sqrt{2} - 1/3}{1 - 1/\sqrt{2}} = PL(1 - \sqrt{2}/3)/3 = 0.1762PL. \tag{30}$$

The optimal total cost of the beam then becomes (Fig. 8d)

$$\Phi_{\min} = \frac{kL}{2}\left[|M_A|\left(1 - \frac{1}{\sqrt{2}}\right) + \frac{M_D}{\sqrt{2}}\right] = \frac{kPL^2}{6}\left[(\sqrt{2} - 1)\left(1 - \frac{1}{\sqrt{2}}\right) + \right.$$

$$\left. +\frac{1}{\sqrt{2}} - \frac{1}{3}\right] = \frac{kPL^2}{6}\left[2(\sqrt{2} - 1) - \frac{1}{3}\right] = 0.08252kPL^2. \tag{31}$$

3.2 Check By Dual Formulation

The same total cost Φ_{opt} can be derived from the dual formula (20) and Fig. 8e. On the basis of the definition of the complementary cost in Fig. 7a and (19), and considering the optimal moment-adjoint curvature relation in Fig. 8c, we obtain the complementary cost (Fig. 8g)

$$\hat{\psi} = 0 \quad (\text{for } \kappa \le k), \quad \hat{\psi} = \infty \quad (\text{for } \kappa > k). \tag{32}$$

This means that in our problem (Fig. 8e) the second term in the integrand of (20) vanishes and hence the optimal total cost is given by

$$\Phi_{\min} = P\overline{u}_D. \tag{33}$$

Using the conjugate beam method described above, the adjoint deflection $\bar{u}_D$ is given by the conjugate moment $\check{M}_D$ at point D for the conjugate beam in Fig. 8f, implying with (33)

$$\Phi_{\min} = kPL^2 \left[\left(1 - \frac{1}{\sqrt{2}}\right) \left(\frac{1 - 1/\sqrt{2}}{2} + \frac{1}{\sqrt{2}} - \frac{1}{3}\right) - \frac{(1/\sqrt{2} - 1/3)^2}{2} \right] =$$

$$= 0.08252kPL^2 . \tag{34}$$

This confirms the result in (31) which was given by the primal formula (8).

3.3 Check by Differentiation

Considering the statically admissible moment diagram in Fig. 8d with an arbitrary value of M_A, we shall introduce the nondimensional quantities m_A, m_D and X with

$$M_A = m_A PL, \quad M_D = m_D PL, \quad x = XL. \tag{35}$$

Then the value of m_D can be calculated from statical considerations

$$m_D = \left(\frac{2}{9} + \frac{m_A}{3}\right), \tag{36}$$

where m_A is negative. The distance x in Fig. 8f can be expressed from similar triangles in Fig. 8d as

$$x = \frac{L}{3} + \frac{\frac{2}{9} + \frac{m_A}{3}}{\frac{2}{9} - \frac{2m_A}{3}} \frac{2L}{3}, \quad X = \frac{1}{3}\left(1 + \frac{2 + 3m_A}{1 - 3m_A}\right), \tag{37}$$

or

$$X = 1/(1 - 3m_A) . \tag{38}$$

The total cost Φ can be represented as

$$\frac{2\Phi}{kPL^2} = -m_A(1 - X) + m_D X . \tag{39}$$

By (36) and (39) we have the nondimensional cost $\tilde{\Phi}$:

$$2\tilde{\Phi} = \frac{2\Phi}{kPL^2} = -m_A\left(1 - \frac{1}{1 - 3m_A}\right) + \left(\frac{2}{9} + \frac{m_A}{3}\right)\frac{1}{1 - 3m_A} \Rightarrow$$

$$\Rightarrow 18\tilde{\Phi} = \frac{2 + 3m_A + 27m_A^2}{(1 - 3m_A)} . \tag{40}$$

Denoting the numerator and denominator of the RHS of (40), respectively, by N and D, we have the minimality condition

$$N'D - D'N = 0 = (3 + 54m_A)(1 - 3m_A) + 3(2 + 3m_A + 27m_A^2), \qquad (41)$$

implying

$$-9m_A^2 + 6m_A + 1 = 0,$$

$$\Rightarrow m_A = \frac{-6 + \sqrt{72}}{-18} = -0.1381, \qquad (42)$$

which confirms the result in (29).

Note. It can be seen that the solution by the optimality criteria method was much simpler algebraically than the solution by differentiation. Moreover, the OC solution, in terms of the location of the point with zero moment in (25), is largely load-independent, i.e. it is valid for all loads that can produce negative (or zero) moments over the left region having a length of $(1 - 1/\sqrt{2})/L$ and positive moments over the right region with a length of $L/\sqrt{2}$ (see, for example, Fig. 8d). This condition is satisfied by any non-negative (downwards) loading on the beam.

4 Conclusions

Some early continuum-based optimality criteria for optimal plastic design, including the Prager-Shield condition, were introduced in this chapter. The latter was illustrated with a simple example, the results of which were also verified by dual formulation and differential calculus.

References

A list of References for all chapters by Rozvany (and co-authors) can be found at the end of Chapter 10.

Chapter 3

COC METHODS FOR A SINGLE GLOBAL CONSTRAINT

G.I.N. Rozvany and M. Zhou
Essen University, Essen, Germany

1 Introduction

In the previous chapter, we discussed optimality criteria methods for optimal plastic design and presented a simple example based on a closed form analytical solution. In optimal elastic design, however, which is to be discussed in this chapter, it is usually necessary to resort to numerical methods because the equations involved are too complicated for an analytical treatment. Moreover, whilst in optimal plastic design (Fig. 6) we had a half real system (involving loads p, stresses $\mathbf{Q}$ and *statical admissibility*) and a half adjoint system (involving adjoint strains $\bar{\mathbf{q}}$, adjoint displacements $\bar{\mathbf{u}}$ and *kinematic admissibility*), in optimal elastic design we have a full real and a full adjoint system.

2 General Formulation

We consider structures for which the cross-sectional geometry is partially prescribed in such a way that the cross-section is fully defined by a finite number of variables

$$\mathbf{z}(\mathbf{x}) = [z_1(\mathbf{x}), \ldots, z_r(\mathbf{x})], \tag{43}$$

termed *cross-sectional parameters*. Examples of such parameters are given in Fig. 10. In earlier publications, it was usual to consider rectangular cross-sections in which either the depth, or the width or the depth/width ratio was kept constant (Fig. 10a).

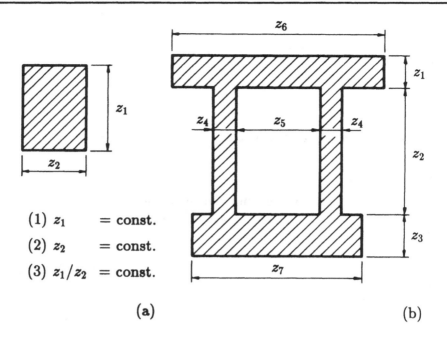

z_1
z_2
z_1/z_2

(1) z_1 = const.
(2) z_2 = const.
(3) z_1/z_2 = const.

(a)   (b)

Fig. 10 Examples of cross-sectional parameters z.

The current formulation can handle much more general cross-sectional geometries, such as the one shown in Fig. 10b.

The specific cost ψ (e.g. the weight, volume or material/construction costs per unit length or unit area) is a function of the cross-sectional parameters:

$$\psi(\mathbf{x}) = \psi[z_1(\mathbf{x}), \ldots, z_r(\mathbf{x})].\qquad(44)$$

The external loads $\mathbf{p}$ equilibrate the generalized stresses $\mathbf{Q}^S$, where the superscript S denotes statical admissibility. The structure is subject to a displacement constraint which can be expressed through the principle of virtual work in the form:

$$\int_D \mathbf{Q}^{S,K} \cdot [\mathbf{F}]\overline{\mathbf{Q}}^S \, dx \le \Delta,\qquad(45)$$

where the superscript K denotes kinematic admissibility, $[\mathbf{F}]$ is the local flexibility matrix converting a generalized stress vector into a generalized strain vector

$$\mathbf{q} = [\mathbf{F}]\mathbf{Q},\qquad(46)$$

and $\overline{\mathbf{Q}}^S$ is the virtual stress vector equilibrating some virtual load $\overline{\mathbf{p}}$. In the case of a displacement constraint at a single point, the virtual load is a unit ("dummy")

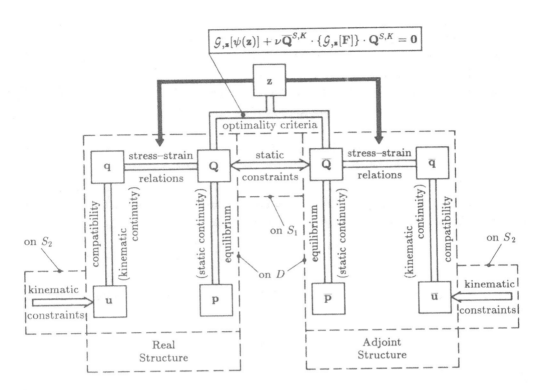

Fig. 11   Fundamental relations of optimal elastic design using the COC approach for freely
varying cross-sections and one deflection constraint.

load at that point in the direction of the prescribed displacement. The displacement
constraint may refer, however, to a weighted combination of several displacements,
in which case the virtual load consists of several forces or couples. Finally, if in the
displacement condition the prescribed quantity is the sum of the products of the
external forces and the corresponding displacements (i.e. the "compliance" C):

$$\Delta = \int_D \mathbf{p} \cdot \mathbf{u} \, dx = C, \tag{47}$$

then the real and virtual loads are the same:

$$\mathbf{p} = \overline{\mathbf{p}}, \quad \mathbf{Q} = \overline{\mathbf{Q}}. \tag{48}$$

3 General Optimality Criteria

The COC approach for a *deflection constraint* is shown schematically in Fig. 11
in which the equilibrium, compatibility and generalized strain-stress relations, as

well as the static constraints (static boundary conditions), are the same for the real and adjoint structures. For rigid and costless supports, the kinematic constraints (kinematic boundary conditions) are also identical, nowever, the latter two differ, if the cost of the supports is non-zero (see Fig. 6 in Chapter 2). For given values of the real and adjoint generalized stresses, the optimality criteria at the top of Fig. 11 furnish the optimal values of the cross-sectional parameters.

4 Application to Bernoulli Beams of Given Depth and an Independent Proof by Variational Methods

4.1 Derivation of the Optimality Criteria from the General Formulae in Fig. 11

In the case of a beam of given depth $h =$ const. but variable width $z = z(x)$, the flexural stiffness s and cross-sectional area ψ become

$$s(x) = \frac{Eh^3}{12} z(x) = rz(x), \quad \psi = hb(x), \tag{49}$$

where E is Young's modulus and $r = Eh^3/12$ is a given constant. The generalized stresses become the moments

$$Q = M, \quad \overline{Q} = \overline{M}, \tag{50}$$

and the flexibility matrix together with its gradient reduce to

$$[\mathbf{F}] = 1/rz, \quad \mathrm{d}[\mathbf{F}]/\mathrm{d}z = -1/rz^2. \tag{51}$$

Then the optimality condition at the top of Fig. 11 furnishes

$$h = \frac{\nu M \overline{M}}{rz^2} \Rightarrow z = \sqrt{\frac{\nu M \overline{M}}{hr}}, \tag{52}$$

or with $\tilde{\nu} = \nu/hr$

$$z = \sqrt{\tilde{\nu} M \overline{M}}, \tag{53}$$

whilst the real and adjoint stress-strain relations in Fig. 11 imply

$$\kappa = M/s = M/rz, \quad \overline{\kappa} = \overline{M}/s = \overline{M}/rz, \tag{54}$$

where $\kappa = -u''$ and $\overline{\kappa} = -\overline{u}''$ are the "curvatures" of the real and adjoint deflections u and $\overline{u}$.

4.2 Derivation of the Optimality Criteria Using the Calculus of Variations

The considered problem can be stated as follows:

$$\min_{z(x)} \Phi = \int_D [hz + \nu M \overline{M}/rz + \overline{u}(M'' + p) + u(\overline{M}'' + \overline{p})] \, \mathrm{d}x - \nu(\Delta - \alpha), \tag{55}$$

where ν is a Lagrangian multiplier (constant), $\overline{u}(x)$ and $u(x)$ are variable Lagrangian multipliers and α is a slack variable. The expression in brackets after $\overline{u}$ and u, respectively, represent the equilibrium conditions for the real and virtual loads. Then for variation of z, M and $\overline{M}$ we obtain the following Euler-Lagrange equations (e.g. Rozvany, 1989, pp. 371-373)

$$\delta z: \quad h - \nu MM/rz^2 = 0, \tag{56}$$

$$\delta M: \quad \nu \overline{M}/rz = -\overline{u}'', \qquad \delta \overline{M}: \quad \nu M/rz = -u''. \tag{57}$$

With the notation $\kappa = -u''/\nu$ and $\overline{\kappa} = -\overline{u}''/\nu$, (56) and (57) imply the optimality conditions in Fig. 11.

Considering the functional $\Phi = \int_D f(s, M, \overline{M})\,dx$ in (55), transversality conditions for variation of M and $\overline{M}$ give the same end conditions for u and $\overline{u}$ as in the case of an elastic beam, provided that the supports are rigid and costless. At a clamped end (B), for example, both the moment M and the shear force $V = -M'$ are variable and hence the following transversality conditions must be satisfied (e.g. Rozvany, 1989, p. 378):

$$\delta M': \qquad (f_{,M''})_B = 0, \tag{58}$$

$$\delta M: \qquad [f_{,M'} - (f_{,M''})']_B = 0, \tag{59}$$

where commas indicate partial derivatives with respect to the symbol in the subscript and primes denote differentiation with respect to x.

Combining (58) and (59) with (55), we have

$$\delta M': \qquad \overline{u}_B = 0, \tag{60}$$

$$\delta M: \qquad \overline{u}_B' = 0, \tag{61}$$

which are the usual end conditions for a clamped end. At a simple support, only the shear force $V = -M'$ is variable and hence only end condition (60) applies, as in the case of an elastic beam. If we also consider the cost Ω of an end reaction $R = V = -M'$ at a point B, then we have a mixed variational problem with a functional

$$\Phi = \int_D f(s, M, \overline{M})\,dx + \Omega[M'(x_B)]. \tag{62}$$

Then the transversality condition for the above mixed problem (e.g. Rozvany, 1989, p. 390) requires

$$\delta M': \qquad \left[f_{,M''} + \Omega_{,(-M')}\right]_B = 0, \tag{63}$$

or by (55) and (63)

$$\overline{u}_B = \Omega_{,R}. \tag{64}$$

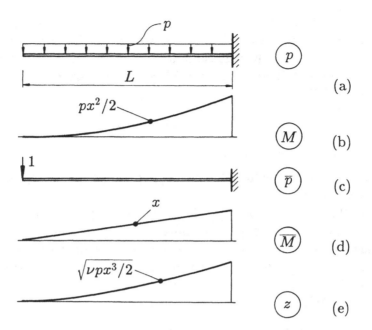

Fig. 12   Illustrative example: cantilever.

This confirms that the boundary conditions for the adjoint displacements can be different from those for the real displacements.

5 A Simple Illustrative Example

The above optimality conditions will be illustrated by a very simple example. We consider a uniformly loaded cantilever (Fig. 12a) of variable width z with a prescribed deflection $u_A = \Delta$ at its free end. The corresponding real moment diagram M, together with the adjoint load $\bar{p}$ and adjoint moment diagram $\overline{M}$, are shown in Figs. 12b, c and d. Using the optimality condition in (53) with $\tilde{\nu} \to \nu$, the optimal width distribution of the beam is given in Fig. 12e.

The value of the Lagrangian can be calculated from the deflection conditions. Using a normalized problem with $\Delta = L = p = d = r = 1$, we have

$$1 = \int_0^1 \frac{M\overline{M}}{\sqrt{\nu M \overline{M}}}\, dx = \int_0^1 \sqrt{\frac{M\overline{M}}{\nu}}\, dx, \tag{65}$$

or

$$\sqrt{2\nu} = \int_0^1 x^{3/2}\, dx = \frac{2}{5} \Rightarrow \nu = \frac{2}{25} = 0.08 . \tag{66}$$

Since in general the relations given in Fig. 11 cannot be solved simultaneously in an analytical form, an iterative procedure is outlined in the next section.

a deflection constraint, both the real and the adjoint systems involve only linear equations.

(b)   We *update* the cross-sectional parameters z using the optimality criteria at the top of Fig. 11.

Apart from the above main steps, some other computations are necessary. This can be seen from Fig. 13 which shows the flowchart of the iterative COC algorithm for a deflection constraint, in the context of Bernoulli-beams of variable width. The same procedure can be used for *any elastic structure* with a deflection constraint. In Fig. 13, the symbol i denotes the i–th element and z_i denotes its flexural stiffness. We stipulate a minimum stiffness constraint

$$z_i \geq z_a \qquad \text{(for all } i). \tag{67}$$

Elements with $z_i > z_a$ shall be termed *active elements* and those with $z_i = z_a$ *passive elements*, as in earlier work by Berke and Khot (1988). The computational steps in Fig. 13 involve the following operations:

(A) The same constant initial stiffness value z_{in} can be adopted for all elements if a better estimate is not available.

(B, C) Any standard FE software can be used for analysing the real and adjoint systems. The adjoint beam is subject to the virtual load associated with the deflection constraint.

(D) The updated value of the Lagrangian ν can be calculated from the discretized work equation

$$\Delta = \sum_i \frac{M_i \overline{M}_i L_i}{z_i}, \tag{68}$$

where M_i and $\overline{M}_i$ are the real and adjoint moment values at the middle of the beam element i, L_i is the length of the element i and $z_i = s_i$ is the normalized beam stiffness. Using the optimality condition (53), the RHS of (68) can be split into active (A) and passive (P) element domains in the form

$$\Delta = \sum_A \frac{M_i \overline{M}_i L_i}{\sqrt{\nu M_i \overline{M}_i}} + \sum_P \frac{M_i \overline{M}_i L_i}{z_a}, \tag{69}$$

which implies the following relation for the updating of the Lagrangian ν:

$$\sqrt{\nu} = \frac{\sum\limits_A L_i \sqrt{M_i \overline{M}_i}}{\Delta - \sum\limits_P \dfrac{M_i \overline{M}_i L_i}{z_a}}. \tag{70}$$

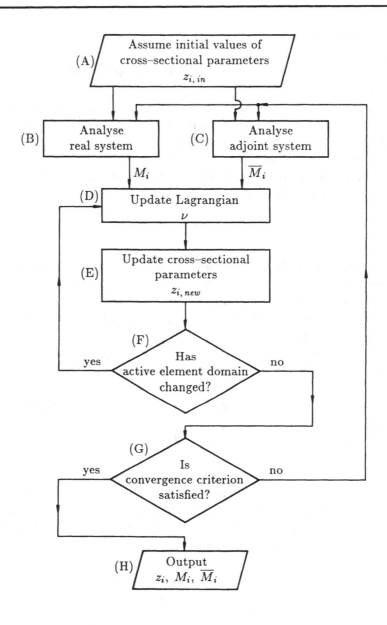

Fig. 13   Iterative COC method for a deflection and prescribed minimum stiffness constraint
in the context of beam optimization.

6 An Iterative COC Algorithm for Large Discretized Systems

The system of equations given in Fig. 11 can be solved iteratively by using the
following two steps in each iterative cycle:

(a) We *analyse* the *real* and *adjoint* systems using some available FE software. For

The relation (70) represents an approximation because the moments in the numerator of the first fraction are from the *current cycle*, while the stiffness $\sqrt{\nu M_i \overline{M}_i}$ is the value in the *forthcoming cycle*. This error becomes negligibly small in late iterations because the change in the moment values becomes insignificant. The fact that in (68) the expression $\sum_i \int_{D_i} [M(x)\overline{M}(x)/z(x)] dx$ is replaced by $\sum_i M_i \overline{M}_i L_i / z_i$ is also an approximation, but the error involved is negligible when a large number of elements are used (Rozvany, Zhou *et al.*, 1989).

(E) The new beam stiffness is calculated from the relations based on an extended version of (53):

$$\text{(for } \nu M_i \overline{M}_i \le z_a^2) \quad z_i = z_a, \quad \text{(for } \nu M_i \overline{M}_i > z_a^2) \quad z_i = \sqrt{\nu M_i \overline{M}_i}. \tag{71}$$

(F) The reason for this step is as follows. If some elements have changed from the active to the passive set or vice versa, then (70) would give an incorrect estimate of ν and hence steps (D) and (E) must be repeated.

(G) The tolerance test can be based on the following criteria:

$$\frac{\Phi_{new} - \Phi_{old}}{\Phi_{new}} \le \overline{E}, \quad \text{(for all } i) \quad \frac{z_{i,new} - z_{i,old}}{z_{i,new}} \le E_1, \tag{72}$$

where $\overline{E}$ and E_1 are given tolerance values.

7 Example of an Ill-Conditioned Problem: Beam with a Deflection Constraint, No Limits on the Cross-Sectional Parameters

The method described in Figs. 11 and 13 is suitable for optimizing any linearly elastic system with a deflection constraint, if the parameters z fully define the cross-sectional geometry and they vary freely along the centroidal axes (subject to minimum values z_a).

In the case of elastic structures, it is useful to know the absolute lower limit of the structural weight when the prescribed minimum cross-sectional area is zero ($z_a = 0$). Since we still want to avoid concentrated rotations at locally vanishing cross-sections, we consider the limiting case of a sequence of solutions in which the prescribed minimum cross-section has first a finite value but then it approaches zero ($z_a \to 0$). This can be evaluated analytically, but in iterative COC calculations it is necessary to specify some small value (for example 10^{-7} times the expected average value of z_i) for z_a.

7.1 Problem Description and Analytical Solution

We consider a clamped, uniformly loaded beam (Fig. 14a) of given depth d but variable width z for which the deflection at midspan (A) has an upper limit

$$u_A \le \Delta. \tag{73}$$

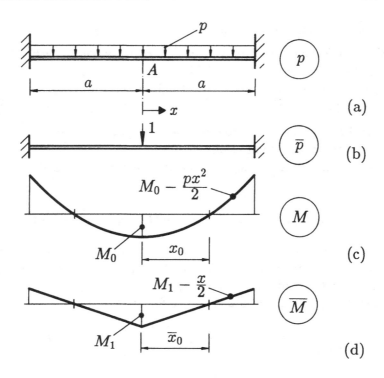

Fig. 14   Beam example — given deflection at midspan and no lower limit on the cross-
sectional area.

The virtual (adjoint) load $\bar{p}$, and the real and adjoint moment diagrams $(M, \overline{M})$
are shown, respectively, in Figs. 14b, c and d.

We introduce the nondimensional notation $\tilde{x} = x/a$, $\tilde{z} = z/a$, $\tilde{M} = M/pa^2$,
$\tilde{\overline{M}} = \overline{M}/a$, $\tilde{u}_A = u_A/\Delta$, $\tilde{s} = 12s/(d^3 aE) = \tilde{z}$, $\tilde{\Phi} = \Phi/(da^2\gamma)$, where x is the
longitudinal coordinate, E is Young's modulus, s is the flexural stiffness, Φ is the
total beam weight, γ is the specific weight of the beam material and other symbols
have been defined earlier. The nondimensional width $\tilde{z}$ and stiffness $\tilde{s}$ have been
made identical since they are linearly interdependent in this problem. The real and
adjoint bending moment diagrams (Figs. 14c and d) are represented by

$$\tilde{M} = \tilde{M}_0 - \tilde{x}^2/2, \quad \tilde{\overline{M}} = \tilde{M}_1 - \tilde{x}/2. \tag{74}$$

The considered problem is to be solved for the case when the prescribed minimum
cross-sectional area approaches zero: $z_a \rightarrow 0$. This means that apart from regions
of infinitesimal length with $z = z_a$, our relevant optimality condition is the nondimensional version of (53). In addition, the compatibility conditions for the real and

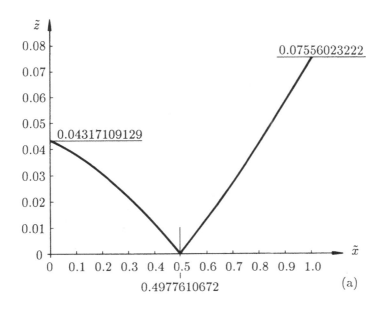

Fig. 15a   Optimal stiffness distribution in the beam example: analytical and COC solutions.

adjoint systems and the nondimensional displacement condition imply:

$$\int_0^1 \frac{\tilde{M}}{\tilde{z}}\,\mathrm{d}\tilde{x} = 0, \quad \int_0^1 \frac{\overline{\tilde{M}}}{\tilde{z}}\,\mathrm{d}\tilde{x} = 0, \quad 2\int_0^1 \frac{\tilde{M}\overline{\tilde{M}}}{\tilde{z}}\,\mathrm{d}\tilde{x} = 1, \tag{75}$$

where $\tilde{M}$, $\overline{\tilde{M}}$ and $\tilde{z}$ are given by (74) and (53). Moreover, (53) gives a non-imaginary (real) value for $\tilde{z}$ only if in Figs. 14c and d

$$x_0 = \overline{x}_0. \tag{76}$$

The relations under (75) and (76), after substitution of (74) and (53), represent four simultaneous equations and the unknown quantities involved are $\tilde{M}_0$, $\tilde{M}_1$ and ν. It can be shown easily that the above equations cannot be satisfied simultaneously. The paradoxical nature of this state of affairs was pointed out by Masur in a discussion of his paper (Masur, 1975) and is considered in detail elsewhere (Rozvany, Rotthaus *et al.*, 1990). As is shown in the above paper, the explanation of this paradox is that in the exact analytical solution for $z_a \rightarrow 0$, the beam stiffness (nondimensionally: $\tilde{s} = \tilde{z}$) takes on a second order infinitesimal stiffness value over a first order infinitesimal beam length with $z = z_a$. The rotations (i.e. integrated curvatures) over such a length tend to finite values $(\theta, \overline{\theta})$ which have been shown to satisfy the relation

$$\theta\overline{V} = \overline{\theta}V, \tag{77}$$

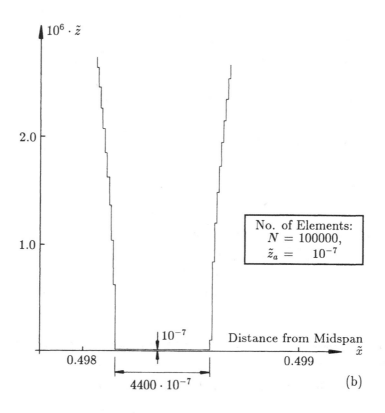

Fig. 15b Enlarged detail of the stiffness distribution in the vicinity of the singularity (COC
solution).

where V and $\overline{V}$ are the shear forces in the real and adjoint beams at the considered
location. It will be seen in Fig. 15b that discretized COC solutions with 100000
elements fully confirm the type of singularity predicted by the analytical solution.

The complete analytical solution (Rozvany, Zhou et $al.$, 1989, Rozvany, Rotthaus
et $al.$, 1990) yields the following optimal stiffness variation for the above problem

$$\tilde{z} = \left[(\tilde{x}_0)^{5/2}(2^{9/2} - 7) - (\tilde{x}_0 + 1)^{3/2}(7\tilde{x}_0 - 3)\right](\tilde{x}_0 - \tilde{x})\sqrt{\tilde{x}_0 + \tilde{x}}/15 , \qquad (78)$$

and the optimal $\tilde{x}_0$ value is furnished by the cubic equation

$$\tilde{x}_0^3\left[2^9 - 7(2)^{11/2}\right] - 63\tilde{x}_0^2 - 15\tilde{x}_0 - 1 = 0 . \qquad (79)$$

Moreover, the optimal total weight becomes

$$\Phi = 4\left[\tilde{x}_0^{5/2}(2^{9/2} - 7) - (\tilde{x}_0 + 1)^{3/2}(7\tilde{x}_0 - 3)\right]^2/225 . \qquad (80)$$

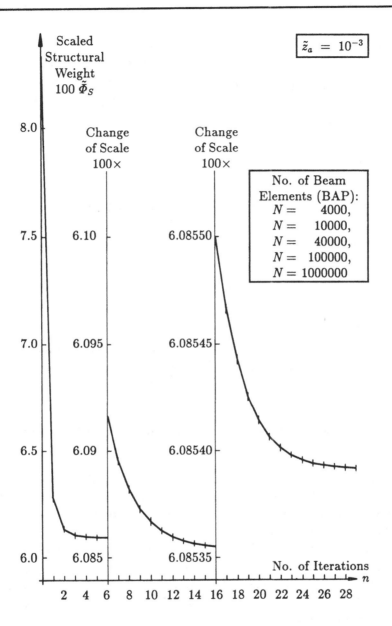

Fig. 16   Scaled total structural weight $\tilde{\Phi}$ as a function of the number of iterations (n).

The relations (79) and (80) furnish the optimal values

$$\tilde{x}_0 = 0.49776107, \qquad \tilde{\Phi}_{opt} = 0.060448186. \tag{81}$$

The optimal variation of the stiffness $\tilde{z}(\tilde{x})$ is shown graphically in Fig. 15a.

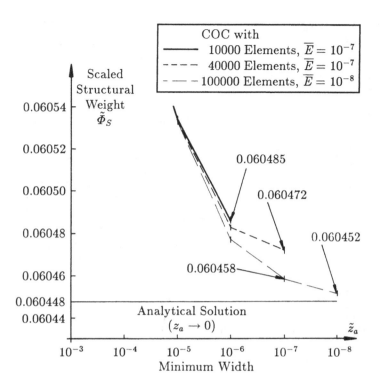

Fig. 17 A comparison of COC results using 10000, 40000 and 100000 elements.

7.2 Iterative COC Solutions

Using the COC procedure described in Section 6 (Figs. 11 and 13), discretized numerical solutions were obtained by using various numbers of elements ranging from fifty to one million. The structural analysis of the real and adjoint systems was carried out using the following programs:

ANSYS (number of elements $N \leq 150$), BAP (number of elements $N \geq 50$).

The special purpose program BAP for beam analysis was found, as expected, to be much more efficient computationally than a general purpose FE program. Both programs yielded identical results for $N = 50, 100$ and 150. Using the COC approach, the following conclusions were reached:

(i) The convergence in the considered example is found to be fully monotonic and almost uniform, and its rate largely independent of the number of elements (N) used (see Fig. 16).

(ii) In spite of an unusual *singularity* in the analytical optimal solution, the total weight in the best COC solution (100000 elements, $\tilde{z}_a = 10^{-8}$, $\tilde{\Phi} = 0.060452$) differs only by *0.006 per cent* from the analytical solution ($\tilde{\Phi} = 0.060448$).

(iii) For problems with near-singularities in the discretized formulation, a very large

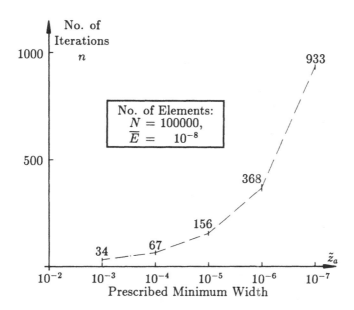

Fig. 18   Number of iterations necessary for the convergence criterion as a function of the prescribed minimum width ($\tilde{z}_a$).

number of elements are required for a reasonably high accuracy (see Fig. 17).

(iv) The type of singularity predicted by the analytical solution (stiffness of higher order infinitesimal over a length of first order infinitesimal) is indicated by the discretized COC solution in the vicinity of the singularity (see Fig. 15b).

(v)   It can be seen from Fig. 18 that with a decreasing value of $\tilde{z}_a$, the problem becomes more and more ill-conditioned due to the near-singularity at $\tilde{z}_i = \tilde{z}_a$ and hence the convergence becomes correspondingly slower.

(vi) The considered problem was also investigated by using mathematical programming (MP) methods. Both Lawo (Karlsruhe, FRG) and Wang (Singapore) have found that with $\tilde{z}_a \to 0$, signs of instability appear when the number of elements (N) exceeds 50. The best MP solution (with $N = 45$) yielded a total weight of $\tilde{\Phi} = 6.088672$ which represents an error of about *0.7 per cent* in comparison with the exact analytical solution (the error of the best MP solution is over *hundred times* higher than that of the best COC solution).

(vii) In solving similar problems, Berke and Khot (1987, 1988) used the recurrence relation

$$(z_i)_{n+1} = (z_i)_n \{\nu M_i \overline{M}_i / z_i^2\}_n^{1/r}, \tag{82}$$

where r is a "step size parameter". For $r = 2$, (82) reduces to the optimality criterion (53) used in this chapter. The number (n) of iterations required to fulfil the convergence criterion for a relatively stable problem (with $\tilde{z}_a = 10^{-3}$)

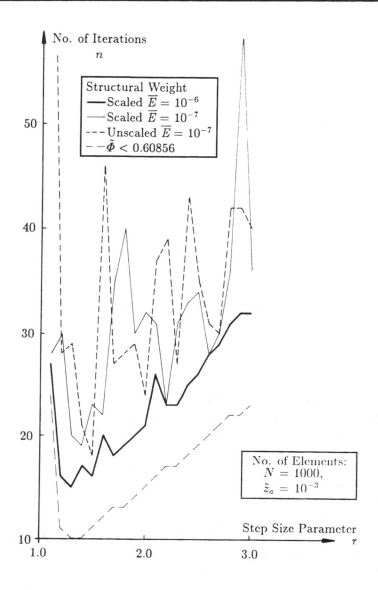

Fig. 19 Beam example: number (n) of iterations required to fulfil the convergence criterion as a function of the Berke-Khot step size parameter.

is shown in Fig. 19 in dependence on the step size parameter in (82). In the latter, either a convergence criterion of the type in (72) is used, or a fixed limit of $\tilde{\Phi} < 0.060856$. The latter type seems to eliminate the "noise" in the convergence polygons. It can be seen from Fig. 19 that for the most stable convergence polygon ($\overline{E} = 10^{-6}$) there is a relatively small difference between the number of iterations for the optimal overrelaxation ($r = 1.3$, $n = 15$) and the "natural" relaxation ($r = 2$, $n = 21$) used herein. Moreover, the r-value for the former is difficult to predict and is very close to an unstable region ($r \leq 1.2$).

3 Concluding Remarks

It was shown in this chapter that, for a single deflection constraint, the COC procedure has an optimization capability of around one million variables on a personal computer and could handle several million variables on a larger computer. The convergence was near-uniform in all test problems right up to the 8-digit accuracy set in the convergence criterion. The treatment of additional design constraints will be discussed in the next chapter.

Chapter 4

COC METHODS FOR ADDITIONAL GEOMETRICAL CONSTRAINTS

G.I.N. Rozvany and M. Zhou
Essen University, Essen, Germany

1 Introduction

In the previous chapter, the COC algorithm was applied to elastic systems with a single deflection constraint and was illustrated with examples involving Bernoulli-beams of variable width. The aim of this exercise was to show that *iterative COC methods not only eliminate but also reverse the existing discrepancy between analysis capability and optimization capability.* Since this study was primarily concerned with an optimizer, the repeated analysis of the structures under consideration was carried out by so-called *FE-simulators* (e.g. BAP) which were shown to generate exactly the same output as standard FE software (e.g. ANSYS), but several orders of magnitude faster. Although FE-simulators can handle only one given set of boundary and loading conditions, they are highly suitable for evaluating the optimization capability in test examples. Whilst present hardware and standard FE software cannot handle such large numbers of elements, the analysis capability of these is expected to keep on increasing in the foreseeable future. The COC algorithm, therefore, represents an optimizer that *can easily handle not only present but also future needs* in the optimization of large structural systems.

Mathematical programming (MP) methods have the *advantage* that they are robust, that is, they can readily handle most problems within and outside the discipline of structural optimization. However, if analytical sensitivities are used in MP methods, then the necessary derivations become strongly problem-oriented.

The disadvantage of COC methods is the necessity to derive the appropriate optimality conditions for the type of structure, cross-sectional topology and design constraints under consideration. This disadvantage could be minimized, however, by developing software for generating automatically the appropriate optimality criteria for a given structural problem. A fairly comprehensive summary of available analytical optimality criteria for various classes of problems is given in the first author's recent book (Rozvany, 1989).

The types of additional design problems to be discussed in this course are summarized in Fig. 20, which also shows in square brackets the chapter and section numbers where they are discussed. These design examples illustrate broader classes of problems which are stated in the section titles.

The last three classes of problems in Fig. 20 will be discussed in the next chapter.

2 Structures with Cross-Sectional Dimensions Freely Variable above a Prescribed Minimum Value

The procedure for this class of problems was already outlined in Section 6 of Chapter 3, except that there the prescribed minimum value of the dimensions was chosen very small (10^{-8} to 10^{-2}) in order to approximate closely an unconstrained variation of the variables. In the present discussion, this lower limit has roughly the same order of magnitude as the average value of the variables concerned. Considering, for example, the problem in Fig. 14 (Chapter 3) with a nondimensional minimum width value of $\tilde{z}_a = 0.01$, the iterative COC method with 10000 elements yielded the width distribution given in Fig. 21. The same solution was verified to six significant digits by a *semi-analytical method* in which the optimality criteria yielded a set of equations containing elliptic integrals. These were evaluated numerically using 50000 intervals and the solution was then obtained by a Newton-Raphson procedure (Rozvany, Zhou *et al.*, 1990). The COC results were also verified by a *sequential quadratic programming (SQP) method* which, however, was restricted to maximum 200 variables. Therefore, for this comparison COC solutions with 50, 100 and 200 elements were obtained. Whilst the agreement between the two sets of results was 8 to 9 significant digits, the SQP method required significantly more CPU time for optimization, as can be seen from Fig. 22. The relative time requirement difference increases with the number of variables, and for 200 variables *the SQP method requires approximately 3000 times more CPU time than the COC procedure.*

3 Structures with Cross-Sectional Dimensions Freely Variable between Prescribed Minimum and Maximum Values

For this class of problems we use essentially the same procedure as in the previous section (see also Section 6 in Chapter 3), except that (71) is replaced by the type of

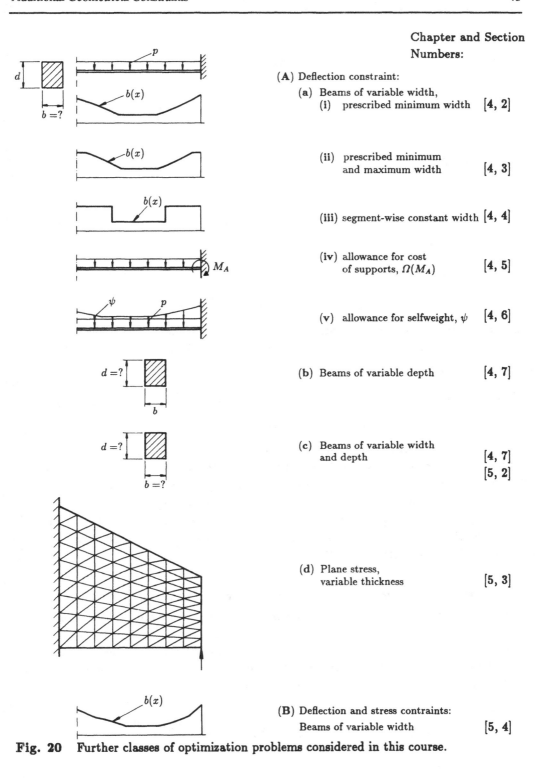

Chapter and Section
Numbers:

(A) Deflection constraint:

(a) Beams of variable width,
    (i)   prescribed minimum width   [4, 2]

    (ii) prescribed minimum
         and maximum width   [4, 3]

    (iii) segment-wise constant width [4, 4]

    (iv) allowance for cost
          of supports, $\Omega(M_A)$ [4, 5]

    (v) allowance for selfweight, ψ [4, 6]

(b) Beams of variable depth   [4, 7]

(c) Beams of variable width
    and depth   [4, 7]
                                     [5, 2]

(d) Plane stress,
    variable thickness   [5, 3]

(B) Deflection and stress contraints:
    Beams of variable width   [5, 4]

Fig. 20 Further classes of optimization problems considered in this course.

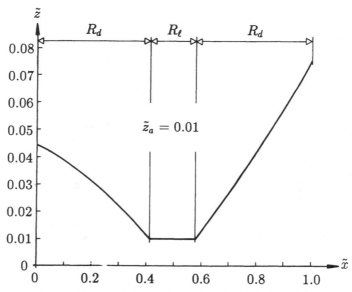

Fig. 21 Optimal distribution of the nondimensional width $\tilde{z}$ with a prescribed minimum value of $\tilde{z}_a = 0.01$ for a clamped beam with a uniformly distributed load.

update formulae:

$$(\text{for } \nu M_i \overline{M}_i < z_a^2) \quad z_i = z_a, \quad (\text{for } z_a^2 < \nu M_i \overline{M}_i < z_b^2) \quad z_i = \sqrt{\nu M_i \overline{M}_i},$$

$$(\text{for } \nu M_i \overline{M}_i > z_b^2) \quad z_i = z_b, \tag{83}$$

where z_a and z_b are the lower and upper limits on the variables z_i.

Moreover, in calculating the Lagrangian ν by (70), the passive set (P) includes all elements governed by the lower and upper limits $(z_i = z_a$ or $z_i = z_b)$.

A solution for the beam shown in Fig. 14 (Chapter 3), but with lower and upper limits of $\tilde{z}_a = 0.01$ and $\tilde{z}_b = 0.05$, is given in Fig. 23. The above solution was verified to a seven significant digits accuracy by the semi-analytical method described in Section 2 (Rozvany, Zhou et al., 1990).

4 Structures with Segmentwise Constant Cross-Sections

In the case of structures with segment-wise constant cross-sections, the structural domain D is divided into segments $D_\alpha (\alpha = 1, 2, \ldots, \omega)$ and each segment may consist of one or several elements. On the basis of a general optimality criterion [Rozvany 1989, (1.98) on p. 72], the optimality condition for beams of constant depth and variable but segment-wise constant width becomes

$$\left[\text{for } \tilde{z}_a^2 L_\alpha > \sum_{i(D_\alpha)} \nu \tilde{M}_i \overline{\tilde{M}}_i \delta_i\right] \quad \tilde{z}_\alpha = \sqrt{\frac{\nu \sum_{i(D_\alpha)} \tilde{M}_i \overline{\tilde{M}}_i \delta_i}{L_\alpha}}, \tag{84}$$

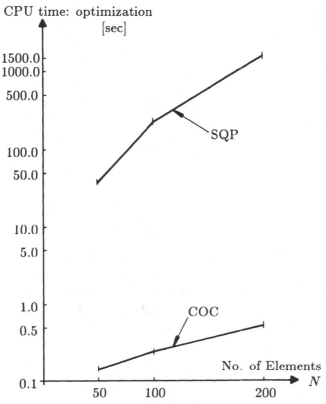

Fig. 22   A comparison of CPU times required for a structural optimization problem solved by the COC and SQP methods.

where L_α is the length of the segment D_α, $i(D_\alpha)$, $(i = 1, 2, \ldots, n)$ are the elements in the segment D_α, δ_i is the length of the element i, $(M_i, \overline{M}_i)$ are the real and adjoint moments at the middle of the same element (i) and ν is a Lagrangian multiplier (constant). The condition (84) replaces (71) in Chapter 3.

An equation for updating the Lagrangian multiplier can be derived as follows. The prescribed deflection Δ is expressed in terms of virtual work as

$$\Delta = \sum_{\alpha(A)} \frac{\displaystyle\sum_{i(D_\alpha)} \tilde{M}_i \tilde{\overline{M}}_i \delta_i}{\sqrt{\dfrac{\nu \sum_{i(D_\alpha)} \tilde{M}_i \tilde{\overline{M}}_i \delta_i}{L_\alpha}}} + \sum_{i(P)} \frac{\tilde{M}_i \tilde{\overline{M}}_i \delta_i}{\tilde{z}_\alpha}, \tag{85}$$

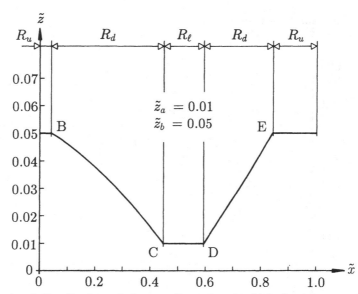

Fig. 23 Optimal distribution of the nondimensional width $\tilde{z}$ with prescribed minimum and maximum values of $\tilde{z}_a = 0.01$ and $\tilde{z}_b = 0.05$ for a clamped beam with a uniformly distributed load.

which implies after simplifications

$$
\sqrt{\nu} = \frac{\displaystyle\sum_{\alpha(A)} \sqrt{L_\alpha \sum_{i(D_\alpha)} \tilde{M}_i \bar{\tilde{M}}_i \delta_i}}{\Delta - \displaystyle\sum_{i(P)} \frac{\tilde{M}_i \bar{\tilde{M}}_i \delta_i}{\tilde{z}_a}}, \tag{86}
$$

where $\alpha(A)$ refers to all active segments with $\tilde{z}_\alpha > \tilde{z}_a$ and $i(P)$ to all elements of the passive segments with $\tilde{z}_\alpha = \tilde{z}_a$. The relation (86) replaces (70) for segmented beams. The COC solution for prescribed segment lengths of $L_1 = 0.3$, $L_2 = 0.4$ and $L_3 = 0.3$ over both half-spans is shown in Fig. 24, which was obtained in 8 iterations. The corresponding nondimensional total weight was $\bar{\Phi} = 0.06757822$.

The above result was verified by using differential calculus to a seven-digit accuracy (Rozvany, Zhou *et al.*, 1990).

In Table 1, the weight of the unconstrained solution (Φ_u) is compared with that of (a) beams having lower and/or upper limits on the width, (b) the optimal segmented beam and (c) the prismatic beam having the same central deflection. It can be seen from Table 1 that the prismatic beam has almost 38% higher weight than the unconstrained optimal solution.

It will be shown in the next section that the COC method can handle allowance for selfweight with a small modification of the optimality criteria.

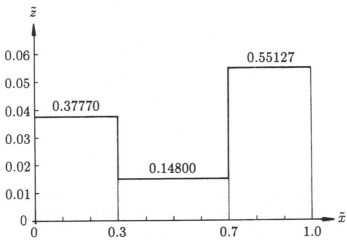

Fig. 24 Optimal distribution of the nondimensional width $\tilde{z}$ for a clamped beam consisting of three given segments of constant width.

Table 1. A comparison of optimal beam weights for various geometrical constraints.

	Optimal weight $\tilde{\Phi}$	$\dfrac{\Phi - \Phi_u}{\Phi_u}$
Unconstrained geometry	0.060448186	0
Prescribed minimum width $z_a = 0.01$	0.062263280	0.0300
Prescribed minimum and maximum widths $z_a = 0.01, z_b = 0.06$ $z_a = 0.01, z_b = 0.05$	0.062529162 0.063833474	0.0344 0.0560
Beam with three prismatic segments	0.06757822	0.1180
Prismatic beam	0.08333333	0.3786

5 Allowance for the Cost of Supports

In the considered class of problems, some subset S of the structural domain D is connected to supports and the total cost of the system becomes

$$\Phi = \int_D \psi[z(x)]\,dx + \int_S \Omega[R(x)]\,dx, \tag{87}$$

where $\Omega[\,]$ is the specific cost of the supports subject to the reactions R, whilst ψ is the specific cost of the structure. On the interior of the structural domain D the

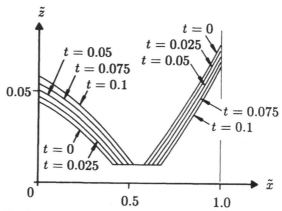

Fig. 25   Optimal distribution of the nondimensional width $\tilde{z}$ for various values of the reaction cost factor t for a clamped beam with a uniformly distributed load.

usual optimality conditions (e.g. Fig. 11) are still valid, but along supports (S) the
adjoint field is subject to non-zero displacements $(\overline{\mathbf{u}})$

$$\text{(on } S) \quad \overline{\mathbf{u}}^K = \mathcal{G}[\Omega(\mathbf{R}^S)]\,, \tag{88}$$

where $\mathcal{G}$ is the G-gradient discussed in Fig. 7 in Chapter 2.

Along the same supports, the real displacements are zero if the supports are
rigid. The above optimality criteria were introduced in the first author's recent book
[Rozvany, 1989, (1.101) on p. 73].

Considering again the beam in Fig. 14 in Chapter 3, but with the reaction costs

$$\Omega(M_C) = t|M_C|\,, \tag{89}$$

where M_C is the clamping moment at the beam ends and t is a given constant, we
have the additional optimality conditions for the beam ends

$$\overline{u}'(-a) = t\,, \quad \overline{u}'(a) = -t\,, \tag{90}$$

where primes denote differentiation with respect to x and hence $\overline{u}'$ represents the
slope of the adjoint displacement field.

The computational procedure was the same as for a beam with a prescribed
minimum width, except that the concentrated rotations t given by (90) must be
taken into consideration in *analysing* the adjoint beam (i.e. in calculating the bending
moments $\overline{M}_i$). Using reaction cost factors of $t = 0$, 0.025, 0.05, 0.075 and 0.1 and a
nondimensional minimum width of $\tilde{z}_a = 0.01$, the width distributions shown in Fig.
25 were obtained for one half of the beam. As expected, with increasing cost of the
clamping moment the stiffness decreases at the beam ends.

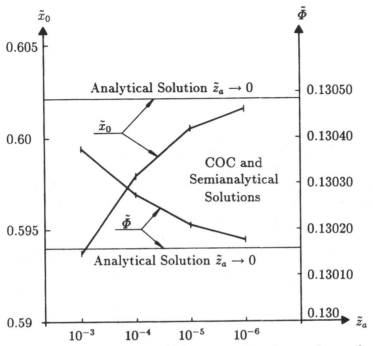

Fig. 26 A comparison of the distances $(\tilde{x}_0)$ of the point of contraflexure from the beam centre and the beam weights $(\tilde{\Phi})$ from the analytical solution with $\tilde{z}_a \to 0$ and from numerical solutions with $\tilde{z}_a > 0$ considering a clamped beam having variable width and a reaction cost factor of $t = 0.1$.

The above solutions were verified by the semi-analytical method described in Section 2 and a nine digits agreement was found. A fully analytical solution was also obtained for $\tilde{z}_a \to 0$ and various t values, and it was found that COC solutions with very small $\tilde{z}_a$–values clearly tend to this analytical solution, see Fig. 26 (Rozvany, Zhou et al., 1990).

6 Allowance for Selfweight

On the basis of the general optimality criteria for elastic structures with selfweight [(1.99) on p. 72 in the first author's recent book, Rozvany, 1989], and beams of variable width (z), the optimality condition of (71) in Chapter 3 must be replaced by

$$\left[\text{for} \ \frac{\nu M(x)\overline{M}(x)}{rc(1+\bar{u})} \geq z_a^2 \right] \quad z = \sqrt{\frac{\nu M \overline{M}}{rc(1+\bar{u})}}, \tag{91}$$

where r and c are given constants and $\bar{u}$ is the adjoint deflection. The same optimality condition can be derived from variational principles. The considered problem can be formulated as

$$\min \Phi = \int_0^L \left\{ cz + \nu \frac{M\overline{M}}{rz} + \lambda(z - z_a - \omega) + \overline{u}(M'' + p + cz) + \right.$$

$$\left. + u(\overline{M}'' + \overline{p}) \right\} dx - \nu(\Delta - \eta) \,, \tag{92}$$

where Φ is the beam weight, L is the beam length, c and r are given constants, z is the beam width, cz is the specific beam weight (weight per unit length), rz is the flexural stiffness, ν is a Lagrangian multiplier (constant), M is the real beam moment, $\overline{M}$ is the adjoint beam moment, $\lambda(x)$, $\overline{u}(x)$ and $u(x)$ are variable Lagrangian multipliers, z_a is the prescribed minimum width, $\omega(x)$ is a slack function, p is the real external load, $\overline{p}$ is the adjoint load, Δ is the prescribed deflection and η is a slack variable. The equilibrium equation $M'' = -p - cz$ includes allowance for the selfweight cz. In the adjoint equilibrium equation $\overline{M}'' = -\overline{p}$ the load $\overline{p}$ is the virtual load (e.g. "unit dummy load"). In (92), $c = h\gamma$ where h is the beam depth and γ is the specific weight of the beam material. Moreover, $r = h^3 E/12$ where E is Young's modulus for the beam material. The usual Euler-Lagrange equations then yield:

Variation of ω :   (for $z > z_a$, $\omega > 0$) $\lambda = 0$. $\qquad$ (93)

Variation of z : $c - \nu M\overline{M}/(rz^2) + c\overline{u} = 0$. $\qquad$ (94)

Variation of M : $\nu \overline{M}/(rz) = \overline{u}''$. $\qquad$ (95)

Variation of $\overline{M}$: $\nu M/(rz) = u''$. $\qquad$ (96)

Relation (94) implies the optimality condition (91).

Since the quantity rz is the flexural beam stiffness, the Lagrangian u becomes a factored (ν) value of the real beam deflection and the Lagrangian $\overline{u}$ represents the factored value of the adjoint deflection associated with the virtual load $\overline{p}$.

Owing to the complexity of nondimensionalization for problems with selfweight, we consider a problem with realistic dimensions. A laminated timber beam has the constant depth $d = 1.0$ m, the specific weight $\gamma = 6.0$ kN/m^3, Young's modulus $E = 10^7$ kN/m^2 and a uniformly distributed load 10 kN/m. The variable width z of the beam is to be optimized. The prescribed minimum width of the beam is 0.1 m and the prescribed deflection at midspan is $\Delta = L/300$ where L is the span length. The optimal width distribution for spans of $L = 22, 30, 50$ and 60 m is shown in Fig. 27.

At $L = 22$ m, most of the design is governed by the minimum width constraint, resulting in an almost prismatic beam. As the span length increases from 30 m to 60 m, the segment with the minimum width shifts gradually towards the centre (note the change of scale between $L = 30$ m and $L = 50$ m). This is to be expected because

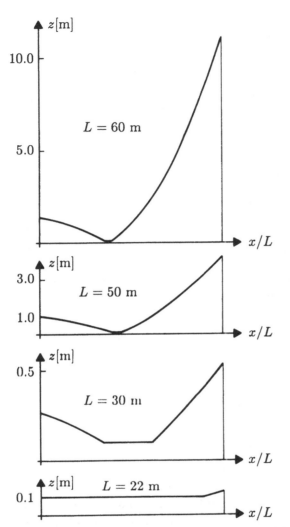

Fig. 27   Optimal width distribution for uniformly loaded clamped beams with allowance
for selfweight, having various span lengths (L), a given depth of 1.0 m, a minimum
width of 0.1 m and a prescribed central deflection of $L/300$.

it is more favourable if at longer spans some of the selfweight moves closer to the
supports.

6.1 A Comparison of Optimal Nonprismatic Designs with Optimal Prismatic Beams

For *prismatic* beams with a deflection constraint and allowance for selfweight, the
optimal design can be determined analytically relatively easily. For the clamped

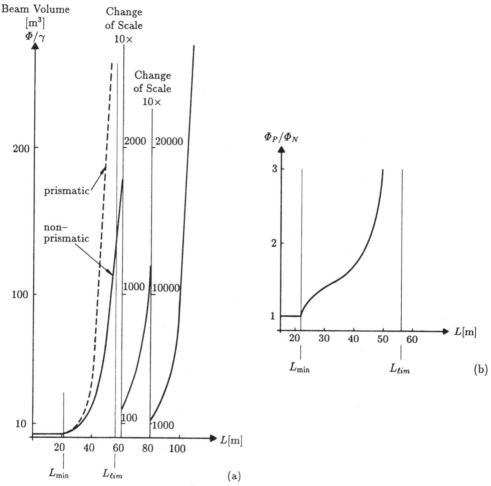

Fig. 28   Beams subject to external load plus selfweight: (a) optimal weights of prismatic and nonprismatic beams of variable width, and (b) ratios of the volumes of prismatic and nonprismatic beams.

beam considered above, we have the deflection constraint

$$\Delta = \frac{(p + bd\gamma)L^4}{384Ebd^3/12} \,, \tag{97}$$

implying

$$b = \frac{pL^4}{32E\Delta d^3 - d\gamma L^4} \,. \tag{98}$$

The upper limit on the span length or spanning capacity L_{lim} of a prismatic beam with the above deflection constraint is reached when the denominator in (98) becomes zero and hence $b \to \infty$. This condition implies

$$L_{\lim} = \sqrt[4]{32E\Delta d^2/\gamma}. \tag{99}$$

In the considered example with $E = 10^7$, $d = 1$, $\gamma = 6$ and $\Delta = L/300$ we have

$$L_{\lim} = \sqrt[3]{32 \cdot 10^7/1800} = 56.2288443 \text{ m}. \tag{100}$$

Considering now the nonprismatic design, below a certain minimum span $L_{\min}$ the beam takes on the prescribed minimum width (0.1 m). Substituting this value for b in (98), we have for the given values for p, E, d, Δ and γ

$$L_{\min} = \sqrt[3]{\frac{8(10^6)}{(10.6)(75)}} = 21.5894189. \tag{101}$$

Figure 28a shows the total volume Φ/γ of optimal prismatic and nonprismatic beams and Fig. 28b shows the ratio Φ_P/Φ_N for various span lengths L where ϕ_P and Φ_N are, respectively, the weights of the prismatic and nonprismatic beams. In the vicinity of the spanning capacity $L_{\lim}$ this ratio approaches infinity: for example for $L = 56.228$ we have $\Phi_P/\Phi_N = 2080185.9/123.19951 = 16884.693$ on the basis of (98) and the COC result for the nonprismatic design.

The advantage of optimal width distribution can be seen from the fact that the spanning capacity (max. feasible span with an infinite beam width requirement) for the *prismatic* beam is around 56 m whilst the optimal *nonprismatic* beam can theoretically span 110 m (but at the cost of an unrealistic volume of 27969.056 m^3). At a more practical level, for a span of 50 m the optimized nonprismatic beam has a volume of 65.596545 m^3, whilst the volume of the prismatic beam is 197.368421 m^3 *which is 200.88% over the optimal nonprismatic volume.*

6.2 A Comparison of Results by the COC and SQP Methods

The results by the iterative COC procedure for beams with external load plus self-weight have also been checked by using a sequential quadratic programming (SQP) method. In the latter, a modified procedure was used for calculating the analytical sensitivities with allowance for selfweight (Zhou, 1990). For a span of $L = 40$ m and hundred elements the COC and SQP methods gave identical weight values ($\Phi = 22.252902$) for the first eight significant digits. The depth values for various cross-sections have shown a similar agreement.

7 Structures with Non-Linear Specific Cost and Stiffness Functions

7.1 Beams of Variable Depth

If the beam has a constant width b but variable depth z then the beam weight per unit length becomes $\psi = cz$ where $c = b\gamma$ and the flexural beam stiffness is rz^3 where

$r = bE/12$. As before, γ denotes the specific weight of the beam material and E is Young's modulus for the beam. Then we can obtain the optimality criteria directly from general formulae [Rozvany, 1989, (1.93) on p. 70], which give

$$(\text{for } \nu M\overline{M} \geq z_a^4) \quad z = \sqrt[4]{\nu M\overline{M}}. \tag{102}$$

The same result can be derived from variational principles by modifying (92) in the form

$$\min \Phi = \int_0^L \left\{ cz + \nu \frac{M\overline{M}}{rz^3} + \lambda(z - z_a - w) + \overline{u}(M'' + p) + \right.$$

$$\left. + u(\overline{M}'' + \overline{p}) \right\} dx - \nu(\Delta - \eta), \tag{103}$$

where most symbols have the same meaning as in (92) unless it is otherwise indicated previously in this section. Then for variation of z we have the Euler equation

$$(\text{for } z > z_a, \ \lambda = 0) \quad c - 3\nu M\overline{M}/(rz^4) = 0 \quad \Rightarrow z = \sqrt[4]{\frac{3\nu M\overline{M}}{cr}}. \tag{104}$$

Redefining $3\nu/(cr)$ as ν, we confirm the optimality condition in (102). For the calculation of the Lagrangian, (70) in Chapter 3 is replaced by

$$\nu^{3/4} = \sum_A \left[(M_i \overline{M}_i)^{1/4} \delta_i \right] / \left(\sum_P \frac{M_i \overline{M}_i \delta_i}{s_a} - \Delta \right), \tag{105}$$

where s_a is the flexural stiffness corresponding to the minimum depth z_a, that is, $s_a = rz_a^3$.

As before, M and $\overline{M}$ denote the moments associated with the real and virtual loads. For a clamped beam with uniformly distributed load, the iterative COC procedure with 10000 elements yielded the nondimensional depth $\tilde{z}$ and flexural stiffness $\tilde{s}$ distributions given in Fig. 29.

In the above calculations, the following nondimensionalization was used: $\tilde{x} = x/a$, $\tilde{z} = z \left[Eb\Delta/(pa^4) \right]^{1/3}$, $\tilde{s} = IE\Delta/(pa^4) = \tilde{z}^3/12$ (where I is the moment of inertia of the cross-section), $\tilde{\Phi} = \Phi \left[E\Delta/(b^2 a^7 p) \right]^{1/3}/\gamma$.

Note. The above expressions for nondimensional quantities may seem to be complicated, but correct nondimensionalization enables us to translate directly our results to optimal solutions for design problems with given dimensions, loading and material properties. Considering, for example, a beam with a load $p = 10$ kN/m, half-span $a = 15$ m, Young's modulus $E = 10^7$ kN/m^2, width $b = 0.1$ m, deflection

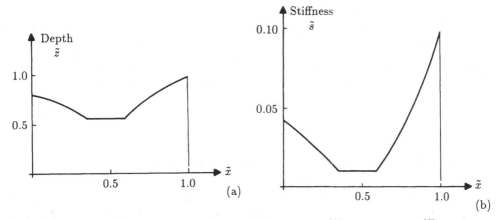

Fig. 29 Optimal variation of the nondimensional depth ($\tilde{z}$) and stiffness ($\tilde{s}$) for beams of variable depth with a prescribed minimum stiffness value of $\tilde{s}_a = 0.01$.

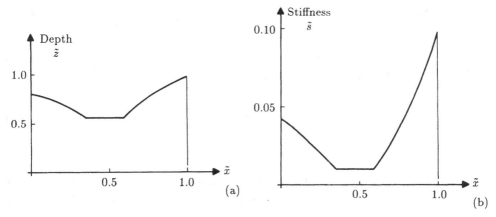

Fig. 30 Optimal variation of the nondimensional depth ($\tilde{z}$) and stiffness ($\tilde{s}$) for beams having a constant depth/width ratio and a minimum stiffness value of $\tilde{s}_a = 0.01$.

$\Delta = 0.1$ m and specific weight $\gamma = 6$ kN/m^2, we obtain the following weights for the optimal and prismatic beams

$$\Phi_{\text{opt}} = (6)(1.370867)\sqrt[3]{\frac{(0.1)^2(15)^7(10)}{(10)^7(0.1)}} = (1.370867)(6)(1.5)^{7/3} = 21.184888 \text{ kN},$$

$$\Phi_P = (1.587401)(6)(1.5)^{7/3} = 24.531127 \text{ kN}. \tag{106}$$

Similarly, the optimal depth values can be calculated for any design data from Fig. 29.

7.2 Beams with a Constant Depth/Width Ratio

Whilst for the class of problems considered in the last section only the specific stiffness function was nonlinear, both the specific cost and specific stiffness functions will be shown to be nonlinear in the problems considered herein. The relevant optimality condition can be readily obtained from general formulae [Rozvany 1989, (1.93) on p. 70] as

$$\text{(for } \nu M \overline{M} \geq z_a^6) \quad z = \sqrt[6]{\nu M \overline{M}}. \tag{107}$$

The same can be derived from a variational formulation by making the beam depth $h = z$ and width $b = cz$, thus having the cross-sectional area as $\psi = cz^2$ and the flexural stiffness as rz^4, where $r = cE/12$ and E is Young's modulus for the beam. Replacing the first two terms in the integrand in (103) by $cz^2 + \nu M \overline{M}/(rz^4)$, the Euler equation for variation of z becomes

$$\text{(for } z > z_a) \quad 2cz - 4\nu M \overline{M}/(rz^5) = 0 \Rightarrow z = \sqrt[6]{\frac{2\nu M \overline{M}}{rc}}. \tag{108}$$

A transformation with $2\nu/(rc) \to \nu$ then yields the result in (107). For the calculation of the Lagrangian, (70) in Chapter 3 is replaced by

$$\nu^{2/3} = \sum_A \left[(M_i \overline{M}_i)^{1/3} \delta_i \right] / \left(\sum_P \frac{M_i \overline{M}_i \delta_i}{s_a} - \Delta \right), \tag{108a}$$

where $s_a = r z_a^4$.

For a clamped beam with a uniformly distributed load the iterative COC method with 10000 elements furnished the nondimensional depth and flexural stiffness distributions in Fig. 30.

The above results are based on the following nondimensionalization:

$$\tilde{x} = x/a, \quad \tilde{s} = IE\Delta/(pa^4), \quad \tilde{z} = (z/a)(Ec\Delta/p)^{1/4},$$

$$\tilde{\Phi} = \left[\Phi/(a^3 \gamma) \right] [E\Delta/(pc)]^{1/2}. \tag{109}$$

8 Conclusions

It has been shown in this chapter that the COC algorithm can easily handle various types of geometrical constraints and nonlinear specific cost and stiffness functions. Further extensions of the COC approach will be discussed in the next chapter.

Chapter 5

COC METHODS: NON-SEPARABILITY, 2D SYSTEMS AND ADDITIONAL LOCAL CONSTRAINTS

G.I.N. Rozvany and M. Zhou
Essen University, Essen, Germany

1 Introduction

It will be shown in this chapter that COC methods are particularly advantageous for problems with nonseparable functions. In addition, applications of this technique to two-dimensional systems will be exemplified through a problem involving plane stress. Finally, the COC algorithm will be extended to a combination of stress and displacement constraints, for which the proposed approach represents major deviation from the traditional DOC methods.

2 Structures with Non-Separable Specific Cost and Stiffness Functions

In all problems discussed in Chapter 4, it was possible to express the specific cost ψ as a sum of element costs ψ_i which depended on only one cross-sectional parameter z_i associated with the element i. Moreover, each element specific stiffness s_i was a function of only one such parameter. In this section, we consider structures for which the element cost ψ_i and the specific element stiffness s_i is a nonseparable function of several cross-sectional parameters (dimensions) z_{ij} $(j = 1, \ldots, r)$.

The above class of problems will be illustrated with examples involving beams of variable width and depth. Other worked examples solved by the authors involved I-sections with variable flange width and web thickness.

2.1 Optimality Criteria for Beams of Independently Variable Width z_1 and Depth z_2 with Bending in Both Horizontal and Vertical Directions

For the above class of problems we have

$$\mathbf{Q} = (M_1, M_2), \quad \overline{\mathbf{Q}} = (\overline{M}_1, \overline{M}_2),$$

$$[\mathbf{F}] = \begin{bmatrix} \dfrac{1}{rz_1^3 z_2} & 0 \\ 0 & \dfrac{1}{rz_1 z_2^3} \end{bmatrix}, \quad \psi = cz_1 z_2, \quad r = E/12, \tag{110}$$

where M_1 and M_2 are the real bending moments in the horizontal and vertical directions, $\overline{M}_1$ and $\overline{M}_2$ are the corresponding virtual moments, $[\mathbf{F}]$ is the specific flexibility matrix, $c = \gamma$ is the specific weight and E is Young's modulus for the beam material. We consider a single deflection constraint which limits the sum of the horizontal and vertical displacements

$$\Delta \geq \int_D \left(\frac{M_1 \overline{M}_1}{rz_1^3 z_2} + \frac{M_2 \overline{M}_2}{rz_1 z_2^3} \right) dx, \tag{111}$$

and the cross-sectional dimensions are constrained from below

$$z_1 \geq z_{1a}, \quad z_2 \geq z_{2a}. \tag{112}$$

For the case when both conditions in (112) are satisfied as strict inequalities, the optimality conditions can be derived directly from general formulae [Rozvany 1989, (193) on p. 70 or Fig. 11 of Chapter 3] which give

$$cz_2 - \nu \left(\frac{3M_1 \overline{M}_1}{rz_1^4 z_2} + \frac{M_2 \overline{M}_2}{rz_1^2 z_2^3} \right) = 0, \tag{113}$$

$$cz_1 - \nu \left(\frac{M_1 \overline{M}_1}{rz_1^3 z_2^2} + \frac{3M_2 \overline{M}_2}{rz_1 z_2^4} \right) = 0. \tag{114}$$

The same result can be obtained from a variational formulation of the particular class of problems:

$$\min \Phi = \int_0^L \left[cz_1 z_2 + \nu \left(\frac{M_1 \overline{M}_1}{rz_1^3 z_2} + \frac{M_2 \overline{M}_2}{rz_1 z_2^3} \right) + +\lambda_1(-z_1 + z_{1a} + w_1) + \right.$$

$$+ \lambda_2(-z_2 + z_{2a} + w_2) + \overline{u}_1(M_1'' + p_1) + \overline{u}_2(M_2'' + p_2) +$$

$$\left. +u_1(\overline{M}_1'' + \overline{p}_1) + u_2(\overline{M}_2'' + \overline{p}_2)\right] dx - \nu(\Delta - \eta), \tag{115}$$

where most symbols have a similar meaning as in (92) in Chapter 4, except that the subscripts 1 and 2 refer to parameters associated with the vertical and horizontal directions. Then the Euler-Lagrange equations for variation of z_1 and z_2 become the ones in (113) and (114), but with the extra term $-\lambda_1$ and $-\lambda_2$, respectively, in the LHSs. The above equations imply after the transformations $\nu/(cr) \to \nu$, $\lambda_i/c \to \lambda_i$:

$$z_1^4 z_2^4 = \nu(3M_1\overline{M}_1 z_2^2 + M_2\overline{M}_2 z_1^2) + z_1^4 z_2^3 \lambda_1,$$

$$z_1^4 z_2^4 = \nu(M_1\overline{M}_1 z_2^2 + 3M_2\overline{M}_2 z_1^2) + z_1^3 z_2^4 \lambda_2. \tag{116}$$

Then we have the following possibilities:

Case A: $z_i > z_{1a}$, $z_2 > z_{2a}$. Then the variation w_1 and w_2 (with $w_1 > 0$, $w_2 > 0$) implies $\lambda_1 = \lambda_2 = 0$ and by (116) we have

$$z_1^2 z_2^4 = 4\nu M_2\overline{M}_2, \quad z_1^4 z_2^2 = 4\nu M_1\overline{M}_1, \tag{117}$$

or

$$z_1^6 = 4\nu\frac{(M_1\overline{M}_1)^2}{M_2\overline{M}_2}, \quad z_2^6 = 4\nu\frac{(M_2\overline{M}_2)^2}{M_1\overline{M}_1}. \tag{118}$$

Since the LHSs of (117) are always positive, these equations cannot have a real root if one of the moment products in the RHS is negative. Hence the condition for the validity of Case A and (118) becomes

$$M_2\overline{M}_2 > 0, \quad M_1\overline{M}_1 > 0, \tag{119}$$

$$z_{1a}^6 < 4\nu\frac{(M_1\overline{M}_1)^2}{M_2\overline{M}_2}, \quad z_{2a}^6 < \frac{(M_2\overline{M}_2)^2}{M_1\overline{M}_1}. \tag{120}$$

Cases B and C: $z_i = z_{ia}$, $z_k > z_{ka}$ ($i = 1$, $k = 2$ and $i = 2$, $k = 1$, respectively).
   Then variation of w_1 and w_2 (with $w_i = 0$, $w_k > 0$) implies $\lambda_i \geq 0$, $\lambda_k = 0$ and by (116) we have

$$z_{ia}^4 z_k^4 = \nu(M_i\overline{M}_i z_k^2 + 3M_k\overline{M}_k z_{ia}^2), \tag{121}$$

$$z_k^2 = \frac{\nu M_i\overline{M}_i}{2z_{ia}^4} + \sqrt{\left(\frac{\nu M_i\overline{M}_i}{2z_{ia}^4}\right)^2 + \frac{3\nu M_k\overline{M}_k}{z_{ia}^2}}, \tag{122}$$

since a negative sign before the square root would give a negative value for z_k^2 if $M_k\overline{M}_k > 0$. Moreover, (116) furnishes the following conditions for the validity of Case B or C and (122):

$$z_{ia}^4 z_k^4 > \nu(3M_i\overline{M}_i z_k^2 + M_k\overline{M}_k z_{ia}^2), \quad M_k\overline{M}_k > 0. \tag{123}$$

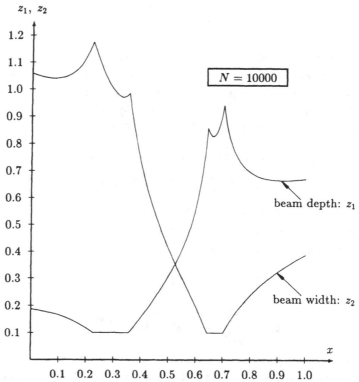

Fig. 31   A beam problem with nonseparable specific cost function: optimal depth z_1 and width z_2 variations obtained by the COC procedure with 10000 elements.

Case D: $z_1 = z_{1a}$, $z_2 = z_{2a}$. Since for this case $\omega_1 = \omega_2 = 0$ and $\lambda_1 \geq 0$, $\lambda_2 \geq 0$, we have by (116)

$$z_{1a}^4 z_{2a}^4 \geq \nu(3M_2\overline{M}_2 z_{1a}^2 + M_1\overline{M}_1 z_{2a}^2), \quad z_{1a}^4 z_{2a}^4 \geq \nu(M_2\overline{M}_2 z_{1a}^2 + 3M_1\overline{M}_1 z_{2a}^2). \quad (124)$$

The calculation of the Lagrangian from the deflection condition in (113) and (114) requires a Newton-Raphson procedure which is described in a research report (Zhou, 1990).

2.2 Solutions by the Iterative COC and SQP Methods

Considering a clamped beam having a span of 2.0 and a uniformly distributed load of unit intensity in the vertical direction and a central unit point load in the horizontal direction (see insert in Fig. 32), a deflection value of $\Delta = 3$ was adopted in (111) and $c = r = 1$ in (115).

Using the COC method with only 50 elements and 100 variables (for a comparison with the SQP method), and a tolerance criterion of $(\Phi_{new} - \Phi_{old})/\Phi_{new} \leq 10^{-8}$, a

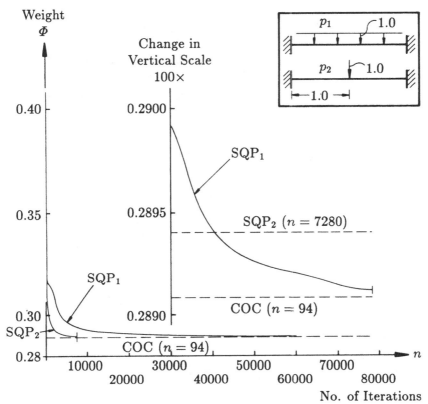

Fig. 32   A beam problem with nonseparable specific cost function: a comparison of weights
Φ obtained for 50 elements by unmodified and modified sequential quadratic methods (SQP$_1$ and SQP$_2$) after various iteration numbers and by the COC procedure
after convergence.

weight value of $\Phi = 0.289081$ was obtained after 94 iterations, with a total CPU
time (analyses plus optimization) of 57 sec. In repeating the COC procedure with
10000 elements and 20000 variables, the above convergence criterion was reached
after 103 iterations with a CPU time of 12096 sec and a weight of $\Phi = 0.288782$. The
corresponding optimal variations of the width and depth are shown in Fig. 31.

The same problem with 50 elements and 100 variables was also computed using
the SQP method with reciprocal variables. The above convergence criterion was
satisfied after 77564 iterations, requiring a CPU time of 1507451 sec (17.45 days).
In Fig. 32, the iteration history of this calculation is compared (see SQP$_1$) with the
result given by the COC method after convergence.

Since the convergence was extremely slow, a modified procedure was used in which
the coupling effect was relaxed through multiplying the mixed second derivatives by
various factors. After testing several such factors, a multiplier of 0.2 was adopted

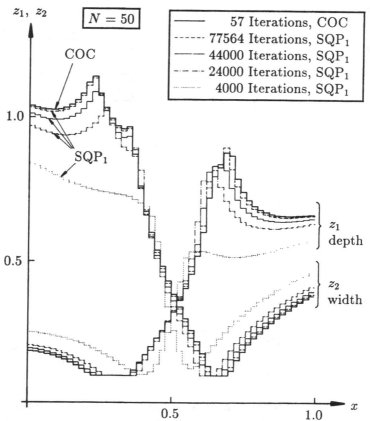

Fig. 33 A comparison of width and depth distributions obtained by the COC and unmodified SQP methods.

which resulted in a much faster convergence (see SQP_2 in Fig. 32). For details of the modified SQP procedure used, refer to a research report (Zhou, 1990).

The SQP_2 procedure reached the convergence criterion after 7280 iterations, requiring a CPU time of "only" 176746 sec (2.05 days). However, due to small instabilities in the convergence, satisfaction of the convergence criterion was somewhat accidental giving a fairly large error in comparison to the COC and SQP_1 results (see the enlarged right-hand part of Fig. 32). The results of the COC method, as well as the unmodified and modified SQP methods are compared in Table 2. The width and depth distributions given by the two SQP methods after various numbers of iterations are compared with those obtained by the COC method in Figs. 33 and 34. It can be concluded from the above diagrams and Table 2 that both the unmodified and modified SQP methods fully confirm the results by the COC algorithm. However, the more accurate SQP method (SQP_1) requires almost 30000 times as much time as the COC procedure.

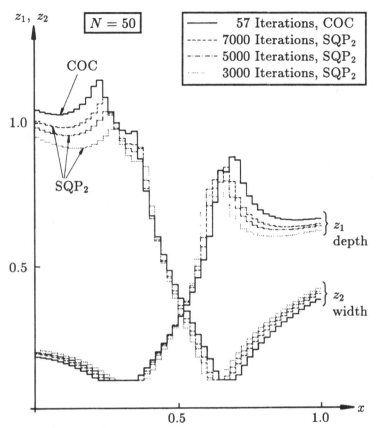

Fig. 34 A comparison of width and depth distributions obtained by the COC and modified SQP methods.

Table 2. A comparison of weight values obtained by the SQP method after various numbers of iterations and by the COC method after convergence ($= 94$ iterations) and percentage differences.

	Weight Φ	$\Delta\%$	No. of Iterations	CPU time sec
COC	0.289081	0	94	57
SQP$_1$	0.289114	0.011	77564	1507451 (17.45 days)
SQP$_2$	0.289408	0.113	7280	176746 (2.05 days)

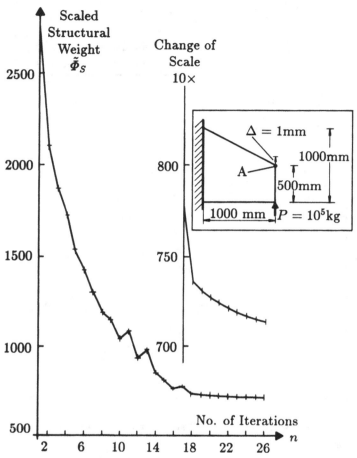

Fig. 35 Scaled structural weight Φ_s as a function of the iteration number for a plate of variable thickness subject to plane stress: optimization by the COC procedure with 1250 elements for a deflection constraint.

3 Two-Dimensional Structural Systems

The proposed COC algorithm can be readily extended to two-dimensional systems (e.g. plates in plane stress or bending, and shells). In the case of a plate of variable thickness z in plane stress with a deflection constraint, we have

$$\mathbf{Q} = (N_x, N_y, N_{xy}), \quad \overline{\mathbf{Q}} = (\overline{N}_x, \overline{N}_y, \overline{N}_{xy}), \quad [\mathbf{F}] = \frac{1}{z}\begin{bmatrix} f_1 & -f_2 & 0 \\ -f_2 & f_1 & 0 \\ 0 & 0 & f_3 \end{bmatrix},$$

$$\psi = cz, \quad f_1 = \frac{1}{E}, \quad f_2 = \frac{v}{E}, \quad f_3 = \frac{1}{G}, \quad c = \gamma, \tag{125}$$

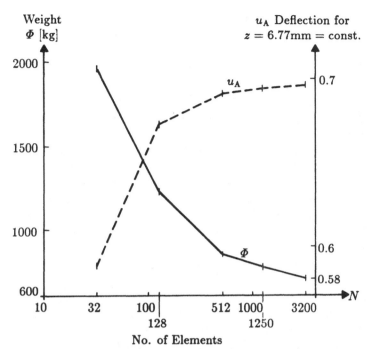

Fig. 36   Variation of the optimal structural weight Φ and of the deflection calculated for a constant thickness as a function of the number of finite elements used.

where N_x and N_y are the axial forces and N_{xy} is the shear force per unit width in the x and y directions, γ is the specific weight, E is Young's modulus, G is the modulus of rigidity and v is Poisson's ratio. Then the general formulae [Rozvany, 1989, (1.93) on p. 70 or Fig. 11 in Chapter 3] imply the optimality criterion:

$$c = (v/z^2)[f_1(N_x\overline{N}_x + N_y\overline{N}_y) - f_2(N_x\overline{N}_y + \overline{N}_x N_y) + f_3 N_{xy}\overline{N}_{xy}],$$

$$z\sqrt{c/v} = \sqrt{[f_1 N_x\overline{N}_x + N_y\overline{N}_y) - f_2(N_x\overline{N}_y + \overline{N}_x N_y) + f_3 N_{xy}\overline{N}_{xy}]}. \qquad (126)$$

For the calculation of the Lagrangian, (70) in Chapter 3 is replaced by

$$v^{1/2} = \sum_A W_i^{1/2} A_i / \left(\sum_P \frac{W_i A_i}{z_a} - \Delta\right), \qquad (127)$$

where W_i is the expression in square brackets in (126), A_i is the area of element i and z_a is the prescribed minimum thickness.

The above optimality criterion was used for a modified version of a test problem suggested to the first author by Dr. G. Kneppe of MBB GmbH (FRG). In the original test problem, deflection and stress conditions were used, but the number of elements was much smaller. For the present test problem, a single deflection constraint was

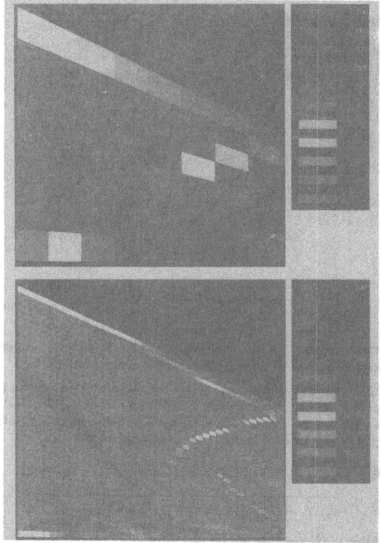

Fig. 37   Optimal distribution of the thickness in mm for a plate in plane stress with a
deflection constraint: (a) 128 elements, (b) 3200 elements.

considered but the above results are being extended to a combination of several
deflection constraints and a permissible stress constraint.

The problem to be investigated is shown in the insert of Fig. 35, with a point load
at the bottom right corner and a prescribed deflection of 1 mm at the top right corner.
For the minimum plate thickness a very low value ($z_a = 0.1$ mm) was prescribed.
The material properties for mild steel have been adopted ($\gamma = 0.785 \times 10^{-4}$ N/mm^3,

$E = 2.1 \times 10^5 \text{ N/mm}^2, v = 0.3).$

Using the iterative COC method, the above problem was solved with discretizations having 32, 128, 512, 1250 and 3200 triangular (constant strain) elements. The thickness of two adjacent elements forming a quadrilateral was linked. The convergence was found not to be quite as smooth as for the beam problems discussed but, apart from small initial fluctuations, it was near-monotonic (see Fig. 35 which was plotted for the system with 1250 elements).

It can be seen from Fig. 36 that, with the type of simple elements used, a large number of elements are required for reasonable accuracy of the deflection. The polygon in broken line gives the calculated deflection values for the same load and same uniform thickness but for systems with different numbers of elements. For 32 elements the computational error compared to the most accurate result (3200 elements) is 15.56% and for 1250 elements it is still 0.26%.

It can also be observed in Fig. 36 that the optimal weight keeps on decreasing with an increase in elements. This can be explained by referring to the optimal thickness distributions in Fig. 37, in which the optimal plate develops rib-like formations. A similar phenomenon was discovered, in the context of solid plates in bending, by Cheng and Olhoff (1981) in numerical solutions and confirmed analytically by Rozvany, Olhoff et al. (1982). Such truss-like solutions were obtained for plate problems more recently by Bendsøe (1989, Fig. 6) who studied homogenization methods.

Table 3. A comparison of computer times for various numbers (N) of elements in the plane stress problem, using SQP and COC methods.

N	CPU time: analysis (sec)		CPU time: optimization (sec)	
	SQP	COC	SQP	COC
32	15.46	14.96	6.52	0.17
128	101.40	88.14	282.46	0.63

In view of the fact that the optimal solution clearly tends to a truss with "concentrated" members, it is understandable that at a coarser finite element net the design is far from the absolute optimal solution (for a continuum), because the material is spread out uniformly along large elements, and thus highly concentrated "ribs" cannot occur in the solution. This can be seen in Fig. 37a in which 128 elements were used. In the solution in Fig. 37b (with 3200 elements) the extent of material concentration along lines is much greater.

For FE nets with 32 and 128 elements, the results were verified independently by using a SQP method. Using a tolerance criterion $|\Phi_{new} - \Phi_{old}|/\Phi_{new} \leq 10^{-8}$, an *eight significant digits* agreement was found between the weight values by SQP and COC

methods (see Table 3). Whereas the difference of analysis time is insignificant, the
time used for optimization becomes increasingly greater with the number of elements
(N) for the SQP method. In both cases, an FE program developed by M. Zhou was
used, but the results were also checked by using ANSYS. M. Zhou's program uses
a variable band width technique with $LD^{-1}L^T$ type decomposition of the stiffness
matrix processed directly in the CPU.

4 Structures with Deflection and Stress Constraints

For a combination of stress and deflection constraints the general form of the optimality criterion [Rozvany, 1989, (1.58b) on p. 62] becomes

$$\mathcal{G}_{,\ast}[\psi(z)] + \sum_{\ell} \lambda_{\ell}(x)\,\mathcal{G}_{,\ast}[S_{\ell}(z,\mathbf{Q})] + \sum_{j} \nu_{j}\overline{\mathbf{Q}}_{j}\{\mathcal{G}_{,\ast}\cdot[\mathbf{F}]\}\mathbf{Q} = 0, \qquad (128)$$

where $\mathcal{G}$ is the G-gradient introduced in Chapter 2 (Fig. 7), $S_{\ell}(z,\mathbf{Q}) = 0$ ($\ell = 1,2,\dots,t$) are stress (strength) constraints, $\int_{D}\overline{\mathbf{Q}}_{j}\cdot[\mathbf{F}]\mathbf{Q}\,dx \leq \Delta_{j}$ the deflection
constraints, $\overline{\mathbf{Q}}_{j}$ the virtual generalized stresses associated with the j-th deflection
constraint, λ_{ℓ} variable Lagrangian multipliers, ν_{j} constant Lagrangian multipliers
and other symbols have been defined earlier. The adjoint generalized strains $\overline{q}$ are
given by [Rozvany, 1989, (1.58a) on p. 62]

$$\overline{q} = \sum_{\ell} \lambda_{\ell}(x)\,\mathcal{G}_{,\mathbf{Q}}[S_{\ell}(z,\mathbf{Q})] + \sum_{j} \nu_{j}[\mathbf{F}]\overline{\mathbf{Q}}_{j}. \qquad (129)$$

For a beam of variable width with a deflection constraint, a flexural stress and a
shear stress constraint (taking the shear deformations into consideration), we have

$$\mathbf{Q} = (M,V), \quad \overline{\mathbf{Q}} = (\overline{M},\overline{V}), \quad [\mathbf{F}] = \begin{bmatrix} f_1 & 0 \\ 0 & f_2 \end{bmatrix}, \quad S_1 = (k_1|M| - z),$$

$$S_2 = (k_2|V| - z), \quad \psi = cz, \quad f_1 = 1/(r_1 z), f_2 = 1/(r_2 z), \quad r_1 = Eh^3/12,$$

$$r_2 = 5hG/6, \qquad (130)$$

where M and V are the bending moment and the shear force on the real beam, whilst
$\overline{M}$ and $\overline{V}$ are the corresponding adjoint stress resultants, E is Young's modulus, h is
the beam depth, f_2V is the generalized shear strain (shear strain for an entire beam
cross-section), G is the modulus of rigidity, $k_1 = 6/(\sigma_p h^2)$ where σ_p is the permissible
flexural stress, $k_2 = 1.5/(\tau_p h)$ where τ_p is the permissible shear stress, z is the beam
width, S_1 and S_2 are the two permissible stress conditions and $c = \gamma h$ where γ is the
specific weight.

In the considered problem, the real and adjoint generalized strain vectors are

$$q = (\kappa, \varsigma), \quad \overline{q} = (\overline{\kappa}, \overline{\varsigma}), \tag{131}$$

where κ is the beam curvature and ς is the generalized shear strain. Then the adjoint displacement $\overline{u}$, for example, is given by

$$\overline{u} = -\int_0^x \int_0^x \overline{\kappa}(x)\, dx\, dx - \int_0^x \overline{\varsigma}(x)\, dx. \tag{132}$$

On the basis of (129) and (130), the adjoint strains become (for $M \neq 0$, $V \neq 0$)

$$\overline{\kappa} = -\lambda_1 k_1 \operatorname{sgn} M + \nu \overline{M}/(r_1 z), \tag{133}$$

$$\overline{\varsigma} = -\lambda_2 k_2 \operatorname{sgn} V + \nu \overline{V}/(r_2 z). \tag{134}$$

Notes. (1) In the case of deflection constraints only, the adjoint strains and displacements depended only on the adjoint generalized stresses (e.g. adjoint moments). For deflection and stress constraints, however, *the adjoint strains and displacements depend on both real and adjoint generalized stresses* which makes their calculation more difficult.

(2) For $M = 0$ or $V = 0$, the extended meaning of the G-gradient (Chapter 2, Fig. 7) gives adjoint strains which are different from those in (133) and (134).

The relations (128) and (130) imply the following optimality criterion:

$$c - \nu\left(\frac{M\overline{M}}{r_1 z^2} + \frac{V\overline{V}}{r_2 z^2}\right) + \lambda_1 + \lambda_2 = 0. \tag{135}$$

Check by variational calculus. With $V = -M'$, the above beam problem can be formulated as

$$\min \Phi = \int_D \left[cz + \nu\left(\frac{M\overline{M}}{r_1 z} + \frac{M'\overline{M}'}{r_2 z}\right) + \lambda_1(-k_1|M| + z + \omega_1) + \right.$$

$$\left. + \lambda_2(-k_2|M'| + z + \omega_2) + \overline{u}(M'' + p) + u(\overline{M}'' + \overline{p}) \right] dx - \nu(\Delta - \eta), \tag{136}$$

where ω_1 and ω_2 are slack functions, Δ is the prescribed maximum deflection value and η is a slack variable.

Then for $M \neq 0$, $V \neq 0$ the Euler-Lagrange equation for variation of z yields (135) and for the variation of M we have

$$-\overline{u}'' = -\lambda_1 k_1 \operatorname{sgn} M - (\lambda_2)' k_2 \operatorname{sgn}(M') + \frac{\nu \overline{M}}{r_1 z} + \left(\frac{\nu \overline{M}'}{r_2 z}\right)', \tag{137}$$

which is implied by (132), (133) and (134).

4.1 Optimality Criteria for Beams of Variable Width with a Deflection Constraint and Flexural Stress Constraints

Neglecting the shear terms in (135) and (137), i.e. with $k_2 \to 0$, $r_2 \to \infty$, we have

$$\lambda_1 = \frac{\nu M \overline{M}}{r_1 z^2} - c, \tag{138}$$

$$-\overline{u}'' = -\lambda_1 k_1 \operatorname{sgn} M + \frac{\nu \overline{M}}{r_1 z}. \tag{139}$$

Then there are two possibilities:

Case A: Stress constraint is active, $k_1|M| = z$.

$$\omega_1 = 0, \quad \lambda_1 \geq 0, \quad (138)(139) \Rightarrow \overline{\kappa} = -\overline{u}'' = -\frac{\nu M \overline{M} \operatorname{sgn} M}{r_1 |M| z} + c k_1 \operatorname{sgn} M +$$

$$+\frac{\nu \overline{M}}{r_1 z} = c k_1 \operatorname{sgn} M. \tag{140}$$

Case B: Stress constraint is inactive $k_1|M| < z$.

$$\omega_1 > 0, \quad \lambda_1 = 0, \quad \overline{\kappa} = -\overline{u}'' = \frac{\nu \overline{M}}{r_1 z}, z = \sqrt{\frac{\nu M \overline{M}}{c r_1}}. \tag{141}$$

Note: In the FE analysis of the adjoint system, a constant curvature given by (140) must be used for stress-controlled regions which can be done by specifying an initial curvature or curvature caused by nonuniform temperature distribution in the cross-section. Since the adjoint moment causes no curvature in stress-controlled regions, for the adjoint analysis a very large flexural stiffness value is adopted for such regions. If a minimum width z_a (with a flexural stiffness of $r_1 z_a$) is specified and the stress constraint is not active, then the adjoint curvature becomes $\nu \overline{M}/(r_1 z_a)$.

4.2 Iterative COC Results for Beams of Variable Width with a Deflection Constraint and Stress Constraints

Using the optimality criteria given in Section 4.1 with $c = 1$, $r_1 = 1$ for a clamped beam with a uniformly distributed load and various values of the stress cost factor k_1, the solutions shown in Fig. 38 were obtained with 10000 elements. The regions R_d are controlled by the deflection constraint, the regions R_ℓ by the minimum width

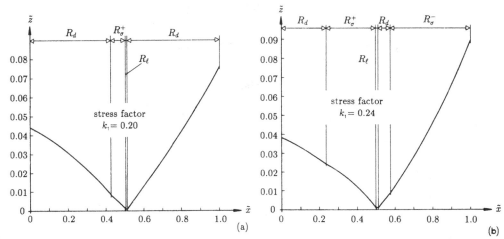

Fig. 38 Optimal variation of the width for a clamped, uniformly loaded beam with a deflection constraint and stress constraints; the value of the stress cost factor is (a) $k_1 = 0.20$, and (b) $k_1 = 0.24$.

(= lower limit) constraint and the regions R_σ^+ and R_σ^- by the stress constraint with a positive and a negative moment, respectively.

In calculating the Lagrangian, the same equation (70) in Chapter 3 can be used as for a deflection constraint without stress constraint, but the passive element set includes all elements in the regions R_ℓ, R_σ^+ and R_σ^-.

The above results were verified by the semi-analytical method described in Chapter 4 (Section 2) to a seven-digit accuracy (Rozvany, Zhou *et al.*, 1990).

4.3 Iterative COC Results for Laminated Timber Beams with a Constraint on the Maximum Deflection and Limits on Both Flexural and Shear Stresses (in Accordance with the German Design Code DIN 1052)

The optimality criteria derived in Section 4 were used for the optimization of long-span laminated timber beams of constant width and variable depth with the following deflection and stress constraints in accordance with DIN 1052:

$$\sigma_p = 11 \, \text{N/mm}, \quad \tau_p = 1.2 \, \text{N/mm}, \quad \Delta = L/300, \quad E = 11000 \, \text{N/mm}^2, \quad (142)$$

where σ_p is the permissible flexural stress, τ_p the permissible shear stress, Δ the permissible deflection, L the beam span and E Young's modulus. The effect of the shear deformations on the above deflection were neglected. An example concerning a proposed cantilever with a uniformly distributed load of $p = 0.01$ MN/m and a span of 20 m is given in Fig. 39, in which the given width of the beam was $b = 0.15$ m. This problem was solved by the COC procedure using up to 100000 elements. As the location of the maximum deflection is not known, this was always based on the

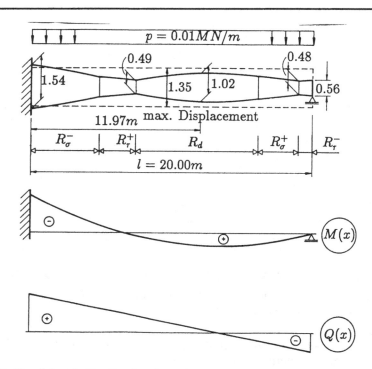

Fig. 39   Optimal depth distribution for a laminated timber beam designed in accordance
with the German Design Code DIN 1052, having constraints on the maximum
deflection, flexural stresses and shear stresses.

analysis of the real beam in the prior iteration. This procedure has resulted in a very
satisfactory convergence, although the equivalent problem with an MP formulation
would have 100000 deflection constraints (i.e. a prescribed deflection at each node),
out of which only one is active.

The optimal variation of the depth for the above problem is shown in Fig. 39,
in which the corresponding moment and shear force diagrams are also indicated.
Regions controlled by the deflection, flexural stress and shear stress constraints, respectively, are denoted by R_d, R_σ and R_τ, and the signs in the superscripts refer
to the sign of the corresponding bending moment or shear force. Figure 39 (broken
lines) also shows the prismatic beam satisfying the same design constraints. The
latter has a 58% higher volume than the optimal one.

The same method was also extended by the authors to frames with additional
axial forces on the cross-sections (Rozvany and Zhou, 1991).

The same procedure was extended to beams with selfweight and external load.
For computational details of this section the reader is referred to a research report
(Zhou, 1990).

5 COC Algorithm: Conclusions

The following conclusions can be drawn from the results reported in Chapters 3-5:

○ Whilst *mathematical programming* (MP) methods have the *advantage* that they are *"robust"*, that is, they can be readily applied to most optimization problems within and outside the discipline of structural optimization, *continuum-type optimality criteria* (COC) methods require *much less storage capacity and CPU time* and *increase the optimization capability* from a few hundred to millions of variables. Owing to the limited optimization capability of MP methods, a direct comparison was carried out only for relatively few variables. In Fig. 22 in Chapter 4 the CPU times for optimization were compared for the sequential quadratic programming (SQP) method and the COC method for 50, 100 and 200 variables; in the last case, SQP requires approx. 3000 times more CPU time. As can be seen from Fig. 22 in Chapter 4, the ratio of the CPU times for the SQP and COC methods increases very rapidly with the number of variables considered.

○ It follows that the COC procedure not only eliminates but also *reverses the discrepancy* that presently exists *between analysis capability and optimization capability* and provides sufficient optimization capability to counter further increases in analysis capability in the foreseeable future.

○ Although COC methods have the *disadvantage* that they require the *analytical derivation of the appropriate optimality criteria* for various types of structures, cross-sectional topologies and design requirements, *general formulae* are available for the latter (Rozvany, 1989) and the development of suitable software is planned for the automatic generation of optimality conditions for specified design problems. One should note, however, that efficient MP methods also require the analytical or semi-analytical derivation of the sensitivities involved.

○ In contrast to discrete optimality criteria (DOC) methods, iterative COC procedures are based on *exact equations representing optimal continua*, which are then reinterpreted in terms of an *"adjoint structure"* and solved in a discretized form using finite element (FE) software. For the above reason, the COC procedure requires a *more specific treatment* of a given problem *than DOC methods* in which optimality criteria are expressed in terms of nodal forces and hence the same formulae can be applied for a wide range of structures.

○ The power and versatility of the iterative COC method was demonstrated by applying it to problems with *various geometrical constraints and design requirements* including lower and upper limits on cross-sectional dimensions, segment-wise constant cross-sections, allowance for the cost of supports and for selfweight, as well as nonlinear and nonseparable specific cost and stiffness functions. The particular advantage of the COC algorithm (also over DOC methods) is the simple handling of *any combination of local and global constraints* (e.g. deflection and stress constraints, see Section 4 of Chapter 5, in which the results are also verified by a semi-analytical method).

○ For a beam problem involving a *nonseparable objective function* with 50 elements and 100 variables, the COC method required a CPU time of 57 sec to satisfy the convergence criterion. For the same problem and convergence criterion, even an improved SQP method needed a CPU time of 176746 sec.

○ The relatively high number of iterations for convergence was due to the *unusually stringent convergence condition* ($|\Phi_{new} - \Phi_{old}|/\Phi_{new} \leq 10^{-8}$ where Φ is the total weight) employed. In practical problems, such an accuracy is not necessary but in the above test problems it was required for a comparison with analytical solutions. In most problems, a *practical optimal design* (with an error well below one per cent) was achieved *after two iterations*.

○ The results of this paper *refute the popular misconceptions* that OC methods (a) result in instabilities in later iterations and hence (b) they provide an approximate solution only. The COC algorithm presented showed an *almost uniform convergence* throughout the iteration histories investigated which resulted in *up to 12 significant digits accuracy* verified by a comparison with the analytical solution.

○ All designs obtained by the COC procedure were *verified by one or two independently derived solutions*, mostly by an analytical or semi-analytical method, but at least by another numerical method (SQP).

○ The COC method was also tested in a problem involving a *plate of variable thickness* subject to plane stress, using up to 3200 elements.

Chapter 6

OPTIMAL LAYOUT THEORY

G.I.N. Rozvany
Essen University, Essen, Germany

1 Introduction

As was indicated in Chapter 1, layout optimization is probably the most difficult class of problems in structural optimization. On the other hand, it is also a very important one, because it results in much higher material savings than cross-section optimization.

To illustrate this point on a simple example, we consider the problem in Fig. 40a in which four point loads P are required to be transmitted by beams of constant depth to simple supports along the boundary of a square domain. In the first solution shown (Fig. 40b), the total "moment area" (area of the moment diagrams), which is a measure of structural weight, is $6Pa^2$. The optimal beam layout in Fig. 40c gives a total moment area of $4Pa^2$. The weight difference of 50% is much greater than the usual savings achieved by cross-section optimization.

An even bigger difference between the weight of optimal and nonoptimal layouts is found for the boundary and loading condition in Fig. 41a, in which the vertical point load P is to be transmitted to two simply supported edges (double lines) of a rectangular domain. The other two edges are unsupported. Fig. 41b shows the optimal beam layout and the corresponding moment diagrams with a total moment area of $8Pa^2$. The nonoptimal layout shown in Fig. 41c has a moment area of $17Pa^2$ which exceeds the optimal one by 112.5%. It will be seen in Chapter 8 that for long span grillages, for which the selfweight is a significant load condition, this difference between optimal and nonoptimal structural weights is often over 1000%.

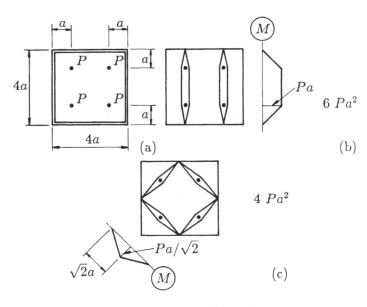

Fig. 40   Comparison of optimal and nonoptimal beam layouts.

The most systematic *analytical* method for layout optimization is the so-called *layout theory*, developed by Prager and the author in the late seventies and extended by the research teams of the latter in the eighties. This theory is a generalization of an approach used around the turn of the century by Michell (1904) (see Chapter 2, Section 2), who extended some ideas by Maxwell (1872). Layout theory is based on two underlying concepts, namely,

(a) the *structural universe*, which is the union of all feasible or potential members, and

(b) *continuum-based* or *static-kinematic optimality criteria*, which are mathematical conditions for the optimality of a structure, reinterpreted in terms of a fictitious *adjoint structure* (see Chapters 2-5).

The adjoint strain-stress relations give a strain requirement, usually in the form of an inequality, also along *vanishing members*, having a zero cross-sectional area and zero generalized stress. This means that in *convex* problems, for which the optimality criteria represent sufficient conditions, their fulfilment for the entire structural universe also represents a *sufficient condition* of optimality *for the structural layout*. The application of the above layout theory will be illustrated with some simple examples.

As explained lucidly by Kirsch (1989), layout optimization problems are solved by the *numerical school* in two stages. First, *topological optimization* is carried out by assuming a "highly connected" *ground structure* (in the author's terminology: structural universe) and then removing non-optimal members. The second stage consists of *geometrical optimization*, in which the topology is assumed to be fixed

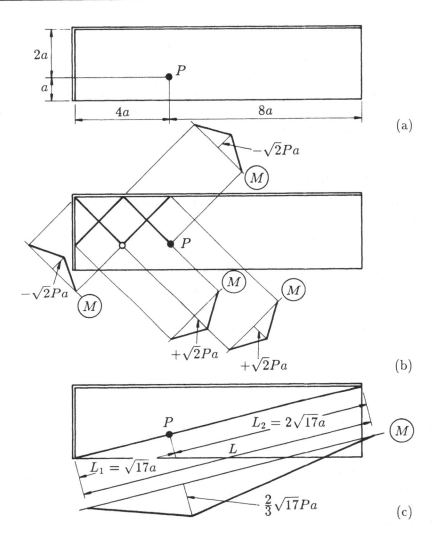

Fig. 41 Another comparison of optimal and nonoptimal beam layouts.

(unless some joints coalesce during the solution process) and the design variables are the coordinates of the joints and the cross-sectional areas. The above design variables are usually optimized by mathematical programming methods.

Topological optimization in the past, even with simplifying assumptions and approximations, was used only for a small number of potential members because of the limited optimization capability of the mathematical programming methods used. Kirsch (1989), for example, uses only 8 potential members in one half of a symmetric truss. Due to the introduction of iterative COC methods, to be described in the next chapter, it has become possible to investigate structural universes with *many thousand potential members.*

Earlier formulations of the layout theory for structural design were introduced by Prager and the author (e.g. Prager, 1974; Rozvany, 1976; Prager and Rozvany, 1977a) and more up-to-date reviews of this field were offered by the author in principal lectures of NATO ASI's (Rozvany, 1981b; Rozvany and Ong, 1987; Rozvany, Gollub and Zhou, 1989b; Rozvany, Zhou and Gollub, 1991), in a book chapter (Rozvany, 1984) and, in particular, in a recent book (Rozvany, 1989).

2 Optimality Criteria Used in This Chapter

The general ideas of layout theory will be explained through examples involving *two types of structures:*

○ beams and beam systems having rectangular cross-sections of given depth and variable width, and

○ trusses or pin-jointed frames.

The two types of *design conditions* to be considered here will be

○ optimal plastic design for a given ultimate load, and

○ optimal elastic design for a given compliance (sum of product of external forces and the corresponding deflections).

2.1 Optimal Plastic Design

For *beams of given depth* [see Chapter 2, Fig. 8 and (21)] and for *trusses*, the specific cost function takes the form

$$\psi = k|Q|, \tag{143}$$

where ψ is the cross-sectional area (volume minimization) or member weight per unit length (weight minimization), Q is a generalized stress (axial force for trusses and bending moment for beams) and k is a given constant. Then the Prager-Shield condition [(14) and Figs. 8b and c in Chapter 2] implies:

$$(\text{for } Q \neq 0) \quad \bar{q} = k \operatorname{sgn} Q, \tag{144}$$

$$(\text{for } Q = 0) \quad |\bar{q}| \leq k, \tag{145}$$

where for trusses $\bar{q} \to \bar{\epsilon}$ is the adjoint axial strain in the members and for beams $\bar{q} \to \bar{\kappa}$ is the curvature $\bar{\kappa} = -\bar{u}''$ of the adjoint deflection $\bar{u}$. For trusses and beams of given depth, respectively, (144) and (145) reduce to Michell's optimality criteria [(1) and (2) in Chapter 2] and Heyman's optimality criteria [(4) and (5) in Chapter 2].

2.2 Optimal Elastic Design for Given Compliance

For the considered class of problems, we have

$$\psi = cz, \quad [\mathbf{F}] = 1/rz, \quad p = \bar{p}, \quad Q = \bar{Q}, \tag{146}$$

where c and r are given constants and z is a cross-sectional variable. Hence the optimality criterion in Fig. 11 (Chapter 3) implies

$$c = \frac{\nu Q^2}{rz^2} = 0 \rightarrow z = \sqrt{\nu/rc}|Q|. \tag{147}$$

If a minimum value for z is prescribed ($z \geq z_a$), then we have [cf. (71) in Chapter 3]:

$$(\text{for } \sqrt{\nu/rc}|Q| \geq z_a) \qquad z = \sqrt{\nu/rc}|Q|, \tag{148}$$
$$(\text{for } \sqrt{\nu/rc}|Q| < z_a) \qquad z = z_a. \tag{149}$$

Since in this class of problems $\bar{q} = \bar{Q}/rz = Q/rz$, (148) and (149) imply

$$(\text{for } \sqrt{\nu/rc}|Q| \geq z_a) \qquad \bar{q} = \sqrt{c/r\nu}\, \mathrm{sgn}\,|Q|, \tag{150}$$
$$(\text{for } \sqrt{\nu/rc}|Q| < z_a) \qquad |\bar{q}| = |Q|/(rz_a) < \sqrt{c/r\nu}. \tag{151}$$

After a limiting process with $z_a \rightarrow 0$ and replacing $\sqrt{c/r\nu}$ with k, we can see that (150) and (151) reduce to (144) and (145). This means that, within a constant multiplier, *the optimality criteria for optimal plastic design and optimal elastic design for compliance are identical* for both trusses and beams of given depth. This confirms earlier observations regarding Michell frames by Hegemier and Prager (1969). These authors have also shown that the same solutions are valid for given natural frequency and for given stiffness in stationary creep.

The value of the multiplier ν can be determined from the work equation and (148) for the limiting case $z_a \rightarrow 0$ as

$$C = \int_D pu\,\mathrm{d}x = \int_D \frac{Q^2}{rz}\,\mathrm{d}x = \int_D \frac{Q^2}{r\sqrt{\nu/rc}|Q|}\,\mathrm{d}x = \int_D \frac{|Q|}{\sqrt{\nu r/c}}\,\mathrm{d}x, \tag{152}$$

furnishing

$$\sqrt{\nu} = \int_D |Q|\,\mathrm{d}x/(C\sqrt{r/c}). \tag{153}$$

Moreover, for the total cost (weight) Φ_{E} in the *elastic* design problem we have by (146), (148) and (153)

$$\Phi_{\mathrm{E}} = \int_D cz\,\mathrm{d}x = \int_D c\sqrt{\nu/rc}|Q|\,\mathrm{d}x = \left(\int_D |Q|\,\mathrm{d}x\right)^2 /C. \tag{154}$$

The same solution is valid, within a constant multiplier, for optimal *plastic* design. In that case, however, on the basis of (143) the total cost Φ_P becomes

$$\Phi_P = \int_D k|Q|\,dx\,. \tag{155}$$

For the normalized problems with $C = k = 1$, we have

$$\Phi_E = (\Phi_P)^2\,. \tag{155a}$$

3 Elementary Examples Illustrating Applications of the Layout Theory

3.1 Two-Span Beam with a Central Point Load Over One Span

This is a very simple layout problem in which the structural universe consists of two beam spans. Over one of the spans the beam cross-section may take on a zero area. Since the optimal deflection diagrams can be subjected to a linear transformation, the constant k in (144) and (145), as well as $\sqrt{c/r\nu}$ in (150) and (151), will be replaced by unity and then we have for $Q \to M$ and $\bar{q} \to \bar{\kappa}$ the optimal beam curvatures

$$\text{(for } |M| > 0)\,, \qquad \bar{\kappa} = \operatorname{sgn} M\,, \tag{156}$$
$$\text{(for } |M| = 0)\,, \qquad |\bar{\kappa}| \le 1\,. \tag{157}$$

Moreover, the optimal values of the width in the elastic problem are given by (148) as

$$z = \sqrt{\nu/cr}|M|\,. \tag{158}$$

It will be shown that the solution depends on the ratio of the loaded and unloaded spans. In Fig. 42a, this ratio is 1:2. Assuming that the beam takes on a zero cross-sectional area over the longer span, we have the moment diagram in Fig. 42c and the adjoint deflection diagram in Fig. 42b satisfies the curvature conditions in (156) and (157). Note that for the vanishing beam region over the longer span we still have an adjoint deflection[*] diagram which plays an important role in determining the optimal solution. By (158) the diagram in Fig. 42c also represents the variation of $z/\sqrt{\nu/rc}$. Prescribing a compliance value of $C = 1$ (corresponding to a unit deflection at the loaded point), the moment area A in Fig. 42c and (154) with $C = 1$, $Q \to M$ and (155) with $k = 1$ furnish the beam weights for the elastic and plastic design problems

$$\Phi_E = (1/8)^2 = 1/64 = 0.015625\,, \qquad \Phi_P = 1/8 = 0.125\,. \tag{159}$$

If we now change the ratio of the loaded and unloaded spans to 2:1 (Fig. 42d) then the type of moment diagrams in Fig. 42f would require by (156) the optimal deflection

---

[*] In Fig. 42, the normalized values of the real and adjoint deflections are the same for nonvanishing cross-sections and hence κ refers to both real and adjoint curvatures.

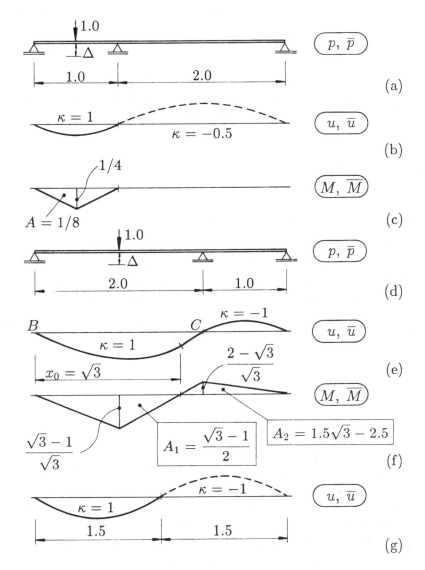

Fig. 42   A simple layout problem: two-span beam.

diagram in Fig. 42e. In the latter, the slope at the support C is $du/dx = 0.5$ and then the end condition $u_B = 0$ requires $2(0.5) + (2 - x_0)(2 + x_0)/2 - x_0^2/2 = 0$, giving $x_0 = \sqrt{3}$. The moment diagram in Fig. 42f, whose absolute value also represents $z/\sqrt{\nu/rc}$, can then be determined from simple statical considerations. By (154), the normalized beam weights become

$$\Phi_E = (A_1 + A_2)^2 = (2\sqrt{3} - 3)^2 = 0.215390309, \quad \Phi_P = 2\sqrt{3} - 3 = 0.464101615.$$
$$(160)$$

The above solutions will be verified by numerical (iterative COC) calculations in Chapter 7.

It is still interesting to know, at which span ratio the elementary topology in Figs. 42a-c changes to the topology in Figs. 42d-f.

Clearly, the topology represented by Figs. 42a-c is valid up to a span ratio of 1:1, at which the curvature in the unloaded span also reaches $\kappa = 1$. Beyond this span ratio, the condition in (157) does not allow a zero cross-sectional area in the unloaded span.

Note that all moment diagrams in this section had to be statically admissible and the corresponding curvatures kinematically admissible in order to fulfill the optimality condition in Fig. 11 in Chapter 3.

3.2 Least-Weight Pin-Jointed Frame for a Point Load Parallel to a Supporting Line

In the case of pin-jointed frames, the generalized stress is the axial force $Q = N$ and the generalized strain is the axial member strain $q = \epsilon$. Then the general optimality conditions (144) and (145) for optimal plastic design reduce to Michell's optimality criteria [(1) and (2) in Chapter 2] and for optimal elastic design with given compliance (150) and (151) give the same solution within a constant multiplier. In this, and the next two examples, therefore, we shall consider plastic design for simplicity. The relevant optimality criteria are repeated here in a normalized form (with $k = 1$) for convenience

$$\text{(for } |N| > 0) \quad \epsilon = \text{sgn } N\,, \tag{161}$$

$$\text{(for } |N| = 0) \quad |\epsilon| \leq 1\,. \tag{162}$$

Figure 43a shows the loading and a structural universe for the considered problem and Fig. 43b the optimal solution which can be proved as follows. One must find a displacement field satisfying (161) and (162) and the kinematic boundary conditions, that is, zero displacements in all directions along the vertical support. This means that along the two bars in Fig. 43b the strains must be $\epsilon = 1$ and $\epsilon = -1$, respectively and along all other members in Fig. 43a the absolute value of ϵ must not exceed unity. Denoting the displacements in the x and y directions in Fig. 43b by u and v, a displacement field satisfying the above conditions is

$$u(x,y) \equiv 0\,, \quad v(x,y) = -2x\,, \tag{163}$$

which clearly satisfies the kinematic boundary conditions $u_x \equiv u_y \equiv 0$ along the support with $x = 0$. Moreover, the strains in the x and y directions are

$$\epsilon_x = \frac{\partial u}{\partial x} = 0\,, \quad \epsilon_y = \frac{\partial v}{\partial y} = 0\,, \quad \gamma_{xy} = \frac{\partial u}{\partial y} + \frac{\partial v}{\partial x} = -2\,. \tag{164}$$

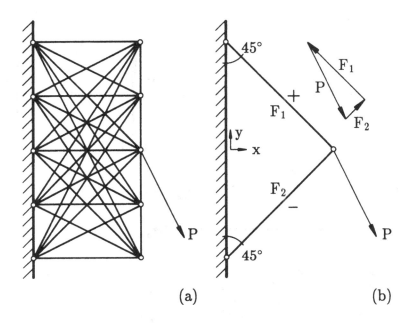

Fig. 43   Optimal layout of a simple pin-jointed frame.

Then in the principal directions at 45° to the vertical the strain values become

$$\epsilon_{1,2} = \frac{\epsilon_x + \epsilon_y}{2} \pm \sqrt{\left(\frac{\epsilon_x - \epsilon_y}{2}\right)^2 + \left(\frac{\gamma_{xy}}{2}\right)^2} = \pm 1 . \tag{165}$$

The above displacement field, therefore, satisfies (161) along the nonvanishing members in Fig. 43a, in which the forces are statically admissible. Moreover, since the principal strains represent the directionally highest absolute values of the strains, (162) is also fulfilled for any vanishing member in Fig. 43a. In fact, the same solution would be still valid, if the structural universe consisted of all possible members contained in the half-plane to the right of the vertical support in Fig. 43a.

3.3 A Simple Beam System

An even simpler structural universe, consisting of two simply supported beams, is shown in Fig. 44a. The beam system is subject to a single pont load P at the intersection of the two beams. In this problem, we shall use an inverse procedure, first assuming a statically admissible stress field and then showing that it satisfies all optimality criteria. The rather obvious optimal solution is given in Figs. 44b and e, in which the entire load P is carried by the short beam AA and the beam BB (with a zero cross-sectional area) is unloaded. The corresponding moment diagrams are shown in Figs. 44c and f.

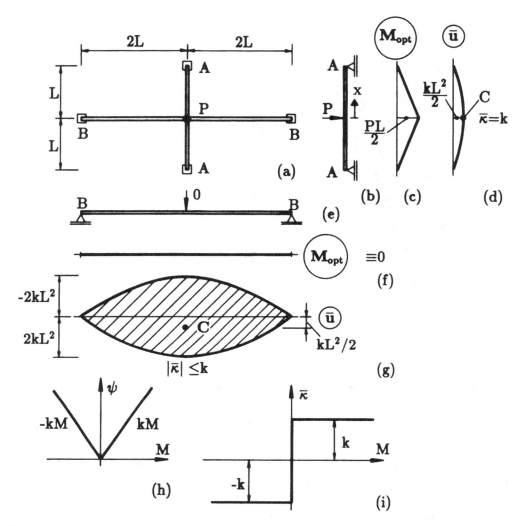

Fig. 44 Another elementary layout problem: grillage consisting of two beams.

The specific cost function and the normalized optimality criteria [(156) and (157)] are represented graphically in Figs. 44h and i (in which $k = 1$).

Since the deflection must be zero at the simple supports of the beams, the adjoint displacement field for the short beam AA becomes

$$\bar{u} = kL^2/2 - kx^2/2 \,, \tag{166}$$

clearly satisfying the boundary conditions $\bar{u}(L) = \bar{u}(-L) = 0$ and the curvature condition (156) $-\bar{u}'' = k$ (see Fig. 44d). For the long beam BB, the bending moment is throughout zero and hence (157) gives a non-unique curvature requirement which

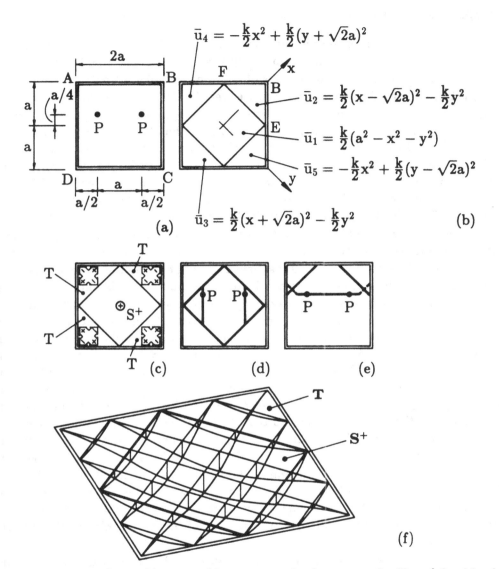

$$\bar{u}_4 = -\frac{k}{2}x^2 + \frac{k}{2}(y + \sqrt{2}a)^2$$

$$\bar{u}_2 = \frac{k}{2}(x - \sqrt{2}a)^2 - \frac{k}{2}y^2$$

$$\bar{u}_1 = \frac{k}{2}(a^2 - x^2 - y^2)$$

$$\bar{u}_5 = -\frac{k}{2}x^2 + \frac{k}{2}(y - \sqrt{2}a)^2$$

(a)

$$\bar{u}_3 = \frac{k}{2}(x + \sqrt{2}a)^2 - \frac{k}{2}y^2$$ (b)

(c)   (d)   (e)

(f)

Fig. 45   A more advanced layout problem: square, simply supported grillage (after Morely, 1966).

furnishes the non-unique adjoint deflection field indicated by the shaded area in Fig. 44g. As the latter does include a central deflection of $kL^2/2$ (Point C in Fig. 44g), kinematic admissibility is also established. Moreover, since the specific cost function is convex (Fig. 44h), necessary and sufficient conditions for optimality have been fulfilled and thus the solution in Figs. 44b-f is optimal.

3.4 Square, Simply Supported Grillage

Figure 45a shows a square, simply supported area $ABCD$ with two point loads (P) which are to be transmitted to the supports by a system of intersecting beams. In this problem, the structural universe consists of an infinite number of potential members in any arbitrary direction, covering the entire area $ABCD$. The specific cost function is again $\psi = k|M|$ (Fig. 44h) and the corresponding optimality criteria are given in Fig. 44i. As the above optimality conditions imply that the directional maximum absolute value of the curvature is k, it follows that such maximal curvatures, and by (156) all optimal beams, must have the same orientation as the *principal* directions of the adjoint displacement field. This means that in order to admit potential beams over the entire area, the latter must be covered by smoothly jointed regions, at all points of which at least in one principal direction the curvature of the adjoint displacement field has an absolute value of k. Figure 45b shows such a displacement field which consists of five distinct regions (u_1 to u_5). It can be checked easily that the kinematic boundary condition $u = 0$ is satisfied along the edges and the displacement field $u(x, y)$ is continuous and slope-continuous along all region boundaries. In the corner regions (in Chapter 8 termed type T) one principal curvature has a value of k and the other one $-k$ (Fig. 45c). In the central region (in Chapter 8 termed type S^+), the curvature takes on a value of k in all directions. It follows then from the optimality conditions (156) and (157) that in the central (S^+) region optimal beams may run in any arbitrary direction, so long as they are subject to positive bending. In the corner (T) regions, optimal beams may only run in two directions; in the direction of the diagonal passing through the considered corner the beams must be in negative bending and in the direction normal to that diagonal the beams must be in positive bending. The above conditions admit an infinite number of optimal beam layouts of equal structural weight. One of these, with only positive moments, is shown in Fig. 46d and another one, with beams in both positive and negative bending, in Fig. 45e. An oblique view of the adjoint field is given in Fig. 45f. The above optimal displacement field, which is also valid for any other non-negative (downward) loading, was first derived by Morley (1966) in the context of optimized reinforced concrete plates.

A detailed treatment of optimal grillage layouts will be given in Chapter 8. The same optimal layouts are valid for the least-weight reinforcement of concrete plates.

4 Concluding Remarks

In this chapter, the optimal layout theory was illustrated with very simple examples for didatic reasons, since in more complicated examples the principles involved could be obscured by computational complexities.

However, it will be shown in Chapters 8, 9 and 10 that closed form analytical solutions for the optimal layout can be obtained for relatively complex boundary and loading conditions, if a systematic approach is used.

Large, real structural systems, on the other hand, are not amenable to analytical treatment. It will be seen from Chapter 7, however, that the iterative COC algorithm is again very efficient for such problems.

Chapter 7

LAYOUT OPTIMIZATION USING THE ITERATIVE COC ALGORITHM

G.I.N. Rozvany and M. Zhou
Essen University, Essen, Germany

1 Introduction

It was explained in Chapter 6 that for *large, real systems* it is difficult, if not impossible to obtain closed form analytical solutions and hence it is necessary to employ *discretization* and *numerical techniques*. As was pointed out in the first five chapters, *mathematical programming* methods have, in general, a very limited optimization capability. However, the *iterative COC algorithm* will be shown to be highly suitable for handling layout (topology) optimization problems with a very large number of potential members.

Since it is easier to find existing FE software and to develop new FE programs for *elastic* analysis than for plastic analysis, we shall consider here elastic structures with a *compliance* constraint. However, the iterative COC algorithm to be discussed can be readily used for elastic systems with *any combination of local and global* (e.g. stress, deflection, stability and natural frequency) *constraints*, although in the case of stress constraints a singular topology (Kirsch and Rozvany, 1992) can cause difficulties. It is mentioned once more that a comprehensive review of optimality criteria for various design constraints was given in a recent book (Rozvany, 1989).

2 Iterative COC Algorithm for Layout Problems

For the layout of all elastic systems with a deflection constraint, basically the same procedure can be adopted as the one described in Chapter 3 (Section 6). The main difference is that for the prescribed minimum value z_{ia} of the parameter z_i the small-est possible value (e.g. 10^{-12} times the average value of that parameter) is used, which can still be handled by a program with double precision and does not cause ill-conditioning.

3 Elementary Examples of Layout Optimization by the Iterative COC Method

3.1 Pin-Jointed Frames

In the problems in Fig. 46, we impose a constraint on the vertical deflection at the point where the load P is applied. In Fig. 46a, we consider a three bar suspension for which the obvious optimal topology consists of a single vertical bar and the sloping members take on a zero cross-section. Fig. 47 shows the iteration history for this problem. The "weight" $\tilde{\Phi}$ here is nondimensional, $\tilde{\Phi} = \Phi Ed/(PL^2\gamma)$, where Φ is the total weight of the structure, E is Young's modulus, d is the prescribed deflection, P is the point load, L is the dimension shown in Fig. 47 and γ is the specific weight of the material used. The nondimensional weight values yielded by the iterative COC procedure are given under "unscaled weight" in Fig. 47. Since the design obtained in each iterative step represents a deflection which differs slightly from the prescribed one ($u_A \neq d$), the cross-sectional areas were scaled to obtain the correct deflection after each iteration. The corresponding weight values are shown under "scaled weight" in Fig. 47. In a practical application, it is not necessary to calculate the scaled weights because the deflection of the unscaled solution converges rapidly to the prescribed value after a few iterations (see Fig. 47 in which the two polygons differ very little after 8 iterations). Fig. 47 shows also an alternative problem in which a prescribed minimum value (0.06) of the nondimensional cross-sectional area $\tilde{z}_i = z_i Ed/(PL)$ is specified. The results in Fig. 47 were obtained both by analytical hand calculations and on the computer using COC software (Rozvany, Zhou *et al.*, 1989).

More complicated layout problems are shown in Figs. 46b and c. For these problems the COC method was combined with either (a) a finite element (FE) program developed by M. Zhou, or (b) ANSYS. Both FE programs yielded identical results. Whilst for the simple example in Fig. 46a this was not necessary, in more complex problems the COC procedure for layout problems requires the specification of a small prescribed minimum cross-sectional area (z_a). For the considered problems the nondimensional value of the latter was $\tilde{z}_a = 10^{-12}$. The variation of the scaled nondimensional weight $\tilde{\Phi}_S$ in dependence on the iteration number is shown in Fig. 46d for the 56-bar truss (Fig. 46b). Using the nondimensionalization given above for $\tilde{\Phi}$, the exact optimal weight for the 56-bar truss is $\tilde{\Phi} = 16$ and the iterative COC procedure yielded after 126 iterations a weight of $\tilde{\Phi}_S = 16.000000000048$ which represents *an agreement of twelve significant digits*. In the COC solution, all non-optimal members took on the prescribed minimum cross-section ($\tilde{z}_a = 10^{-12}$), except members a and b

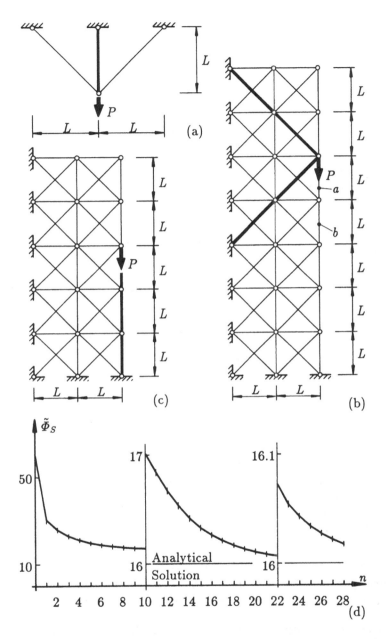

Fig. 46 (a)-(c) Various simple layout optimization problems solved by iterative COC methods; and (d) nondimensional scaled structural weight $\tilde{\Phi}_s$ vs. number of iterations n for the problem under (b).

(Fig. 46b) which had a cross-sectional area of $\tilde{z}_i = 2.88 \cdot 10^{-12}$ and $\tilde{z}_i = 1.74 \cdot 10^{-12}$.

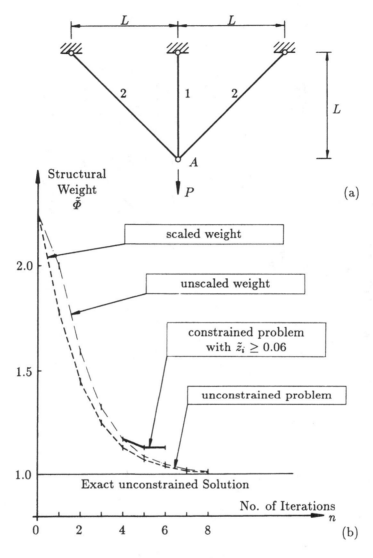

Fig. 47 Elementary layout problem with a structural universe consisting of three bars: (a) geometry and loading, and (b) convergence history for scaled and unscaled weight with and without a minimum cross-sectional area constraint.

This indicates that all non-optimal members vanish when $\tilde{z}_a \to 0$, as in the analytical solution. The cross-sections of the optimal members (thick lines in Fig. 46b) agreed with the analytical solution for the first 12 significant digits.

In the 40-bar problem (Fig. 46c), *all* non-optimal members took on the value of $\tilde{z}_a = 10^{-12}$ in the iterative COC solution and the cross-sectional area for the optimal

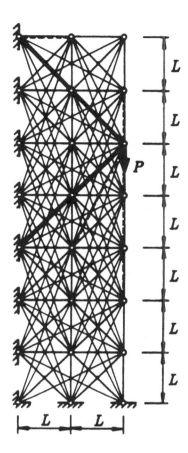

Fig. 48   A more complex structural universe with 114 members for the example in Fig.
46b.

members agreed again with the analytical solution.

The problem in Fig. 46b (56-bar truss) was also run on the COC program with a
structural universe having 114 members (Fig. 48). The optimal members (thick lines)
again took on a cross-sectional area of 2.8284271 ($\approx 2\sqrt{2}$) and all other members took
on the prescribed minimum value of 10^{-12}, except the ones shown in broken lines,
which were all under 10^{-11}. The above run yielded an optimal total cost value of
16.00000000016 after 231 iterations. The larger error compared to the analytical
solution is due to (i) the larger number of members and (ii) the greater average
length of the members.

The same problem was investigated with a structural universe having 804 members (Fig. 49). The optimal members are again shown in thick line (Fig. 50) and
the members having an area slightly greater than the prescribed minimum value
($10^{-12} < z_i < z^{-11}$) in broken line. The unusually stringent convergence criterion of

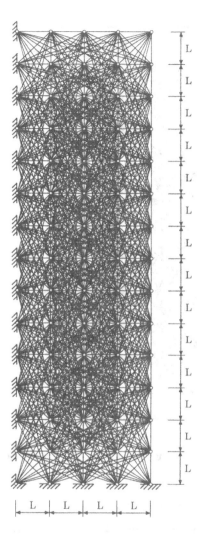

Fig. 49 A structural universe with 804 members for the problem in Fig. 46b.

$(\Phi_{new} - \Phi_{old})/\Phi_{new} < 10^{-14}$ was satisfied after 627 iterations and gave a total cost value of 16.00000000081.

At the time of these investigations, the analytical solution was not available. It was obtained somewhat later and is given in Chapter 10.

3.2 Two-Span Beams

The problem solved analytically in Section 3.1 of Chapter 6 was also computed by the iterative COC procedure using a prescribed minimum width value of $z_a = 10^{-8}$ with $N = 300$ and $N = 1800$ elements. The span ratio in Fig. 42a (Chapter 6) yielded

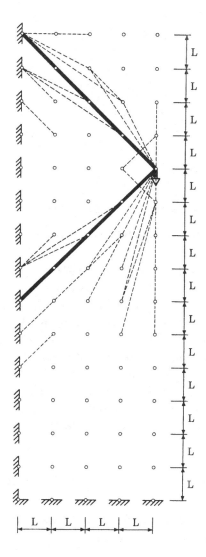

Fig. 50 Optimal members for the structural universe in Fig. 49.

total cost values of $\Phi_{300} = 0.015625018$ (22 iterations) and $\Phi_{1800} = 0.015625013$ (21 iterations), both showing a 7-digit agreement with the analytical result of $(1/8)^2$ in (159). For the span ratio in Fig. 42d the COC method gave $\Phi_{300} = 0.21540290$ (22 iterations) and $\Phi_{1800} = 0.21539237$ (36 iterations) which represent four and five digit agreements, respectively, with the analytical solution of $(2\sqrt{3} - 3)^2$ in (160). The above iteration numbers were necessary for a tolerance value of $\overline{E} = 10^{-6}$ in (72), but a Φ-value with an error of less than 2 per cent was obtained already after 3 iterations.

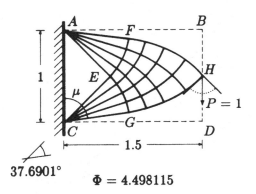

$$\Phi = 4.498115$$

37.6901°

Fig. 51 Optimal layout consisting partly of a Hencky net.

4 More Advanced Examples of Layout Optimization by the Iterative COC Method

4.1 Truss-Like Continuum Containing a Hencky Net

It was pointed out by Prager (e.g. 1974) that many least-weight truss layouts consist of an infinite number of members having an infinitesimal spacing and hence they should be termed "truss-like continua". One such optimal layout is shown in Fig. 51 in which the truss is restricted to the rectangle $ABCD$ and a point load $P = 1$ is to be transmitted to the shorter side at the opposite end. The triangle ACE contains no members on its interior and the regions AEF and CEG consist of straight radial members. The region $EFGH$ is a Hencky net. For obvious reasons, only a finite number of members are indicated. The analytical solution of the above problem is discussed by Hemp (1973, pp. 97-99). For a side ratio of 1.5 to 1.0, Hemp's equation (4.120) gives

$$1.5 = \frac{1}{2} \int_0^{2\mu} [I_0(t) + I_1(t)]\, \mathrm{d}t = \frac{1}{2}\left[I_0(2\mu) - 1 + 2 \sum_{n=0}^{\infty} (-1)^n I_{2n+1}(2\mu) \right], \qquad (167)$$

furnishing the angle $\mu = 82.690133°$. Then the total truss volume can be calculated from Hemp's equations (4.123) for a unit load and $k = 1$ (in Hemp's notation $\sqrt{2}FR/\sigma = 1$) as

$$\Phi = (1 + 2\mu)I_0(2\mu) + 2\mu I_1(2\mu), \qquad (168)$$

where I_0 and I_1 are modified Bessel functions. For the above μ-value, (167) furnishes the optimal truss volume

$$\Phi_{\text{opt, plastic}} = 4.498115. \qquad (169)$$

The above value represents an optimal *plastic design* [(155) in Chapter 6 with $k = 1$]. For *elastic design with a compliance constraint* ($C = 1$) we have by (154)

$$\Phi_{\text{opt, compliance}} = 4.498115^2 = 20.233042. \qquad (170)$$

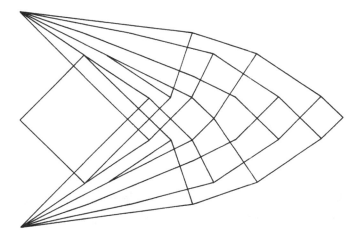

Fig. 52   Iterative COC solution corresponding to the analytical solution in Fig. 51 with 5055 potential members.

Using the iterative COC algorithm and structural universes with 5055 and 12992 members, respectively, truss volumes of $\Phi = 20.540807$ and $\Phi = 20.419699$ were obtained for elastic design with a compliance constraint. For plastic design with $k = 1$, by (155a), the square root of the above values must be taken, giving $\Phi = 4.532197$ and $\Phi = 4.518816$, which represent 0.76% and 0.46% errors compared to the analytical solution. In the above procedure, a minimum prescribed cross-sectional area of 10^{-12} was used. The system with 5055 members required over 3500 iterations with a convergence tolerance value of $\overline{E} = 10^{-8}$ [see (72) in Chapter 3], but a Φ-value of 20.77 was reached already after 100 iterations (1% error compared to the value after satisfaction of the convergence criterion). The layout of members having a cross-sectional area over $z = 0.1$ in elastic compliance design is shown in Fig. 52 which exhibits clear similarities with the analytical solution in Fig. 51.

The above cross-sectional area [by (153) and (154), with $c = r = C = 1$] corresponds to $z = 0.1/\Phi_{\text{opt, plastic}} = 0.1/4.498115 \approx 0.0222$ in plastic design with $k = 1$.

4.2 Single Point Load Parallel to a Supporting Line, Triangular Structural Domain
For the above problem, in which the structural members are restricted to the triangle ABC in Fig. 53, with a support along AB and free edges along AC and BC, the fairly obvious analytical solution consists of members running along the free edges of the domain (Fig. 53). Before a rather complicated proof for the above solution was found (see Chapter 10), the same problem was investigated by the iterative COC procedure, using a structural universe with 720 members (Fig. 54). The analytical solution for the above problem yields $\Phi = 25$ and the iterative COC solution with a prescribed minimum cross-sectional area of $z_a = 10^{-12}$ and a convergence tolerance value [see (72) in Chapter 3] of $\overline{E} = 10^{-14}$ gave after 268 iterations $\Phi = 25.000000001168$.

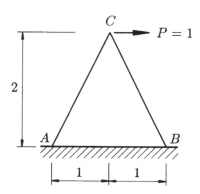

Fig. 53   Loading and analytical solution for another layout problem.

5 Layout Problems with Alternative Loading Conditions

5.1 Plastic Design

If the structure is subjected to several alternative loading conditions ($p_1, p_2, \ldots, p_j$, $\ldots, p_n$) equilibrating the statically admissible stress fields Q_j ($j = 1, 2, \ldots, n$), our optimization problem in *plastic design* becomes

$$\min_{Q_j^S} \Phi = \int \overline{\psi} \, dx,$$

subject to

$$(\text{for all } x \in D) \quad \overline{\psi}(x) = \max_j \psi \left[Q_j^S(x) \right], \tag{171}$$

where ψ is the specific cost requirement for a given statically admissible stress vector Q_j^S and $\overline{\psi}$ is the design value of the specific cost (e.g. cross-sectional area).

For the above problem, the optimality criteria are (Rozvany, 1989, p. 47, Eq. 1.24):

$$(\text{for all } x \in D) \quad q_j^K = \lambda_j(x) \, \mathcal{G} \left\{ \psi \left[Q_j^S(x) \right] \right\},$$

$$\lambda_j \geq 0, \quad \lambda_j > 0 \quad \text{only if} \quad \overline{\psi} = \psi(Q^S), \quad \sum_j \lambda_j = 1. \tag{172}$$

Considering now the class of problems with (a) a specific cost function $\psi = k|Q_j|$ where Q_j is a scalar, and (b) *two* loading conditions, the optimality criteria become

$$(\text{for } k|Q_1| = \overline{\psi}, \quad k|Q_2| < \overline{\psi}, \quad |Q_1| > 0) \quad q_1 = k \operatorname{sgn} Q_1, \quad q_2 = 0, \tag{173}$$

$$(\text{for } k|Q_2| = \overline{\psi}, \quad k|Q_1| < \overline{\psi}, \quad |Q_2| > 0) \quad q_2 = k \operatorname{sgn} Q_2, \quad q_1 = 0, \tag{174}$$

$$(\text{for } k|Q_1| = k|Q_2| = \overline{\psi} > 0), \quad q_1 = \lambda k \operatorname{sgn} Q_1, \quad q_2 = (1 - \lambda)k \operatorname{sgn} Q_2,$$

$$1 \geq \lambda \geq 0, \tag{175}$$

$$(\text{for } Q_1 = Q_2 = \overline{\psi} = 0) \quad |q_1| + |q_2| \leq k. \tag{176}$$

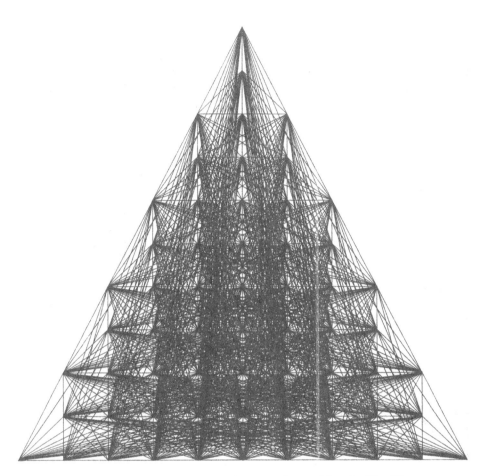

Fig. 54   Structural universe with 720 members for the problem in Fig. 53.

It is relatively difficult to find a solution satisfying the above optimality conditions. However, it was shown independently by Nagtegaal and Prager (1973), Spillers and Lev (1971), and Hemp (1973) that one can employ a *superposition principle* consisting of the following steps. First, construct the *component load systems*

$$\mathbf{p}_1^* = (\mathbf{p}_1 + \mathbf{p}_2)/2\,, \quad \mathbf{p}_2^* = (\mathbf{p}_1 - \mathbf{p}_2)/2\,. \tag{177}$$

Then optimize the structure for $\mathbf{p}_1^*$ and $\mathbf{p}_2^*$ separately and add the corresponding specific costs ψ_1^* and ψ_2^*,

$$\psi = \psi_1^* + \psi_2^*\,. \tag{178}$$

The above specific cost (ψ) values represent the optimal solution for the two alternate loading conditions ($\mathbf{p}_1$ and $\mathbf{p}_2$). Denoting the adjoint strains of the optimal solutions

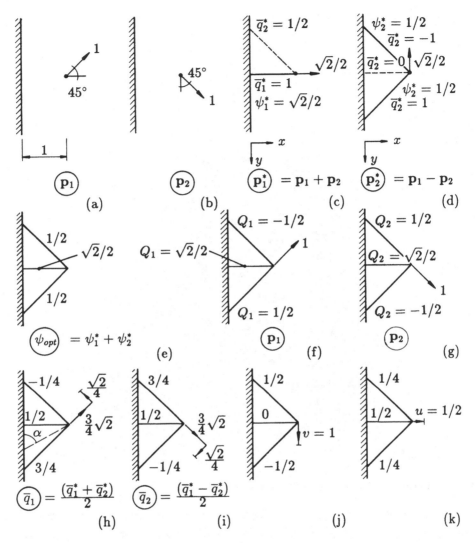

Fig. 55 Optimal layout for plastic design: two alternative loads.

for the component loads by $\bar{q}_1^*$ and $\bar{q}_2^*$, the adjoint strains for the original two loading conditions, satisfying (173)-(175), are given by

$$\bar{q}_1 = \bar{q}_1^* + \bar{q}_2^*, \quad \bar{q}_2 = \bar{q}_1^* - \bar{q}_2^*. \tag{179}$$

The above superposition principle will be illustrated with a simple example. Determine the optimal truss layout for the two load conditions in Figs. 55a and b. The component loads and the corresponding optimal solutions are given in Figs. 55c and

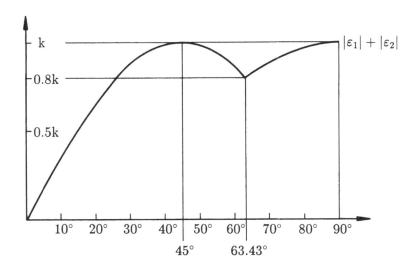

Fig. 56   Check on optimality condition (176).

d, in which the adjoint displacements, respectively, are

$$\bar{u}_1^* = kx, \quad \bar{v}_1^* \equiv 0; \quad \bar{u}_2^* = 0, \quad \bar{v}_2^* = -2kx, \qquad (180)$$

where $\bar{u}$ and $\bar{v}$ denote displacements in the x and y directions. It can be readily checked that the displacements in (180) give the strains in Figs. 55c and d and that the latter satisfy the optimality conditions in (161) and (162) (Chapter 6). The superimposed optimal solution for the two alternate loads is shown in Fig. 55e and the member forces for the two load conditions in Figs. 55f and g.* Note that only *statical* admissibility is required in plastic design. It can be seen from the above figures that all members are fully stressed $(k|Q_j| = \bar{\psi})$ for *both* loading conditions and hence optimality criterion (175) applies. The adjoint strains for the alternate loads are given in Figs. 55h and i, calculated on the basis of (179) and Figs. 55c and d. Kinematic admissibility of the strains in Fig. 55i, for example, can be checked by superimposing the simple displacement fields in Figs. 55j and k. The strain fields in Figs. 55h and i satisfy the optimality criterion (175) with $\lambda = 1/4$, $\lambda = 1/2$ and $\lambda = 3/4$ for the top, middle and bottom members, respectively and hence optimality of the considered solution is confirmed.

It is still necessary to show that the strains for nonoptimal directions also satisfy conditions (176). For an arbitrary angle α, the strains become (Fig. 55h)

$$\epsilon_1 = \frac{\sqrt{2}}{4}\sin\alpha[3\cos(135° - \alpha) - \cos(45° - \alpha)],$$

---

* Since this is plastic design, compatibility need not be satisfied by the corresponding elastic strains.

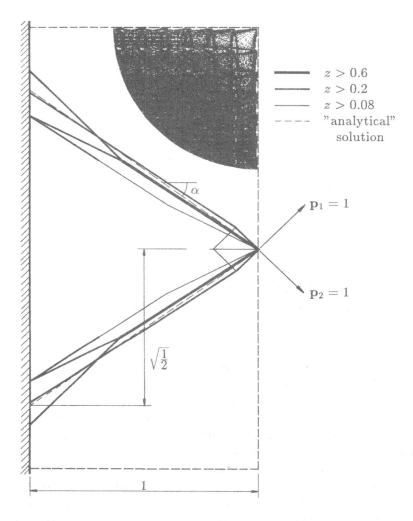

Fig. 57 Iterative COC solution for a compliance constraint with two alternative loading conditions and part of the structural universe.

$$\epsilon_2 = \frac{\sqrt{2}}{4} \sin \alpha [\cos(135° - \alpha) + 3\cos(45° - \alpha)]. \tag{181}$$

By condition (176),

$$|\epsilon_1| + |\epsilon_2| \leq k, \tag{182}$$

for any value of α in an optimal solution. Indeed, relations (181) and (182) imply

$$\frac{\sin \alpha}{4}[|3(\sin \alpha - \cos \alpha) - (\sin \alpha + \cos \alpha)| +$$

$$+|3(\sin\alpha+\cos\alpha)-(\sin\alpha-\cos\alpha)|]\leq k\,,\qquad(183)$$

which can readily be shown to be correct. The actual variation of the left-hand side of relation (183) is given in Fig. 56. It can be seen that $(|\epsilon_1|+|\epsilon_2|)$ only takes on the value k in the optimal directions and condition (182) is also satisfied for all other directions. The above problem was investigated in an earlier publication (Rozvany, 1974).

The superposition principle discussed here has been extended to an arbitrary number (2^n) of loading conditions by Rozvany and Hill (1978a).

5.2 Elastic Design for Compliance

For several loading conditions $(k=1,2,\ldots,s)$ and several displacement constraints $(j=1,2,\ldots v)$, the optimality criterion in Fig. 11 (Chapter 3) changes to (Rozvany, 1989, p. 64):

$$\overline{q}_k^K=\sum_j\nu_{jk}[\mathbf{F}]\overline{Q}_j^S\,,\qquad(184)$$

$$\mathcal{G}_{,\mathbf{z}}[\psi(z)]+\sum_j\sum_k\nu_{jk}\overline{Q}_j^{S,K}\cdot\{\mathcal{G}[\mathbf{F}]\}Q_k^{S,K}=0\,.\qquad(185)$$

Considering the class of simple problems with $\psi=cz$, $[\mathbf{F}]=1/rz$, and two compliance constraints [each associated with a different load condition, with scalar stresses $\overline{Q}_k=Q_k]$ $\int_D(Q_k^2/rz)\,dx=C$ $(k=1,2)$, (184) and (185) reduce to

$$\overline{q}_k=\nu_{kk}Q_k/rz\quad(k=1,2)\,,\qquad(186)$$

$$c-\sum_{k=1}^2\nu_{kk}Q_k^2/rz=0\quad\Rightarrow\quad z=\sqrt{(\nu_{11}Q_1^2+\nu_{22}Q_2^2)/rc}\,.\qquad(187)$$

The above optimality criterion is difficult to handle in analytical derivations, but is highly suitable for an iterative COC procedure.

For a discretized formulation, the values of the Lagrangians ν_{11} and ν_{22} can be calculated from the compliance constraints:

$$C=\sum_A\frac{Q_{ik}^2\delta}{r\sqrt{(\nu_{11}Q_{i1}^2+\nu_{22}Q_{i2}^2)/rc}}+\sum_P\frac{Q_{ik}^2\delta}{rz_a}\quad(k=1,2)\,,\qquad(188)$$

where δ is the element length. Equations (188) can, in general, only be solved iteratively.

In the case of a symmetric boundary and two anti-symmetric load systems (as in Fig. 55), we have $\nu_{11}=\nu_{22}=\nu$. Moreover, we can carry out the analysis for one load condition only and denote by Q and $\hat{Q}$ the strains in the corresponding elements of the two symmetric half-structures. Then (187) reduces to

$$z = \sqrt{\nu(Q^2 + \hat{Q}^2)/rc}.\tag{189}$$

If the entire structure has $2n$ elements and they are numbered in the same sequence for the two halves, then the discretized equivalent of (185) becomes

$$z_i = z_{i+n} = \sqrt{\nu(Q_i^2 + Q_{i+n}^2)/rc}\quad(i = 1, 2, \ldots, n).\tag{190}$$

Moreover, for the considered symmetric problems, (188) reduces to

$$\sqrt{\nu} = \sum_A \frac{Q_i^2 \delta}{\sqrt{r(Q_i^2 + Q_i^2 \pm n)/c}} \Big/ \left(C - \sum_P \frac{Q_i^2 \delta}{r z_a}\right),\tag{191}$$

where in the subscript "$\pm$", we have "$+$" for $1 \le i \le n$ and "$-$" for $n + 1 \le i \le 2n$.

The problem in Figs. 55a and b was solved for an elastic compliance constraint using the above iterative COC procedure with 7170 potential bars. After 1500 iterations a total weight value of $\Phi = 3.49295726$ was obtained. The plot of bars having a cross-sectional area over $z = 0.08$ is shown in Fig. 57 (continuous lines), in which three different line thicknesses show various ranges of member sizes ($4z$). The "structural universe" consisted of 11×21 grid-points and the connecting members were restricted to slopes of $0, 1 : 1, 1 : 2, \ldots, 1 : 10, 2 : 3, 2 : 5, \ldots, 2 : 9$ and their reciprocals (see part of the structural universe in the top right corner of Fig. 57).

The exact analytical solution for the above problem is not known to the authors. However, if we assume a symmetric two-bar system (broken lines in Fig. 57), then the optimal solution within this topology can be determined easily. It can be shown from statical considerations that for $p_1 = 1$ the member forces are

$$N_1 = \frac{1}{2\sqrt{2}}\left(\frac{1}{\cos\alpha} + \frac{1}{\sin\alpha}\right), \quad N_2 = \frac{1}{2\sqrt{2}}\left(\frac{1}{\cos\alpha} - \frac{1}{\sin\alpha}\right),\tag{192}$$

and due to symmetry of the solution we have

$$C = 1 = \frac{N_1^2}{A}L + \frac{N_2^2}{A}L = \frac{1}{A\cos\alpha\sin^2(2\alpha)},\tag{193}$$

where A is the cross-sectional area and L is the member length. The relation (193) implies:

$$A = \frac{1}{\cos\alpha\sin^2(2\alpha)}, \quad \Phi = 2AL = \frac{2}{\cos^2\alpha\sin^2(2\alpha)}.\tag{194}$$

Then the usual stationarity condition $(d\Phi/d\alpha = 0)$ implies

$$2\cos\alpha\sin\alpha\sin^2(2\alpha) - 4\cos^2\alpha\sin(2\alpha)\cos(2\alpha) = 0, \qquad \tan^2\alpha = 1/2,$$

$$\Phi = \frac{27}{8} = 3.375, \quad A = \frac{27}{16\sqrt{1.5}} = 1.37783798, \quad \alpha = 35.26438968°. \qquad (195)$$

The optimal two-bar system, having a slope of $\tan\alpha = 1/\sqrt{2}$ is shown in broken lines in Fig. 56. Since this system has 3.495% lower weight than the COC solution with 7170 potential members, it could be the absolute optimal layout. It can be seen from Fig. 57, that the COC procedure, within the limited range of admissible member slopes, is trying to achieve this solution.

Another iterative COC calculation with 12202 members gave a weight of 3.375668 which is only 0.0198% above the assumed analytical solution.

Note. It can be seen from Sections 5.1 and 5.2 that, whilst the elastic compliance and plastic designs were identical (within a constant factor) for a single load condition, *the solutions for plastic design and elastic compliance design differ significantly if several alternate loads are considered.*

6 Concluding Remarks

It was shown in this chapter that the iterative COC procedure is highly suitable for the simultaneous optimization of the topology and geometry of structural systems. The solution is extremely accurate if members of the structural universe include the optimal members (see Section 3) and represents an excellent approximation if the structural universe does not contain the optimal members but has a sufficiently large number of elements.

Chapter 8

OPTIMAL GRILLAGE LAYOUTS

G.I.N. Rozvany

Essen University, Essen, Germany

1 Introduction

The theory of optimal structural layouts was discussed in Chapter 6 and its iterative
applications to large systems in Chapter 7. In order to check the validity and accuracy of such numerical methods, however, it is necessary to obtain *exact, analytical
solutions* for relatively complex layout problems. One of the most successful applications of the exact layout theory was the optimization of grillage layouts, as can
be seen from the following remark by Prager: "Although the literature on Michell
trusses" "(i.e. least-weight trusses)" "is quite extensive, the mathematically similar
theory of grillages of least-weight was only developed during the last decade. Despite
its late start, this theory advanced farther than that of optimal trusses. In fact, grillages of least-weight constitute the first class of plane structural systems for which
the problem of optimal layout can be solved for almost all loadings and boundary
conditions" (Prager and Rozvany, 1977b). The problem of grillage optimization can
be described as follows (Fig. 58): A structural domain D, bound by two horizontal
planes and some vertical surfaces, is subject to a system of vertical loads which are
to be transmitted to given supports by beams of rectangular cross-section having a
given depth and variable width. The beams are to be contained in the structural
domain and are to take on a minimum weight (or volume). The beam system is to
be designed plastically (see Chapter 2, Fig. 1).

As can be seen from the quotation above, Prager regarded the grillage optimization problem as particularly important because of the following unique features:

(a) Grillages constitute the first class of truly two-dimensional structural optimization problems for which closed form analytical solutions are available for most boundary and loading conditions.

(b) Optimal grillages are more practical than Michell structures (least-weight trusses), because the latter are subject to instability which is ignored in the formulation.

(c) The optimal rib layout of least-weight plates has been found similar to that of minimum weight grillages (see Chapter 10, Fig. 102).

(d) A computer algorithm is available for generating analytically and plotting optimal beam layouts for a wide range of boundary conditions (Rozvany and Hill, 1978b; Hill and Rozvany, 1985).

(e) It has been shown that the same grillage layout is optimal for plastic design and elastic design with a stress or compliance or natural frequency constraint (Rozvany, 1976; Olhoff and Rozvany, 1982).

(f) The optimal grillage layout is independent of the (non-negative) load distribution if no internal simple supports are present.

(g) The adjoint displacement field can be readily generated and it provides an *influence surface* for any (non-negative) loading (the total structural weight equals the integral of the product of loads and deflections).

(h) A number of additional refinements have been added to the optimal grillage theory.

2 Optimality Criteria and Their Implications

The specific cost functions ψ for grillages of given depth is $\psi = k|M|$, where k is a given constant and M is the beam bending moment, and the optimality conditions are basically those of Heyman (1959), see (4) and (5) in Chapter 2. The latter imply (see Section 3.4 of Chapter 6) that all optimal beams must have the same orientation as the principal directions with a directional curvature $-\bar{u}'' = k$ of the adjoint displacement field $\bar{u}(x,y)$.

This means that the optimal solution must consist of the following types of regions:

$$
\left.
\begin{array}{llll}
R^+ : & M_1 > 0, & M_2 = 0, & \bar{\kappa}_1 = k, \ |\bar{\kappa}_2| \le k, \\
R^- : & M_1 < 0, & M_2 = 0, & \bar{\kappa}_1 = -k, \ |\bar{\kappa}_2| \le k, \\
S^+ : & M_1 > 0, & M_2 > 0, & \bar{\kappa}_1 = \bar{\kappa}_2 = k, \\
S^- : & M_1 < 0, & M_2 < 0, & \bar{\kappa}_1 = \bar{\kappa}_2 = -k, \\
T : & M_1 > 0, & M_2 < 0, & \bar{\kappa}_1 = -\bar{\kappa}_2 = k,
\end{array}
\right\}
\quad (|\bar{\kappa}_1| \ge |\bar{\kappa}_2|), \ (|M_1| \ge |M_2|).
$$

$$(196)$$

The above optimal regions are shown in Fig. 59 in which continuous and broken lines, respectively, indicate optimal beams with positive and negative bending moments. Arrows indicate the directions of principal adjoint curvatures. In the case of

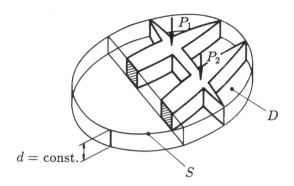

Fig. 58   The grillage layout problem.

circles enclosing a sign all directions are equally optimal (the curvature is the same in all directions).

The optimal regions are also represented graphically, in terms of moment-curvature relations, in Fig. 60. Shaded regions of the latter are invalid if we observe the inequalities on the right-hand side of (196).

The above optimality conditions mean that we have replaced a complicated variational problem with a relatively simple geometrical task which requires that (i) the structural domain is covered with the types of optimal regions given in Fig. 59, (ii) the above field is continuous and slope-continuous at region boundaries and satisfies the kinematic boundary conditions, and (iii) beams in the optimal directions given by such an adjoint field are capable of transmitting the loads to the supports (i.e. a statically admissible moment field is associated with the adjoint displacement field).

3 Properties of Optimal Regions

3.1 Properties of S and T Type Regions

In S type regions both principal adjoint curvatures take on the same value (k or $-k$) and, therefore, the curvature is the same in all directions. Then, by (4) and (5) in Chapter 2, the directions of nonzero moments are indeterminate and only their sign is prescribed. If we replace $\bar{\kappa}_1$ and $\bar{\kappa}_2$ with $-\bar{u}_{,xx}$ and $-\bar{u}_{,yy}$ in (196), then the curvature conditions for S type regions furnish, after integration, the following general equations for the adjoint field $\bar{u}(x,y)$:

$$S^+: \quad \bar{u} = -\frac{1}{2}k(x^2 + y^2) + a + bx + cy, \tag{197}$$

$$S^-: \quad \bar{u} = \frac{1}{2}k(x^2 + y^2) + a + bx + cy, \tag{198}$$

where a, b and c are constants.

Next, it is shown that in R and T type regions, the centroidal axis of beams having a nonzero capacity is always straight.

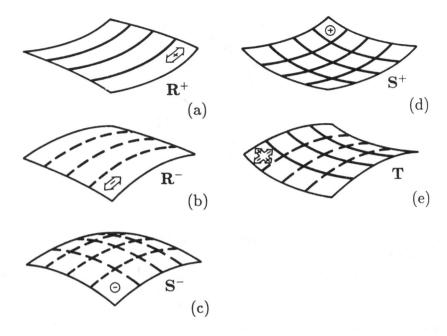

Fig. 59   Optimal regions for least-weight grillages.

Shield (1960) obtained the following relation for principal curvatures in plates:

$$\partial \bar{\kappa}_1 / \partial s_2 = (1/\rho_1)(\bar{\kappa}_2 - \bar{\kappa}_1) \ , \tag{199}$$

where s_2 is a curvilinear coordinate measured in the direction of the line of principal curvature $\bar{\kappa}_1$ and ρ_1 is the radius of *in-plane* curvature of the principal line "1".

If $M_1 \neq 0$, then by (4), $\bar{\kappa}_1 =$ const. (i.e. k or $-k$). The derivative in (199) therefore takes on a zero value, and hence $\rho_1 = \infty$ for $\bar{\kappa}_1 \neq \bar{\kappa}_2$ Q.E.D.

This means that T type regions consist of principal lines in constant directions at right angles, and the general equation for their adjoint displacement field is

$$T: \quad \bar{u} = \frac{1}{2}k(y^2 - x^2) + a + bx + cy \ , \tag{200}$$

$$T: \quad \bar{u} = kxy + a + bx + cy \ , \tag{201}$$

respectively, for principal directions parallel and at 45° to the coordinate axes x and y. The symbols a, b and c denote constants.

In T type regions, the direction and signs of the principal moments are prescribed. However, for both T and S type regions, the foregoing conditions, in general, admit an infinite number of moment fields, which all yield the same optimal cost.

3.2 Properties of R Type Regions

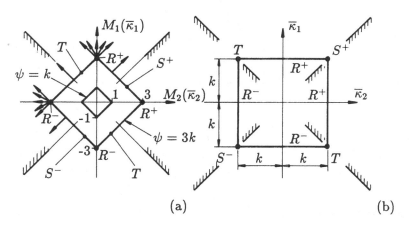

Fig. 60 Optimal moment-adjoint curvature relations for least-weight grillages.

The geometrical properties of R type regions are rather complicated and are discussed elsewhere (Rozvany and Hill, 1976). Two of the simplest properties are reproduced here; for proofs the reader is referred to the above publication.

Proposition 1. A principal line ℓ of an R region with $\bar{\kappa} = k$ intersects the contour lines K_1 and K_2 with $\bar{u} = $ const. at the same angle (α, see Fig. 61).

Proposition 2. The slope $\bar{\omega} = \partial \bar{u}/\partial x$ of the adjoint field $\bar{u}$ normal to a principal line is constant and equals $k(FW)$ in Fig. 61.

4 The Optimal Topology of Grillage Layouts

In Section 3 we discussed the geometrical properties of various types of optimal regions without taking the boundary conditions into consideration. Additional geometrical properties of optimal adjoint fields can be derived by making an allowance for kinematic constraints on the adjoint field $\bar{u}(x,y)$.

Along lines of simple supports (S_1) and clamped supports (S_2), for example, the following constraints must be observed:

$$\text{on } S_1: \quad \bar{u} = 0 \,,$$
$$\text{on } S_2: \quad \bar{u} = 0, \ u_{,x} = 0, \ u_{,y} = 0 \,, \tag{202}$$

where x and y are Cartesian coordinates along the middle plane of the grillage.

Subsequent discussion will be restricted to the following class of problems:
(i) The load is nonnegative $p(x,y) \leq 0$, i.e. downward on the domain D.
(ii) All boundaries of the domain D are either simply supported or clamped.

In order to examine various characteristics of optimal adjoint fields in an example, a relatively complex solution is shown in Fig. 62, in which the system has a clamped support along the external boundary (thick lines) and at the point P.

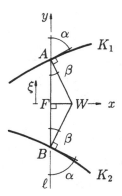

Fig. 61   Properties of R regions.

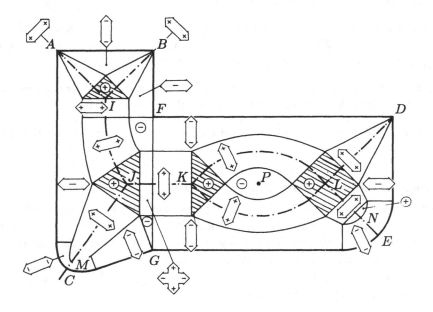

Fig. 62   Optimal layout illustrating the topology of least-weight grillages.

The domain D in Fig. 62 is divided into two subsets, which shall be termed *branches* (unshaded areas) and *junctions* (shaded area). The so-called *centrelines* of branches are shown in dash-dot lines. Considering the adjoint field $\overline{u}(x,y)$ along one principal line only, the former takes on its maximum value at the centreline. Each branch is divided by the centreline into two subsets, which may fall into one of four categoeries termed α, β, γ and δ *fields*. It follows that according to the two fields they contain, branches are of ten different types termed as $\alpha\alpha$, $\alpha\beta$, $\alpha\gamma$, $\alpha\delta$, $\beta\beta$, $\beta\gamma$, $\beta\phi$, $\gamma\gamma$, $\gamma\delta$ and $\delta\delta$ type. β fields are always associated with *clamped boundaries* and

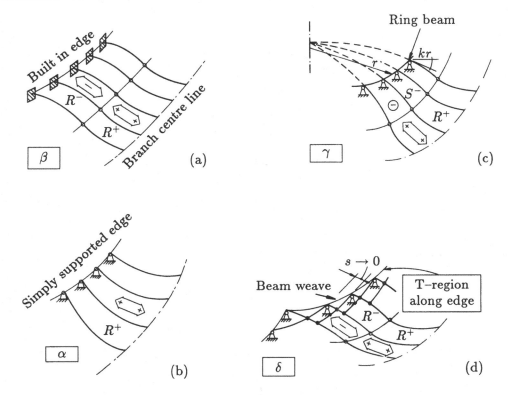

Fig. 63   Optimal fields associated with simply supported and clamped edges.

α, γ, and δ fields with *simply supported boundaries.*

The difference between optimal regions (R^+, S^+, etc.) and optimal fields (α, β, etc.) is that the former are based on the adjoint curvature conditions [(4) and (5)] only and the latter take boundary and kinematic continuity conditions also into consideration. Only clamped supports ($\beta\beta$ branches) are considered in Section 5 and mixed support conditions are discussed in Section 6.

An overview of optimal fields is given in Fig. 63. β *fields* are associated with a *clamped support* and consist of an R^- and an R^+ region, whilst α, γ and δ fields are associated with *simple supports.* Moreover, α *fields* consist of an R^+ region, while γ *fields* contain an R^+ and an R^- field and a "concentrated" beam along the support, which resists a circumferential moment impulse. Finally, δ *fields* consist of an R^- and an R^+ region, as well as of a T region of infinitesimal width along the supporting line. The latter transmits the moment along the support, usually to a corner.

5 Properties of Solutions for Clamped Boundaries

Topological properties of grillage layouts were discussed in Section 4. The particular case of clamped boundaries ($\beta\beta$ branches) will be considered first because of the

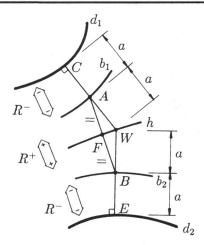

Fig. 64   Property 1 of $\beta\beta$ branches.

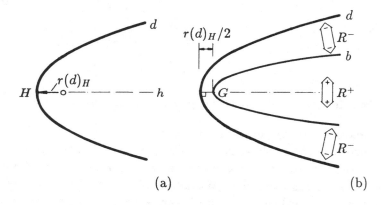

(a)   (b)

Fig. 65   Property 2 of $\beta\beta$ branches.

relative simplicity of their geometry. *$\beta\beta$ branches are bounded by two clamped edges* (d_1 and d_2 in Fig. 64) and have the following properties.

Property 1. A $\beta\beta$ branch consists of one inner and two outer regions. In general, the outer regions are of R^- type and the inner region of R^+ type (which may degenerate into S^- and T regions, respectively, or the inner region into an S^+ type region, see Properties 4 and 5). The region boundaries (b_1 and b_2 in Fig. 64) between the inner and outer regions and the centreline h may be determined on the basis of the construction in Fig. 64 in which $CA = AW = EB = BW; AF = BF; CW \perp d_1; EW \perp d_2; A \in$

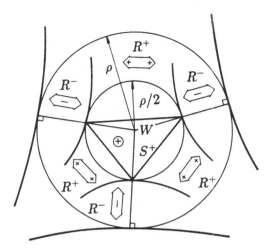

Fig. 66   Property 3 of $\beta\beta$ branches.

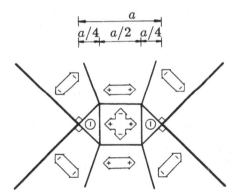

Fig. 67   Property 4 of $\beta\beta$ branches.

$b_1; B \in b_2; F \in h; \bar\kappa_2 = -k$ on CA and BE; and $\bar\kappa_1 = k$ on AB. W is termed the *intercept point*.

Property 2. Denoting the curvature of the boundary d by $r(d)$, branch centreline (h in Fig. 65) and the region boundary (b in Fig. 65) may only have their intersection G at the distance $r(d)_H/2$ from the point H having locally maximum boundary curvature, measured at right angles to the domain boundary.

Property 3. If the intercept point (W in Figs. 64 and 66) is at an equal distance from n boundary points with $n > 2$, then the inner region changes into an n-sided S^+ type region in the vicinity of such an intercept point, and its boundaries can be constructed on the basis of Fig. 66 (which shows the specific case $n = 3$).

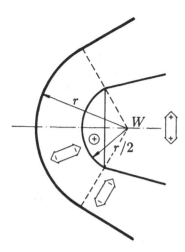

Fig. 68 Property 5 of $\beta\beta$ branches.

The foregoing S^+ type region is termed a *junction* of n $\beta\beta$ type branches and the point W is called the *centre* of the junction. If W is contained in the junction, then it represents a local maximum of the adjoint displacement field $\bar{u}(x,y)$.

Property 4. The outer regions degenerate into S^- type regions around reentrant corners and clamped-point supports, and the inner region degenerates into a T type region between two such S^- type regions. The foregoing regions can be constructed on the basis of Fig. 67. Boundaries between S^- and R^- type regions are normal to the domain boundary segments adjacent to the reentrant corners.

Property 5. When an intercept point (W in Fig. 68) is at a constant distance r from a circular domain boundary segment, then the inner region degenerates into an S^+ type region (Fig. 68) and the centreline intersects the circular domain boundary segment at its midpoint (E in Fig. 68).

Property 6. If a $\beta\beta$ branch is associated with a clamped point or reentrant corner and a straight boundary segment at a distance a (Fig. 69) from each other, then the region boundaries are parabolas:

$$y_B = a/4 + x_B^2/4a \,, \tag{203}$$

$$y_A = 3a/4 + x_A^2/a \,, \tag{204}$$

and the centreline (dash-dot line), intercept line (broken line), and envelope of the principal lines in the R^+ region (dash-double dot line), respectively, are given by

$$y_F = a/2 + 4x_F^2/9a \,, \quad y_W = a/2 + x_W^2/2a \,, \quad y = 1.25a - 3/(\bar{x}/4)^{2/3}a^{1/3} \,, \tag{205}$$

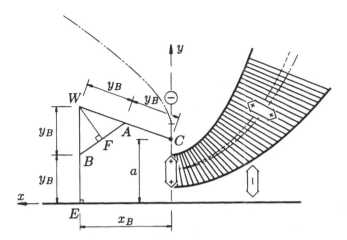

Fig. 69   Property 6 of $\beta\beta$ branches.

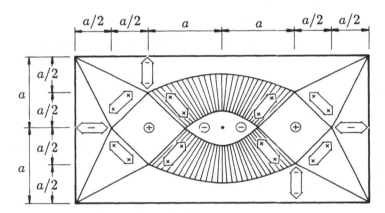

Fig. 70   An example of an optimal grillage layout for clamped supports.

where x and y are the Cartesian coordinates shown in Fig. 69. The principal lines in the inner (R^+) regions, are normal to the parabola given by (204).

For proofs of the above properties, refer to the paper by Rozvany and Hill (1976).

An example of an optimal grillage layout for clamped supports (rectangular boundary with a point support) is given in Fig. 70.

6 Properties of Solutions for a Combination of Simply Supported and Clamped Boundaries

6.1 Branches Containing α and β Fields Only

Along simply supported boundary segments, three classes of optimal fields termed α, γ and δ types may occur in the solution. Because of their relative geometrical

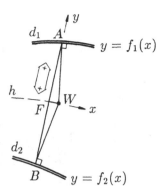

Fig. 71   Property 7 of $\alpha\alpha$ branches.

simplicity, $\alpha\alpha$ and $\alpha\beta$ type branches will be discussed first and branches containing γ or δ type fields will be dealt with in Section 6.2.

Basic properties of α fields are:

(i) The field consists of an R^+ type region.

(ii) Both principal moments take on a zero value along the domain boundary.

(iii) The curvature is $\overline{\kappa}_1 = k$ along principal lines.

It will be seen later that γ fields consist of an S^- type region and in δ fields the two principal curvatures take on values of $\overline{\kappa}_1 = k$ and $\overline{\kappa}_2 = -k$ along the domain boundary with $\overline{\kappa}_2 = -k$ on the interior of the field.

$\alpha\alpha$ *branches* are bounded by two simply supported edges and $\alpha\beta$ *branches* by one simply supported edge and one clamped edge. Their properties are as follows.

Property 7. The principal lines of an $\alpha\alpha$ branch may be constructed on the basis of Fig. 71, in which $AW = BW, AW \perp d_1, BW \perp d_2, AF = BF, F \in h$, and $\overline{\kappa}_1 = k$ on AB, where d_1 and d_2 are simply supported domain boundaries and h is the branch centreline. The foregoing construction is admissible only if

$$|f_{i,x}^2(0) + f_i(0)f_{i,xx}(0)| \leq 1, \quad i = 1, 2 , \tag{206}$$

where $y = f_1(x)$ and $y = f_2(x)$ represent the boundaries d_1 and d_2 in the coordinates shown in Fig. 71.

Property 8. If (206) is satisfied as an inequality then the corresponding part of an $\alpha\alpha$ branch is an R^+ type region. If $f_1(x) = f_2(x) = f(x)$ and $(f_{,x}^2 + ff_{,xx})|_{x=0} = 1, -1$, respectively, then the $\alpha\alpha$ branch degenerates into an S^+ and a T type region. The second principal curvature at domain boundaries is furnished by

$$\overline{\kappa}_2 = -k(f_{i,x}^2 + f_i f_{i,xx})|_{x=0}, \quad i = 1, 2 , \tag{207}$$

Property 9. The centreline h of a branch may only intersect the domain boundary d at a local maximum of the curvature of d.

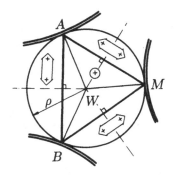

Fig. 72   Property 10 of $\alpha\alpha$ branches.

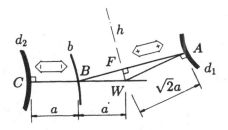

Fig. 73   Property 11 of $\alpha\beta$ branches.

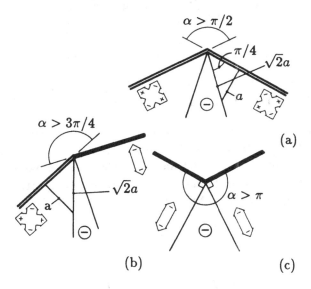

Fig. 74 S^- type regions at corners.

Property 10. If the intercept point W in Fig. 71 is at an equal distance from n domain

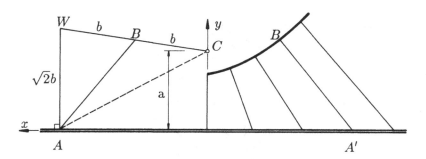

Fig. 75   Property 14 of $\alpha\beta$ branches.

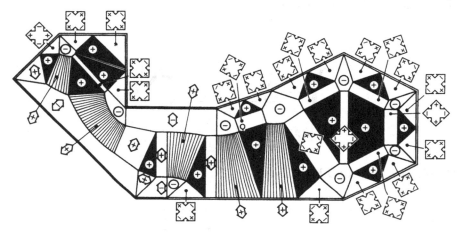

Fig. 76   Example of an optimal grillage layout for mixed support conditions.

boundary points with $n > 2$, then an n-sided S^+ type region occurs in the solution, which can be constructed on the basis of Fig. 72 (in which $n = 3$).

Property 11. The principal lines of $\alpha\beta$ branches may be constructed on the basis of Fig. 73, in which $AW = \sqrt{2}BW = \sqrt{2}CB, AW \perp d_1, BC \perp d_2, FW \perp AB, F \in h$, and $B \in b$, where d_1 is a simply supported and d_2 a clamped domain boundary, h the branch centreline, and b a region boundary. AB and BC, respectively, are principal lines in R^+ and R^- type regions. The foregoing construction is admissible only if

$$|f_{,x}^2 + ff_{,xx}|_{x=0} \leq 1 , \tag{208}$$

and

$$CW < r_c(d_2) , \qquad \text{for} \quad r_c(d_2) > 0 , \tag{209}$$

where $y = f(x)$ represents d_1, the axes x and y are given by FW and FA (Fig. 73), and $r_c(d_2)$ is the radius of curvature of d_2 at C.

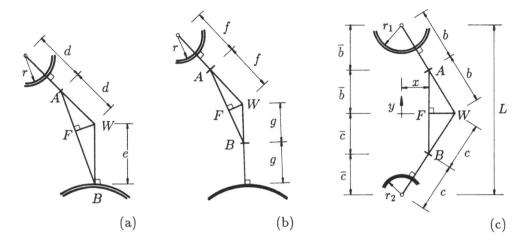

(a)   (b)   (c)

Fig. 77   Construction of $\alpha\gamma$, $\beta\gamma$ and $\gamma\gamma$ branches.

Property 12. The centreline of an $\alpha\beta$ branch may only intersect the domain boundary at a point where a line of simple support changes to a line of clamped support.

Property 13. An S^- type region occurs at a corner of a domain if the angle between two adjacent sides is
   (i) $> \pi/2$ for two simply supported edges (Fig. 74a),
   (ii) $> 3\pi/4$ for one simply supported and one clamped edge (Fig. 74b),
and
   (iii) $> \pi$ for two clamped edges (Fig. 74c).

Property 14. Let an $\alpha\beta$ branch be associated with a clamped-point support or reentrant corner C and a straight simply supported line support AA' at a distance a from each other. Then the region boundary is given by (Fig. 75)

$$y_B^2 = x_B^2 + a^2/2 \,. \tag{210}$$

The optimal grillage layout for a domain with mixed support conditions is given in Fig. 76, in which double lines indicate simple supports and thick lines a clamped edge.

6.2 Branches Containing γ and δ Fields
γ fields have the following properties:
(i) They consist of an S^- type region and an R^+ type region.
(ii) Along the boundary, the slope of the adjoint field in the normal direction is

$$\overline{\omega} = kr_0 \,, \tag{211}$$

where r is the radius of the boundary.

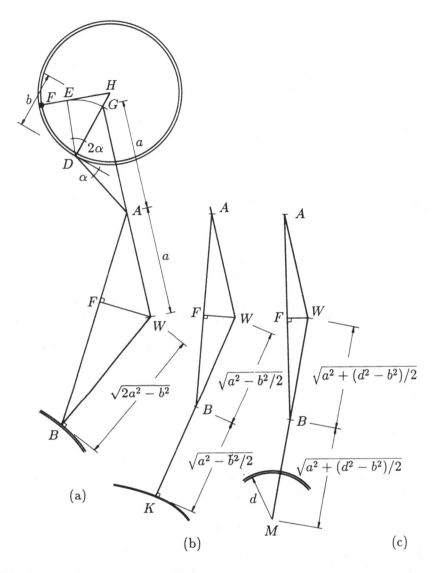

Fig. 78   Construction of $\alpha\delta$, $\beta\delta$ and $\delta\delta$ branches.

(iii) γ fields may occur only along simply supported boundaries of constant radius.
The circular boundary must be convex towards the domain.
(iv) If the radial moment along the boundary is M_r then we have a "concentrated" circumferential moment M_θ represented by

$$M_\theta = r_0 M_r \delta(r - r_0),\qquad (212)$$

where $\delta(\ )$ is a Dirac distribution (impulse).

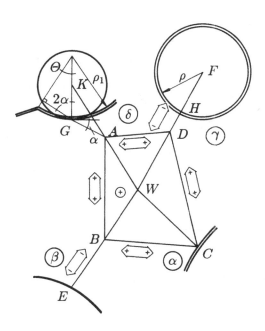

Fig. 79   Construction of a junction associated with α, β, γ and δ fields.

δ fields consist of an R^- type region and an R^+ type region. The R^- type region must fulfill the following:

$$\bar{\kappa}_2 = -k, \quad |\bar{\kappa}_1| \leq k, \quad \text{on Int } D, \tag{213}$$

$$\bar{\kappa}_2 = -k, \quad \bar{\kappa}_1 = k, \quad \text{on Bd } D. \tag{214}$$

Using the foregoing relations, various properties of δ fields have been determined (Rozvany and Hill, 1976).

The construction of $\alpha\gamma$, $\beta\gamma$ and $\gamma\gamma$ branches is given in Fig. 77 and that of $\alpha\delta$, $\beta\delta$, $\gamma\delta$ and $\delta\delta$ branches in Fig. 78 in which point F is a clamped point along a simply supported circular boundary. Moreover, construction of a junction associated with α, β, γ and δ fields is given in Fig. 79.

7 Optimal Grillage Layouts for Domains with Free Edges

By about 1974, the optimal grillage theory could handle all boundary conditions with the exception of free edges, which defied all attempts to find a solution. Finally, the so-called "free-edge paradox" was cleared up by Hill and Rozvany (1977) and Prager and Rozvany (1977b), who found that, in general, a "beam-weave" (T region of infinitesimal width) occurs along the free edge in optimal solutions (Fig. 80a). The solution for a triangular domain with two free edges and a simply supported side was

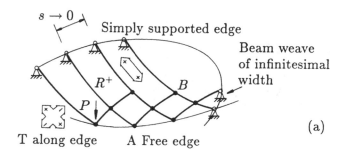

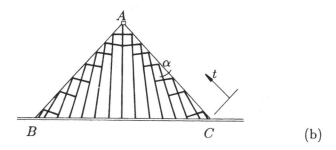

Fig. 80   Optimal layouts for grillages with free edges.

also derived by the above authors, see Fig. 80b in which

$$t = L\left(1 + \frac{1}{\sin 2\alpha}\right)^{1/2} e^{(1-\cot\alpha)/2}. \tag{215}$$

Closed form analytical solutions for other boundary conditions, including free edges parallel, at right angles and at an arbitrary angle to a simply supported edge, were given by Rozvany (1981b).

8 Optimal Grillage Layouts Obtained by a Non-Numeric Computer Algorithm

As was mentioned in Section 1, a computer algorithm was also developed for generating and plotting grillage layouts by using purely analytical operations (Rozvany and Hill, 1978b, Hill and Rozvany, 1985). One computer-produced optimal solution is shown in Fig. 81. An extended version of the above computer algorithm has been developed by D. Gerdes and W. Gollub at Essen University, and can handle circular boundary segments as well as straight ones. In addition, allowance for the cost of supports is included in this more recent algorithm.

Fig. 81   Optimal grillage layout generated on the computer by purely analytical operations.

9 Various Extensions of the Optimal Grillage Theory

Further extensions of the optimal grillage theory included *plastic grillages* with deeper edge-stiffeners and non-uniform depth (e.g. Rozvany, 1981b), solutions for up to four alternate loads (Rozvany and Hill, 1978a), partial discretization (Rozvany and Prager, 1976), allowance for the cost of supports (Rozvany, 1977), bending and shear dependent cost (Rozvany, 1979), upper constraint on the beam density (Rozvany and Wang, 1983a), allowance for selfweight with bending-dependent cost (Rozvany and Wang, 1984) and bending and shear-dependent cost (Rozvany, Yep and Sandler, 1984, 1986), as well as *elastic grillages* with deflection (Rozvany and Ong, 1986a), deflection and stress (Rozvany, Booz and Ong, 1987), and natural frequency constraints (Olhoff and Rozvany, 1982).

In Fig. 82, for example, the weight of optimal (continuous lines) and non-optimal (broken and dash-dot lines) solutions for long-span circular truss grids is compared, for which both bending and shear on a truss have an effect on the specific cost.

The parameter "c" represents the relative magnitude of the cost of shear in comparison to the cost of bending and R is the nondimensional radius of the boundary. It will be seen that a purely circumferential chord-system is optimal up to a critical radius of the boundary. At longer spans, there are circumferential chords in an

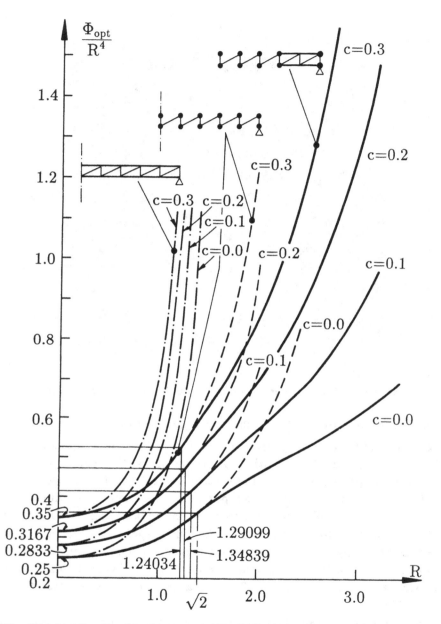

Fig. 82 Weight of optimal and non-optimal solutions for long-span circular truss grids.

inner region, radial chords in the outer region and there is a pair of "concentrated" circumferential chords along the simple supports (resisting the radial moments). To give the reader some idea of the mathematical complexity of this solution, the radial moments M_r in the outer region are represented by the function (Rozvany, Yep and

Sandler, 1986):

$$rM_r = e^{\alpha r}[A\cosh(r\beta) + B\sinh(r\beta)] - cR + r, \qquad (216)$$

with

$$A = -\sinh(a\beta)e^{a(\alpha+a/2)}\left[e^{2\alpha^2}\sqrt{\frac{\pi}{2}}2\alpha\left(\operatorname{erf}\frac{a+cR}{\sqrt{2}} - \operatorname{erf}\frac{cR}{\sqrt{2}}\right) - 1\right]/\beta +$$

$$+(cR-a)e^{-\alpha a}[\cosh(a\beta) + \alpha\sinh(a\beta)]/\beta, \qquad (217)$$

$$B = \cosh(a\beta)\{\sqrt{\frac{\pi}{2}}2\alpha e^{[(a^2/2)+2\alpha^2+\alpha a]}\left(\operatorname{erf}\frac{a+cR}{\sqrt{2}} - \operatorname{erf}\frac{cR}{\sqrt{2}}\right) - e^{[a^2/2+\alpha a]} -$$

$$-(cR-a)e^{-\alpha a}[\beta\tanh(a\beta) + \alpha]\}/\beta, \quad \alpha = cR/2, \quad \beta = (1+\alpha^2)^{1/2}, \qquad (218)$$

where

$$\operatorname{erf}(r) = \frac{2}{\sqrt{\pi}}\int_0^r e^{-t^2}\, dt \qquad (219)$$

is the "error function". It can be seen from Fig. 82 that there is often a several hundred per cent difference between the structural weight of optimal and non-optimal solutions.

10 Concluding Remarks

The rather extensive research concerning exact optimal grillage layouts was summarized briefly in this chapter. The next chapter will give an overview of the theory of the optimal layout of trusses.

Chapter 9

OPTIMAL LAYOUT OF TRUSSES: SIMPLE SOLUTIONS

G.I.N. Rozvany and W. Gollub
Essen University, Essen, Germany

1 Introduction

As was mentioned in Chapter 2, this class of optimal layouts was pioneered around
the turn of the century by an Australian scientist, A.G.M. Michell (1904). The
specific cost function ψ for the considered structures is $\psi = k|N|$, where $k = 1/\sigma_p$
is a given constant, σ_p is the permissible stress and N is the member force. The
corresponding optimality criteria were given in (1) and (2) in Chapter 2. They imply
that least-weight trusses *in the plane* must consist of the following types of optimal
regions for points with non-vanishing members:

$$
\left.
\begin{array}{lllll}
R^+ : & N_1 > 0, & N_2 = 0, & \bar{\epsilon}_1 = k, & |\bar{\epsilon}_2| \leq k, \\
R^- : & N_1 < 0, & N_2 = 0, & \bar{\epsilon}_1 = -k, & |\bar{\epsilon}_2| \leq k, \\
S^+ : & N_1 > 0, & N_2 > 0, & \bar{\epsilon}_1 = \bar{\epsilon}_2 = k, \\
S^- : & N_1 < 0, & N_2 < 0, & \bar{\epsilon}_1 = \bar{\epsilon}_2 = -k, \\
T : & N_1 > 0, & N_2 < 0, & \bar{\epsilon}_1 = -\bar{\epsilon}_2 = k,
\end{array}
\right\} \quad (|\bar{\epsilon}_1| \geq |\bar{\epsilon}_2|), \ (|N_1| \geq |N_2|),
$$

$$(220)$$

where N_1 and N_2 are the principal forces whilst $\bar{\epsilon}_1$ and $\bar{\epsilon}_2$ are the principal strains.

If we have different permissible stresses for compression (σ_{pc}) and tension (σ_{pt}),
then the corresponding cost factors become $k_c = 1/\sigma_{pc}$ and $k_t = 1/\sigma_{pt}$ and the
optimality conditions in (1) and (2) are replaced by

$$\bar{\epsilon} = k_t \quad \text{(for } N > 0\text{)} , \tag{221}$$

$$\bar{\epsilon} = -k_c \quad \text{(for } N < 0\text{)} , \tag{222}$$

$$-k_c \leq \bar{\epsilon} \leq k_t \quad \text{(for } N = 0\text{)} . \tag{223}$$

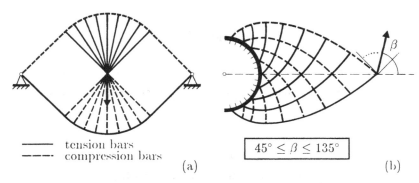

$$45° \leq \beta \leq 135°$$

(a)   (b)

Fig. 83   Some known Michell layouts.

R^+ R^- S^+ S^- T

Fig. 84   Symbols representing various types of optimal regions for grillage and Michell layouts.

For different permissible stresses in tension and compression, the regions in (220) must be modified in accordance with (221)-(223).

In constructing optimal topologies for Michell frameworks, the experience with grillage layouts is utilized. For this reason, the above two classes of problems are compared subsequently and a possible explanation for the scarcity of existing Michell-solutions is given.

Most of the known solutions for Michell frameworks have highly restrictive statical boundary conditions, such as point supports (Fig. 83a) or a support along a small circle (Fig. 83b). It will be shown in this paper, that the solution becomes relatively simple if the structural domain is bounded, at least partially, by line supports.

The classes of problems to be considered are *less restricted from a statical point of view*, because the external loads can be transmitted to a greater number of points. On the other hand, the so-called *adjoint field is more restricted kinematically*, since a zero displacement is prescribed along a larger subset of the structural domain (i.e. along the supporting lines).

Obvious similarities of least-weight truss and grillage theories can be seen from (196) and (220). The notation for various regions, established in the grillage literature and to be used herein, is shown in Fig. 84.

It has been found that, in the case of grillage layouts, *the topology and geometry*

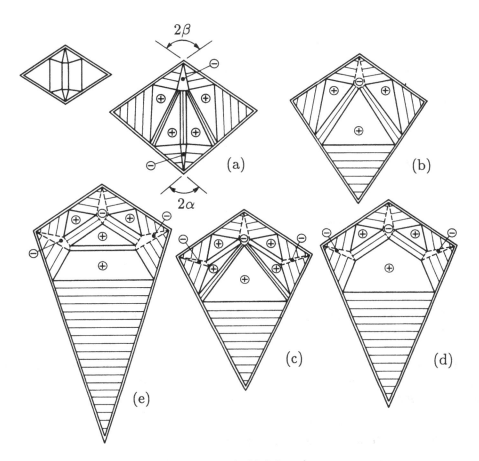

Fig. 85 Optimal grillage layouts for rhomboidal domains.

become more complicated if the kinematic boundary conditions are less restrictive. For clamped boundaries, which require zero deflection and zero slope in all directions at boundary points, the topology is relatively simple. For simply supported boundaries, requiring only zero deflection, the topology is potentially much more complicated (see Fig. 85). If the grillage has free (unsupported) edges, then the geometry becomes even more complex than the ones shown in Fig. 85.

In a very broad class of grillage problems (clamped or simply supported boundaries, clamped inner supports, non-negative loading), the adjoint displacement field is *independent of the load distribution* or the location of point loads and hence the optimal solution for complicated load conditions can be obtained by *superposition*. While the same can be observed for some Michell structures, in general the latter are highly load-dependent, which makes their derivation for complex force systems difficult.

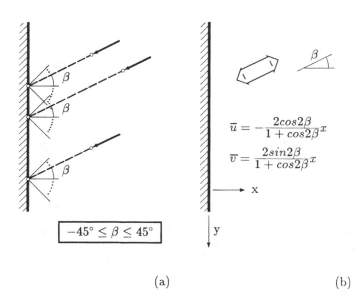

$$\bar{u} = -\frac{2\cos 2\beta}{1 + \cos 2\beta}x$$

$$\bar{v} = \frac{2\sin 2\beta}{1 + \cos 2\beta}x$$

$-45° \leq \beta \leq 45°$

(a)   (b)

Fig. 86   Optimal one-bar topology for a straight supporting line and parallel loads (a) and adjoint field (b).

As in the development of the theory of least-weight grillages, we shall start here also with simple boundary components (such as straight sides or corners) and then proceed onto complete domains surrounded by supports.

2 Solutions for a Straight Supporting Line

Note: Throughout this chapter, the range of valid load directions are indicated by dotted circular lines (see, for example, Fig. 87a). If the direction of the force must be the same as the direction of the bar, then the dotted circular arc shows the range of valid bar directions (e.g. Fig. 86a).

2.1 Parallel Loads Enclosing an Angle of $|\beta| \leq 45°$ to the Normal of the Boundary
In this chapter, the term "parallel" is used in the stricter sense, requiring all vectors to point in the same direction (and not turned around by 180°), see Fig. 86a.
   For the above geometry, the adjoint displacement components $(\bar{u}, \bar{v})$ in the x and y directions are given in Fig. 86b. It can be checked easily that we have $\bar{u} \equiv \bar{v} \equiv 0$ along the support, while the principal strains are $\bar{\epsilon}_1 = -1$ in the direction of the force and $|\bar{\epsilon}_2| \leq 1$ for $|\beta| \leq 45°$. The above strain field, therefore, satisfies the optimality conditions (1) and (2) in a normalized form (i.e. with $k = 1$, which will be used throughout this paper). Since the considered adjoint field contains a single R^- type region only, the optimal layout consists of parallel bars in the same direction (Fig. 86a) for any system of parallel forces.

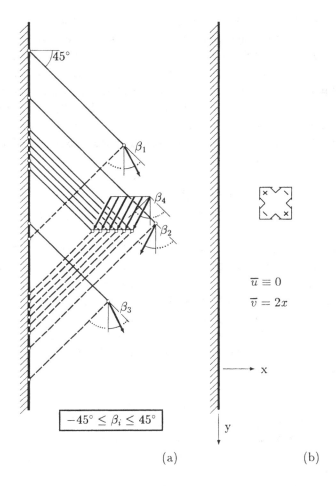

Fig. 87 Optimal two-bar topology for a straight supporting line and a single point load
(a) and adjoint field (b).

2.2 Loads Enclosing an Angle of $|\beta| \leq 45°$ with the Boundary

Whilst the class of loads discussed in Section 2.1 was limited to parallel forces, this
restriction is no longer necessary within the above range of directions, as can be seen
from the solutions in Fig. 87a. The adjoint field for this class of loads is given in Fig.
87b and represents a simple T type region. The optimal layouts in Fig. 87a consist of
two-bar trusses and the loading includes point loads and a distributed load (see the
angle β_4). The latter is transmitted by a "distributed" truss, consisting of a dense
system of two-bar systems of theoretically infinitesimal spacing. For both loading
conditions considered above, the solution is valid only if *all* forces point partially
downwards, i.e. the solution would not be valid if some forces were turned around by
180 degrees.

2.3 Two Symmetrically Positioned Sloping Point Loads Pointing Inwards

Until now we discussed load systems in which, within the restrictions given, the optimal topology could be determined by *superposition* of the solutions for individual forces, that is, there was *no interaction* between individual point loads. This is not necessarily the case if two point loads are oriented in such a way that one points partially upwards and the other one partially downwards. Clearly, if the considered loads are far enough from each other, then there is no interaction and superposition can be used. Considering, for simplicity, two symmetrically positioned sloping forces, the optimal truss topology depends on

(a) the orientation of the forces (β in Fig. 88) and

(b) their distance from each other (A in Fig. 88).

It will be shown subsequently that the optimal topology can be determined as follows.

(i) If the *forces are within 45° to the normal of the boundary* (i.e. $45° \leq \beta \leq 90°$ in Fig. 88) then each force is transmitted by a single bar of the same direction, provided that the distance (A in Fig. 88) of the forces is less then twice the distance (B in Fig. 88a) of the intersections of the lines of action with the supporting line. If this inequality is reversed, then we have a three-bar topology, all member forces having the same sign (Fig. 88b). The geometry of this three-bar layout is characterized by the property that the distance of the forces (A) is twice the distance ($A/2$) of the intersection of the bars with the supporting line.

(ii) If the *forces are within 45° to the direction of the boundary* (i.e. $0 \leq \beta \leq 45°$) then we have two two-bar systems (with one bar in tension and the other one in compression, Fig. 88c), provided that the distance (A) of the forces exceeds four times their distance (C) from the supporting line. If this condition is not fulfilled, then we again have a three-bar topology (Fig. 88d, all member forces have the same sign).

In the case of the limiting angle $\beta = 45°$, both conditions can be applied since for that angle $A = 2B$ implies $A = 4C$.

The conclusions under (i) above can be derived from the adjoint field in Fig. 89b consisting of three R^- type regions. The above adjoint field admits the topologies shown in Fig. 89a for the *simultaneous application* of point loads at a constant orientation (η) in the outer regions and symmetric pairs of forces (having any orientation within the limits $0 \leq \beta \leq 90° - \eta$) on the region boundary.

It is shown elsewhere (Rozvany and Gollub, 1990) that compatibility is satisfied by the above displacement fields and the principal direction in the outer regions is given by the equation for η in Fig. 89a. The adjoint field in Fig. 89b is valid only up to a central angle of $\alpha = 63.4349°$ corresponding to $\tan \alpha = 2.0$. Beyond that limiting value the optimal topologies in Fig. 89d take over with $\eta = 45°$. If the loads act at points lying on the region boundaries with $\tan \alpha = 2.0$, then the optimal topology consists of a five-bar truss (Fig. 89d) which allows an infinite number of

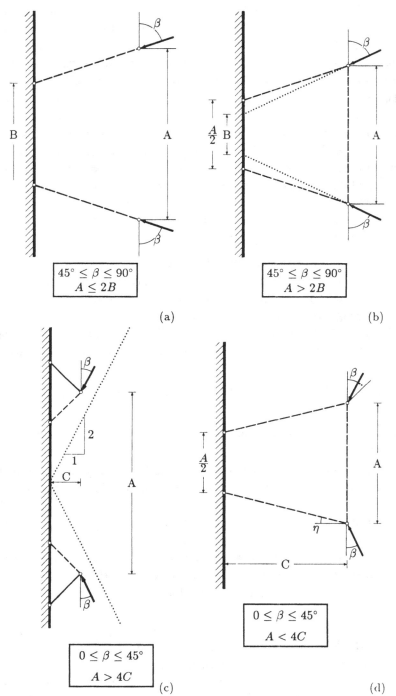

Fig. 88 Optimal topologies for a straight supporting line and two symmetrically placed point loads.

equally optimal force distributions in those bars, some of which may take on a zero cross-sectional area. The adjoint field for this solution is similar to the one in Fig. 90b and will be explained in Section 3.1. The solution in Fig. 89d is also valid for $\beta_1 \neq \beta_2$ and $\beta_3 \neq \beta_4$ within $|\beta_i| \leq 45°$; moreover, for an optimal topology, forces lying in the outer regions do not need to be located symmetrically.

Check by statical considerations and differentiation. Although the kinematic method used in this paper ensures both optimal topology and geometry, we show here on a simple example that the geometry within a given topology is indeed optimal. Let in Fig. 88d $A = 2$, $C = 1$ and $\beta = 45°$. Denoting the orientation of the near-horizontal bars by η, for unit loads the bar forces will be $1/(\sqrt{2}\cos\eta)$ and $(1 - \tan\eta)/\sqrt{2}$, while the corresponding lengths are $1/\cos\eta$ (twice) and 2. Then the total truss volume (with $k = 1$) becomes $\Phi/\sqrt{2} = (1/\cos^2\eta) + 1 - \tan\eta$. The usual stationarity condition gives $d\Phi/d\eta = (2\sin\eta/\cos^3\eta) - 1/\cos^2\eta = 0$ or $\tan\eta = 0.5$ which confirms the solution in Fig. 88d.

2.4 Two Symmetrically Positioned Point Loads Pointing Outwards

For this problem, the adjoint field is similar to that given in Fig. 89b, except that in the inner region we have an R^+ type region (with tension) and the other two principal directions are sloping outwards (instead of inwards) at an angle η, the absolute value of which is still given by the equation in Fig. 89a. The corresponding truss topologies are given in Fig. 89c. For $\tan\alpha \geq 2$ and $|\beta| \leq 45°$, the layouts in Fig. 89d are optimal.

3 A Combination of a Vertical and a Horizontal Supporting Line Forming a Corner

3.1 Equal Permissible Stress in Compression and Tension, Loads at an Angle of $|\beta| \leq 45°$ to the Vertical

The optimal topologies for this case are shown in Fig. 90a and the corresponding adjoint field in Fig. 90b. Above the region boundary, which has a slope of 2:1, two-bar trusses are optimal and below that line a single vertical member minimizes the weight or volume for vertical loads. In the case of forces along the boundary, the optimal topology consists of a three-bar truss (see the force with the angle β_4 in Fig. 90a), which allows an infinite number of optimal solutions including three different two-bar trusses. In the latter solutions, however, the direction of the force is restricted further: if, for example, the top bar vanishes, then the angle of force must be $-45° \leq \beta_4 \leq 0°$, sloping downwards from right to left.

Figure 90 verifies the optimality of the solutions in Figs. 46b and c in Chapter 7. Moreover, for the load in Fig. 46b, we could also adopt a region boundary with a slope of 2.5:1 (instead of 2:1, as in Fig. 90), with principal strains of $\bar{\epsilon}_y = -0.8$ and $\bar{\epsilon}_x = 0$. This would be a valid adjoint field, confirming the same optimal solution, although it contains an "understressed" region. This example confirms, however, our earlier observation about non-uniqueness of the adjoint.

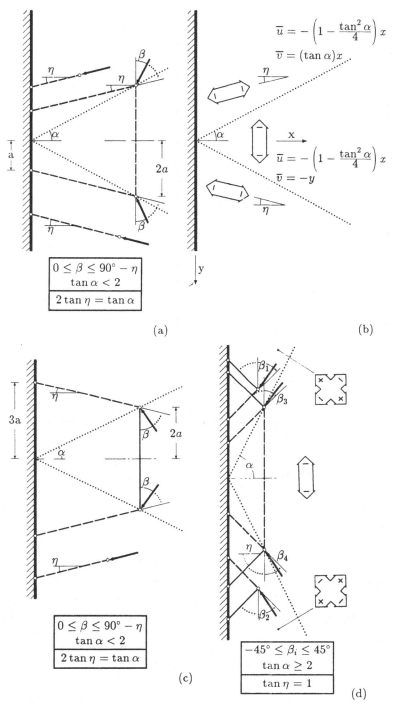

$$\overline{u} = -\left(1 - \frac{\tan^2 \alpha}{4}\right) x$$

$$\overline{v} = (\tan \alpha) x$$

$$\overline{u} = -\left(1 - \frac{\tan^2 \alpha}{4}\right) x$$

$$\overline{v} = -y$$

$$0 \le \beta \le 90° - \eta$$
$$\tan \alpha < 2$$
$$2 \tan \eta = \tan \alpha$$

(a)

(b)

$$0 \le \beta \le 90° - \eta$$
$$\tan \alpha < 2$$
$$2 \tan \eta = \tan \alpha$$

(c)

$$-45° \le \beta_i \le 45°$$
$$\tan \alpha \ge 2$$
$$\tan \eta = 1$$

(d)

Fig. 89   The adjoint fields and corresponding optimal layouts for a straight supporting line with two symmetrically placed point loads.

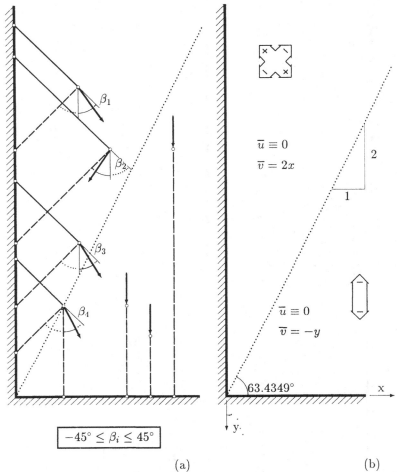

$$\boxed{-45° \leq \beta_i \leq 45°}$$

(a)   (b)

Fig. 90   Adjoint field and optimal layout for a corner region with vertical (and partly near-vertical) point loads.

3.2 Different Permissible Stresses in Compression and Tension

For cost factors of $k_t = 1$ in tension and $k_c = 3$ in compression, the optimal layouts in Fig. 91a and the adjoint field in Fig. 91b replace those in Figs. 90a and 90b. In general, for the values $k_t = 1$ and $k_c = c$, we have the following relations for optimality:

$$\cos(2\alpha) = \frac{c-1}{c+1}, \quad \tan \alpha_1 = \sqrt{c}, \tag{224}$$

where $\alpha_1 = 90° - \alpha$,

$$\tan(2\eta) = \frac{c-1}{2\sqrt{c}}, \tag{225}$$

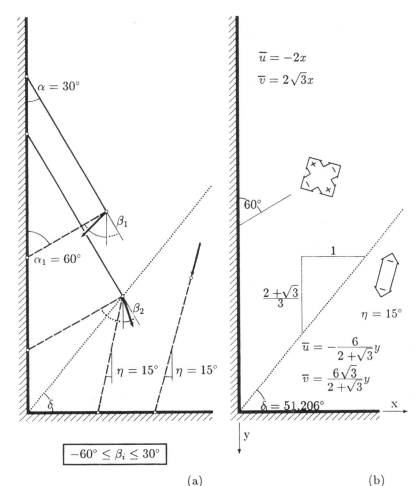

(a)   (b)

Fig. 91 Adjoint field and optimal layouts for a corner region with a point load: different permissible stresses in tension and compression $(\sigma_{pt}/\sigma_{pc} = 3)$.

$$\tan \delta = \frac{(1 + \sqrt{c})^2}{2c} = \frac{1}{2} + \frac{1}{\sqrt{c}} + \frac{1}{2c} \ . \tag{226}$$

For $c = 3$ and $c = 1$ the optimal values of α, α_1, γ and δ reduce to those in Fig. 91 and Fig. 90, respectively. For the general case considered above, the adjoint displacement fields are

$$\bar{u} = -(c-1)x \ , \quad \bar{v} = 2\sqrt{c}x \tag{227}$$

in the upper region and

$$\bar{u} = -\frac{2(c-1)c}{(1+\sqrt{c})^2}y \ , \quad \bar{v} = \frac{4c\sqrt{c}}{(1+\sqrt{c})^2}y \tag{228}$$

in the lower region.

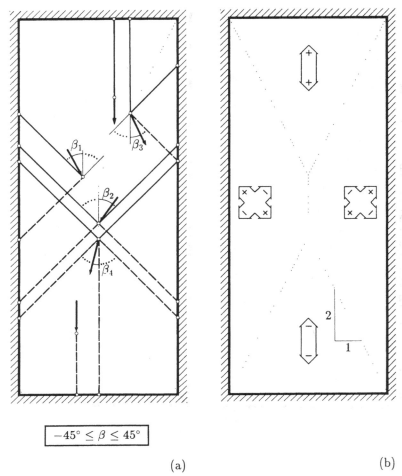

$$-45° \leq \beta \leq 45°$$

(a)   (b)

Fig. 92 Adjoint field and optimal layouts for a rectangular domain with a vertical (and partly near-vertical) point load.

4 Rectangular Domains Bounded by Two Supporting Lines: Loads Parallel to One Side

On the basis of the adjoint field in Fig. 90b, optimal topologies and the corresponding adjoint displacement regions for rectangular domains are shown in Figs. 92a and b, respectively. In the R type regions, all forces must be vertical and are transmitted by a single bar. In the T type regions, the loads may enclose an angle of $|\beta| \leq 45°$ with the vertical and they are resisted by a two-bar truss. Finally, for forces along the region boundaries the optimal layout consists of three, four or five bars; however, all these bars transmit forces only if the direction of the load allows the appropriate sign (Fig. 92) of statically admissible forces in the bars concerned. In Fig. 93, the optimal topologies and the corresponding regions of the adjoint field are shown for a

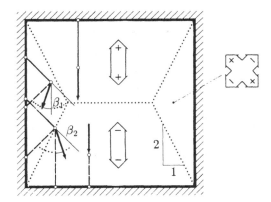

Fig. 93   Adjoint field and optimal layouts for a square domain with a vertical (and partly near-vertical) point load.

square domain.

5 Square Domains with Supports Along all Edges: Loads of Arbitrary Direction Along the Diagonals

The optimal topology for this class of layout problems depends on the direction of the load. This direction will be characterized by the angle (β_i) between the loads and one of the diagonals (Fig. 94a, bottom).

5.1 Load Orientation: $-22.5° \leq \beta_i \leq 22.5°$
For this case the optimal topologies and the corresponding adjoint field are given in Figs. 94a and b. This solution was originally described by Lagache (1981). All trusses consist of either two compression bars or two tension bars, except that for a load at the centre the optimal truss may have four members.

5.2 Load Orientation: $67.5° \leq \beta_i \leq 112.5°$
This class of topologies differs from the last one, in so far as all optimal trusses consist of a tension and a compression bar, see Figs. 94c and d. The only exception is the central load for which we have again four bars in the optimal layout.

5.3 Load Orientation: $22.5° \leq \beta \leq 67.5°$
In this case we denote the relevant angle by β (instead of β_i) because all forces along the diagonal must have the same orientation. Moreover, η denotes the angle between the horizontal and the (parallel) forces. For this loading all optimal layouts for loads along the diagonal consist of a single bar, except that for the load at the centre we have two bars of the same direction. Figure 94f shows the adjoint field and Fig. 94e the optimal topologies; the latter includes layouts for loads away from the diagonals. The derivation of the adjoint field, giving the limiting line at which the topology changes from one bar to two bars, is outlined in a paper by Rozvany and Gollub

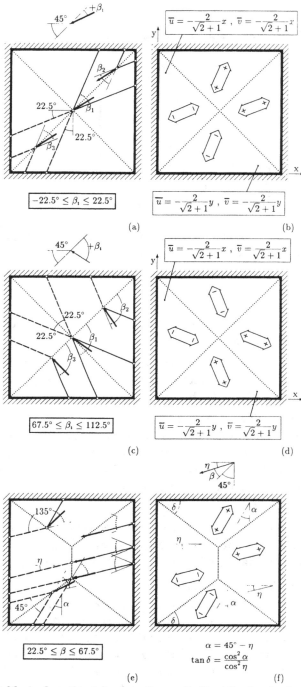

Fig. 94 Adjoint fields and optimal layouts for a single point load of arbitrary direction along the diagonal of a square domain.

(1990). Alternative formulae for the calculation of the boundary direction δ are the following:

$$\tan \delta = \frac{\cos^2(45° - \eta)}{\cos^2 \eta} = \frac{(1 + \tan \eta)^2}{2} \Rightarrow \tan \eta = \sqrt{2 \tan \delta} - 1. \qquad (229)$$

Important Note: In Figs. 94a and c the solutions for individual forces *can be superimposed* and the forces along the diagonal may have *different directions* (β_i) within the limitations given. However, in Fig. 94e all forces acting along the diagonal must have the *same direction* (β) if we want to superimpose them.

In the case of two-bar topologies in Fig. 94e, corresponding to loads along the region boundaries, the loads may take on any orientation within the directions of the bars involved, so long as the bar forces have the appropriate sign (given by Fig. 94f).

For the case of $\beta = 45°$ (horizontal forces) with $\eta = 0$, the equations in Fig. 94f give $\alpha = 45° \tan \delta = 1/2$. Hence the solution and the appropriate adjoint field reduce to those in Fig. 93 (which is to be turned around by 90°).

6 Rectangular Corner with a Single Point Load Having an Arbitrary Location and Direction

For a rectangular corner, the optimal topology may consist of two compression or two tension bars (Fig. 95a), a compression and a tension bar (Fig. 95b), or a single bar in either compression or tension (Fig. 95c). The adjoint fields for these solutions are similar to the one in Section 5.3 (see also Fig. 94f).

The optimal directions of the bars in Fig. 95a have two surprising geometrical properties:
(a) the angle between the two bars has a constant value of 45°, independently of the load location (represented by the angle δ); and
(b) the distance of the bottom end of the bars from the corner is the same on both sides (see the distance b and the equation for it in terms of δ in Fig. 95a).

The symbol d denotes the distance of the point load from the corner. The dotted line in Fig. 95a shows the distance d for the limiting case with $\eta = 45°$ and $\alpha = 0°$. The equation for b in Fig. 95a follows from geometrical considerations (Rozvany and Gollub, 1990).

Similarly, the angle between the two bars in Fig. 95b has a constant value of 135°.

The topology with two bars having forces of the same sign (Fig. 95a) can only be optimal for forces of certain orientation in the region inside the dotted line with a slope of 2:1.

7 Concluding Remarks

Optimal truss layouts were derived for some simple boundary and loading conditions in this chapter. Solutions are also available for much more complicated boundary

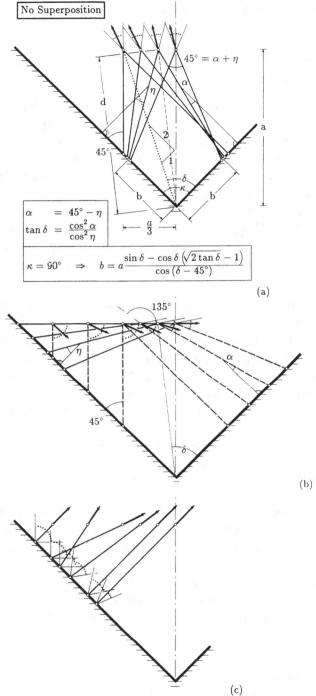

$\alpha = 45° - \eta$

$\tan \delta = \dfrac{\cos^2 \alpha}{\cos^2 \eta}$

$\kappa = 90° \quad \Rightarrow \quad b = a \dfrac{\sin \delta - \cos \delta \left(\sqrt{2 \tan \delta} - 1\right)}{\cos(\delta - 45°)}$

(a)

(b)

(c)

Fig. 95 Optimal layouts for an orthogonal corner region with a single point load in arbitrary location and direction.

conditions which, however, require a lengthy treatment (see Rozvany and Gollub, 1990; Rozvany, Gollub and Zhou, 1992). For this reason, only a few end results will be presented in the next chapter, together with some other advanced developments.

Chapter 10

OPTIMAL LAYOUT THEORY: AN OVERVIEW
OF ADVANCED DEVELOPMENTS

G.I.N. Rozvany, W. Gollub, M. Zhou and D. Gerdes
Essen University, Essen, Germany

1 Some Advanced Truss Layouts

In the last chapter, relatively simple least-weight truss topologies were discussed. Optimal truss layouts have also been determined for *non-reentrant corner regions* with a point load having an arbitrary location and direction (Rozvany and Gollub, 1990). Further developments concern *reentrant corners*, domains with free edges, convex domains bounded by supporting lines and symmetric supports consisting of two curves (Rozvany, Gollub and Zhou, 1992). Moreover, Mr. D. Gerdes has developed a *computer program for the automatic generation of least-weight trusses* for various boundary conditions. Figure 96, for example, shows the optimal layout for two different locations of a point load, which is to be transmitted to a quadrilateral boundary. The corresponding adjoint fields are also shown. The above layouts were derived analytically and fully confirmed by Mr. D. Gerdes' computer program which uses only analytical operations.

Another relatively complex problem is the verification of the solution for a *triangular domain* shown in Fig. 97b which was already derived by the iterative COC method (see Figs. 53 and 54 in Chapter 7). The adjoint field, confirming the above solution is shown in Fig. 97a. It is an interesting feature of this solution that for a very simple optimal topology, the adjoint field turns out to be rather complicated. The "O" Regions in the above field are rigid (with $\epsilon_1 = \epsilon_2 = 0$). It can be shown that the above solution is also valid for the boundary conditions in Figs. 97c and d

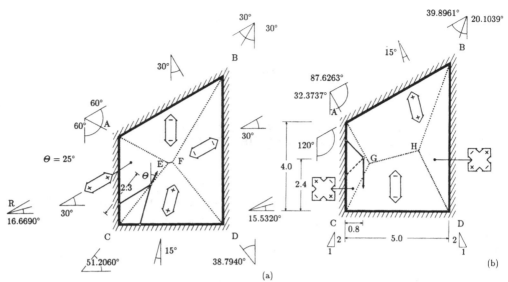

Fig. 96 Optimal layout of trusses for transmitting a point load to a quadrilateral boundary; analytical results confirmed by a non-numeric computer program.

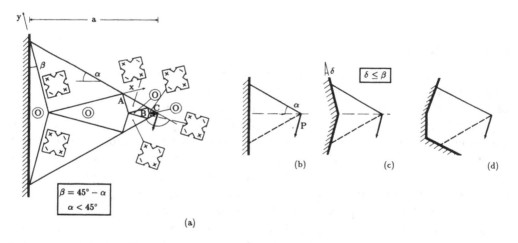

Fig. 97 Adjoint field for a layout problem with free edges.

(Rozvany, Gollub and Zhou, 1992).

In the same paper, various optimal topologies were derived rigorously for *two circular supports* and a central point load, using the analytical COC method (see Fig. 98). In Fig. 98a, for example, the layout consists of two Hencky nets, two fans and some concentrated members. If the force is sufficiently far from the supports in the vertical direction, then first the fan disappears from the solution (Fig. 98c) and at even greater distances the optimal topology consists of two bars (Figs. 98d and e).

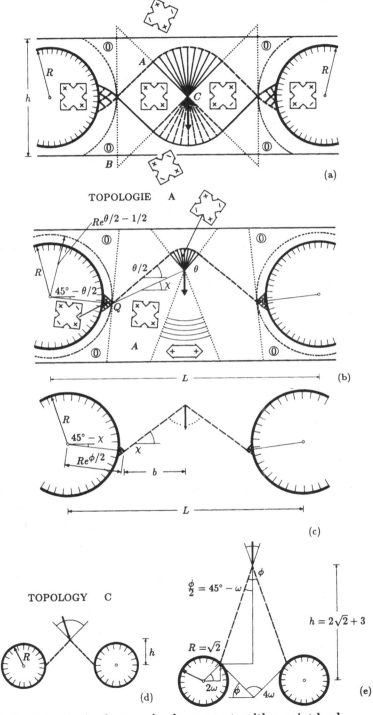

Fig. 98 Optimal topologies for two circular supports with a point load.

	WITHOUT SELFWEIGHT	WITH SELFWEIGHT
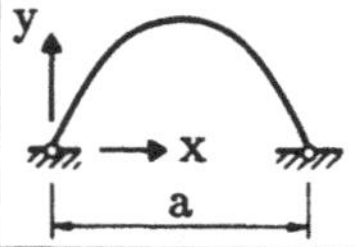	$$\dfrac{\int_0^a (dy/dx)^2\,dx}{a} = 1$$	$$\dfrac{\int_0^a (dy/dx)^2 e^{2ky}\,dx}{\int_0^a e^{2ky}\,dx} = 1$$
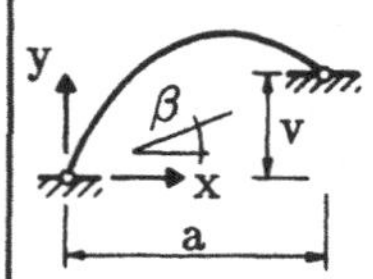	$$\dfrac{\int_0^a (dy/dx)^2\,dx}{a}$$ $$= 1 + 2\tan^2\beta$$	$$\int_0^a ke^{2ky}\left[1 - (dy/dx)^2\right]\left[1 -\right.$$ $$\left. C\int_0^x e^{-2ky}\,dx\right]dx - Cv = 0$$ where $$C = (e^{2kv} - 1)/(e^{2kv}\int_0^a e^{-2ky}\,dx)$$

Optimal Geometry   (a)

	WITHOUT SELFWEIGHT	WITH SELFWEIGHT
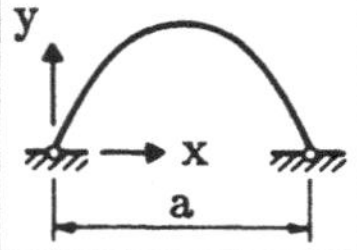	$u_y = 2ky$	$u_y = e^{2ky} - 1$
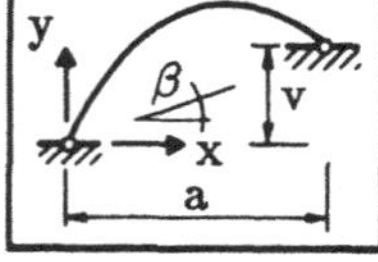	$u_y = 2k(y - x\tan\beta)$	$u_y = e^{2ky}\left[1 - \dfrac{(e^{2kv} - 1)\int_0^x e^{-2ky}\,dx}{e^{2kv}\int_0^a e^{-2ky}\,dx}\right] - 1$

Influence Lines   (b)

Fig. 99   Optimal geometry and influence lines for plane Prager structures.

2 Archgrids and Cable Nets of Optimal Layout (Prager Structures)

A Prager structure can be defined as a surface structure consisting of intersecting arches or cables for which the shape of the middle surface and the member layout are to be optimized. Moreover, the (usually vertical) loads are movable along their line of action. Alternatively, a Prager structure can be regarded as a special class of Michell frames for which (a) either the compressive or the tensile permissible stress tends to zero and (b) the position of (usually vertical) loads is unspecified and to be optimized. This special class of Michell structures has been shown to reduce always to a surface structure in 3D space (or a line structure in plane). On the basis of (221)–(223) in Chapter 9 the following optimality conditions apply to Prager structures:

$$\epsilon = k \ (\text{for } N > 0), \quad -\infty < \epsilon \le k \quad (\text{for } N = 0),$$

$$N \ge 0, \quad \epsilon_{\text{vertical}} \equiv 0. \tag{230}$$

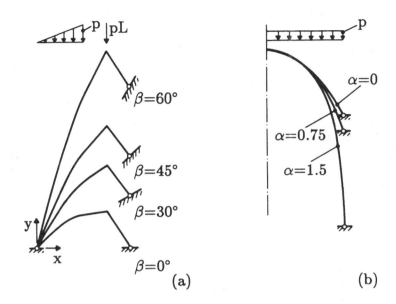

Fig. 100   Examples of plane Prager structures.

The last condition is due to the fact that loads are "movable" vertically which is equivalent to having weightless (i.e. costless) members in that direction. Closed form analytical solutions are now available for any vertical axisymmetric load in three-dimensional space and for any vertical load system in the plane and also for additional selfweight [in which case k is replaced by $k(1 + \bar{u})$ in (230) where $\bar{u}$ is the adjoint vertical displacement]. Moreover, the above solutions are extended to "quasi-axisymmetric" loads (concentrated loads distributed in the circumferential direction at equal angular intervals and axisymmetric support conditions). Fig. 99a shows the general form of the optimal solution for plane Prager structures with two supports at the same level and at different levels (at an angle β), both with or without selfweight. It can be seen that for two supports at the same level, the optimal form of the Prager structure is given by a funicular whose *mean square slope is unity*. The above optimality criteria are also valid for several supports and for axisymmetric and quasi-axisymmetric systems.

Moreover, Fig. 99b gives optimal cost influence lines u_y for various types of plane Prager structures in the sense that the total cost for any vertical load system is given by $\int_D p u_y \, dx$.

It can be seen from Fig. 99b that for the simplest case (two supports at equal elevation, no selfweight) the optimal cost (structural weight) is simply the *sum* (or integral) *of the products of loads p(x) and their elevation y(x)* multiplied by a constant $(2k)$. Using Maxwell's (1872) theorem, an alternative optimality condition for that simple case can be stated as follows: *the sum of the products of vertical forces and their elevation must equal the product of the horizontal reaction component and the distance between the two supports* (Rozvany and Wang, 1983b).

Fig. 100a shows Prager structures *without selfweight* for two supports at various angles in the vertical plane and a linearly varying load plus point load while Fig. 100b gives solutions *with selfweight* for a uniform external load and supports at the same level. The parameter α is the nondimensional span ($\alpha = kL$ where k is the cost factor and L is the span). For further information on Prager structures, the reader is referred to papers by Rozvany and Prager (1979), Rozvany, Nakamura and Kuhnell (1980), Rozvany, Wang and Dow (1982), Rozvany and Wang (1983b), Wang and Rozvany (1983) and Rozvany (1984). Quite recently, Wang derived optimality conditions for Prager structures with movable internal supports (Wang, 1987).

3 Classical and Advanced Layout Theories

All examples up to this point were based on the so-called "classical" layout theory, a generalization of Michell's (1904) theorem, which has been used for the optimization of "low-density" structural systems whose structural material occupies only a small proportion of the feasible space. This theory has two fundamental features: (a) at any point of the structural domain potential members may run in any number of directions (Fig. 101a), but (b) the effect of the member intersections on both the cost and strength (or stiffness) is neglected. It follows that the specific cost function ψ is the sum of several terms each of which depends on a stiffness (or stress resultant) value s_i,

$$\psi = \psi_1(s_1) + \psi_2(s_2) + ... + \psi_n(s_n). \tag{231}$$

"Advanced" layout theory is used for "high density" structures in which material occupies a high proportion of the feasible space or the structure consists of several materials whose interfaces are to be optimized. In this case, the microstructure of a perforated or composite structure is first optimized locally by minimizing, for given stiffnesses or stress resultants in the principal directions, the specific cost ψ (e.g. material volume per unit area or volume of the structural domain for perforated structures and some factored combination of the material volumes per unit area or volume of the structural domain for composite structures). This means that the specific cost function, e.g. $\psi(s_1, s_2)$ in Fig. 101b, is in general a non-separable function of the principal stiffnesses or stress resultants.

Advanced layout theory results in substantial extra savings for "high density" structures, but the optimal solutions given by this theory tend to those of classical layout theory if the material volume/feasible volume ratio approaches zero.

Whilst applications of classical layout theory were discussed in detail in Chapters 6-9 and in the first two section of this chapter, some applications of the advanced layout theory will be briefly reviewed in the next section. Other advanced layout problems will be discussed in Chapter 12 by Kikuchi.

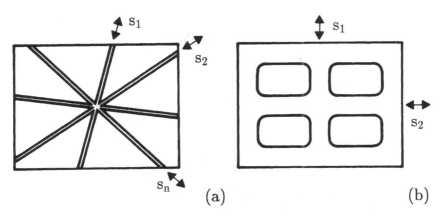

Fig. 101   Classical and advanced layout theories.

4 Applications of the Advanced Layout Theory

4.1 Optimal Plastic Design of Solid Plates

It was established already in the late sixties (e.g. Kozłowski and Mróz, 1969) and confirmed more rigorously later (Rozvany, Olhoff, Cheng and Taylor, 1982) that the weight of solid plates can be reduced to an arbitrarily small value by employing a system of sufficiently high and thin ribs. Naturally, the above solution is of purely theoretical interest because it ignores the lateral instability of such ribs. A finite value for the structural weight can be ensured, however, by introducing an *upper constraint* on the plate thickness. As can be seen from Fig. 102, Cheng and Olhoff (1981) have discovered through numerical solutions that stiffener-like formations occur in the optimal solution for such systems. Complete solutions were obtained for plastically designed solid Tresca-plates with an upper limit on the plate thickness by Rozvany, Olhoff, Cheng and Taylor (1982) and extended to other boundary conditions by Wang, Rozvany and Olhoff (1984). For simply supported circular plates, for example, the optimal solutions are compared with some non-optimal ones in Fig. 103 for various levels of the nondimensional load ν. The above solutions were obtained by establishing a specific cost function $\psi(M_1, M_2)$ through local optimization of the rib/plate configuration for given values of the principal moments (M_1, M_2) and then employing the Prager-Shield (1967) condition [(14) in Chapter 2] together with principles of the (advanced) optimal layout theory. The nondimensionalized specific cost function τ as a function of the nondimensional moments (μ_1, μ_2) is represented graphically in Fig. 104. All least-weight solutions have been found to consist of the following types of regions:

(α): solid plate of non-maximum thickness with $M_1 \equiv M_2$;

(β): ribs of maximum depth in one principal direction only, infinitesimal plate thickness in between ribs;

(γ) and (δ): solid plates of maximum feasible thickness.

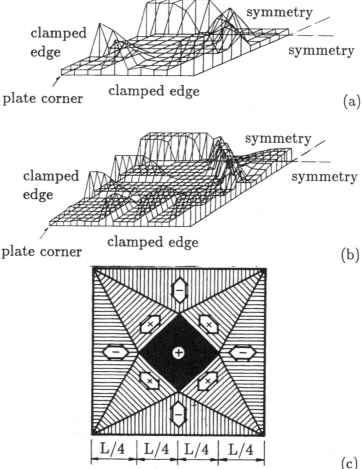

Fig. 102   (a and b) Numerical solutions by Cheng and Olhoff (1981) showing rib-like formations and (c) optimal grillage of similar layout (Rozvany and Adidam, 1972d).

It has also been found that at very low levels of the nondimensional load (ν in Fig. 103), the solution tends to the *optimal grillage layout* with type (β) regions only. This confirms Prager's intuitive insight, communicated to Olhoff shortly before his untimely death, that the rib layout for optimal solid plates is similar to the layout of least-weight grillages (see Fig. 102).

4.2 Optimal Plastic Design of Perforated Plates

A "perforated plate" may only have two thicknesses, a prescribed (maximum) thickness or zero thickness. The latter occurs over "perforations" whose in-plane dimensions are assumed to be sufficiently small so that the load over areas of zero thickness can be transmitted to the adjacent plate segments by some secondary system of negligibly small volume. In minimizing the total material volume of perforated plates,

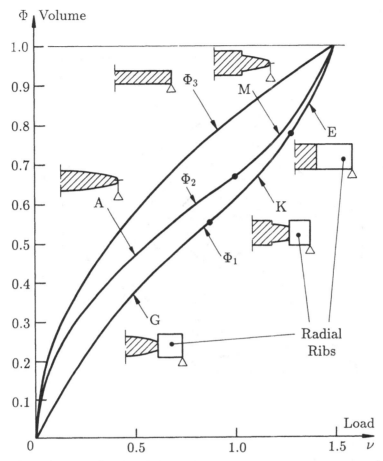

Fig. 103 Plastic plate design with a prescribed maximum thickness: a comparison of the
weight of the absolute optimal (ribbed) solution Φ_1 with that of smooth Φ_2 and
a constant thickness Φ_3 optimal solutions.

it was assumed (Rozvany and Ong, 1986b and c; Ong, 1987) that the plate material
obeys Tresca's yield condition. In perforated regions, the optimal microstructure can
be shown to consist of ribs running in the directions of the principal moments (M_1,
M_2). As stresses of the same sign do not influence the yield value of the major stress
in Tresca's yield condition, a saving can be achieved at rib intersections with sgn
$M_1 = $ sgn M_2 and the nondimensional specific cost ψ becomes

$$\psi \equiv |M_1| + |M_2| - M_1 M_2. \tag{232}$$

However, such a saving is not possible if the stresses are of opposite sign (sgn $M_1 \neq$
sgn M_2):

$$\psi = |M_1| + |M_2|. \tag{233}$$

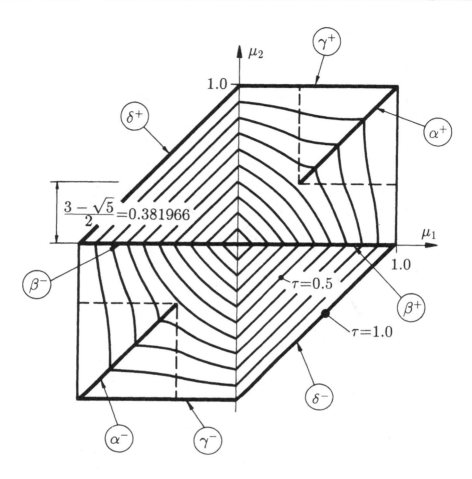

Fig. 104 Specific cost functions in terms of principal moments for solid plastic plates.

The above specific cost function (ψ) is represented graphically in Fig. 105. Making use of the Prager-Shield (1967) condition [(14) in Chapter 2], Ong (1987) has shown that for plastic axisymmetric perforated plates the least-weight solution may only consist of:

(α) unperforated regions (stress regimes α and γ in Fig. 105); and

(β) ribs in the radial direction ($M_\theta \equiv 0$, stress regime β in Fig. 105).

Introducing the nondimensional notation $r = \bar{r}/R$, $p = \bar{p}\overline{R}^2/\overline{M}$, $M_i = \overline{M}_i/\overline{M}$ ($i = r, \theta$) where $\bar{r}$ is the radial coordinate, $\overline{R}$ is the plate radius and $\overline{M}$ is the maximum feasible moment capacity, the optimal solution for simply supported uniformly loaded circular plates turns out to be the following:

$$(0 \leq r \leq g) \quad M_\theta = 1, \ M_r = 1 - pr^2/6 \quad \text{(Regime } \alpha\text{)},$$

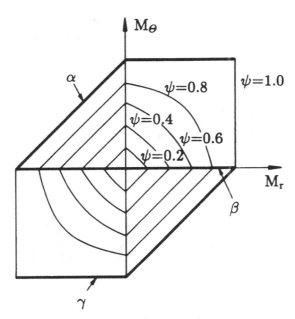

Fig. 105   Specific cost function for perforated plastic plates.

$$(g \leq r \leq 1) \quad M_\theta = 0, \ M_r = p(1 - r^3)/6r \qquad (\text{Regime } \beta),$$

$$g = p/6. \qquad\qquad (234)$$

To demonstrate the validity of the above conclusion, the volumes of various intuitively selected designs are compared in Fig. 106. Design A consists of circumferential ribs only ($M_r \equiv 0$, the shear transmission is assumed to be costless), Design B of radial and circumferential ribs of equal width ($M_\theta \equiv M_r$) throughout, Design C with $M_r \equiv M_\theta$ in an inner region and $M_\theta \equiv 0$ (only radial ribs) in the outer region and Design D (optimal design) with an unperforated plate in an inner region and $M_\theta \equiv 0$ (only radial ribs) in the outer region. Although even at small values of the nondimensional load p the optimal Design D gives a lower volume than Design C, the difference is too small for showing it graphically in Fig. 106.

4.3 Optimal Elastic Design of Perforated Plates with a Compliance Constraint

It was shown in recent papers by Murat and Tartar (1985), Lurie, Cherkaev and Fedorov (1982), Lurie, Fedorov and Cherkaev (1984), Lurie and Cherkaev (1983, 1984a, b and c), as well as Kohn and Strang (1986) and Strang and Kohn (1986), that optimal solutions in plane systems often contain regions with two sets of intersecting ribs (strips of material) at right angles: one such set has a first order infinitesimal spacing [of O(δ) with $\delta \to 0$] and the other set a second order infinitesimal spacing [of O(δ^2)]. The implications of these results for plate optimization were investigated

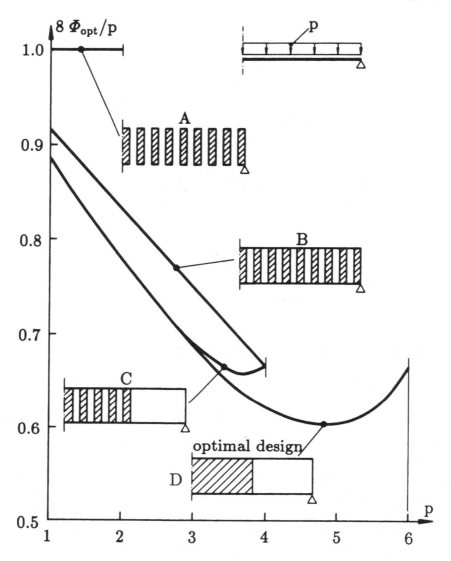

Fig. 106 A comparison of the volume of optimal and non-optimal solutions for plastic perforated circular plates.

by Rozvany, Ong, Olhoff, Bendsøe, Szeto and Sandler (1987) who arrived at the following conclusions:

(A) Given a *horizontal system of intersecting first and second order ribs* whose depth is significantly smaller than its span, it is reasonable to assume that *under a distributed vertical load* the normal stresses in the ribs are proportional to the distance from the middle surface. As a consequence of St. Venant's principle, (and a detailed

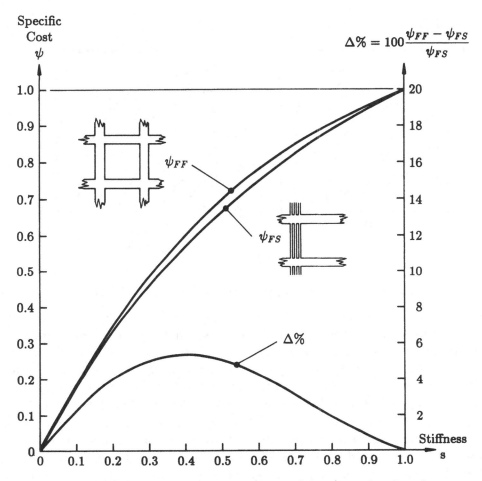

Fig. 107   Comparison of the economy of first/first and first/second order microstructures for given equal stiffnesses.

finite difference/finite element analysis), any horizontal slice of a rib of second order infinitesimal width is subjected on its interior to stresses in the direction of the rib middle plane only. The same conclusion can be obtained for a *first/second order rib system in plane stress*. Hence *second order ribs do not contribute to the stiffness in the direction normal to their middle plane*. In this respect, the formulation by Rozvany, Ong *et al.* (1987) differs from some of the mathematical studies of the authors listed above.

(B) In minimizing the weight of a perforated plate for a *given compliance*, it was shown by Rozvany, Ong *et al.* (1987) on the basis of an intuitive argument that at *low rib-densities* the material consumption is smaller for a first/second order ribbed microstructure than for a first/first order system. Moreover, detailed finite

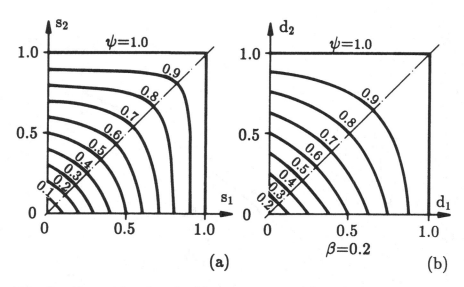

Fig. 108   Specific cost functions for (a) perforated and (b) composite plates.

element/finite difference analyses at Monash and Essen universities by Ong, Szeto, Booz, Menkenhagen and Spengemann have shown that the *first/second order microstructure is more economical than a prismatic first/first order one at all rib densities* [see Fig. 107 in which ψ_{FF} and ψ_{FS} denote, respectively, the volume of first/first and first/second order microstructures for given equal principal stiffnesses (s) and Δ is the percentage difference between the volumes of the two microstructures].

(C) Assuming a first/second order microstructure at all rib densities, a *specific cost function* was derived for perforated plates in bending or plane stress, using the simplifying feature mentioned under (A). *For a zero value of Poisson's ratio*, the latter gives a relationship between the stiffnesses (s_1, s_2) in the principal directions and the material volume ψ per unit area of the middle surface (Fig. 108a)

$$\psi = \frac{s_1 - 2s_1 s_2 + s_2}{1 - s_1 s_2}. \tag{235}$$

(D) The above specific cost function was then used for examining the design of *least-weight axisymmetric transversely loaded elastic perforated plates* of given compliance.

(E) It was found that the optimal design for the above problem reduces to that of *grillages* if the average rib density approaches zero (i.e. at very low load or high compliance levels).

(F) *Optimality criteria* were derived from a variational analysis, using the proposed microstructure.

(G) The above conditions together with static/kinematic admissibility indicated that for transversely loaded axially symmetric plates only the following *two types of regions* may occur in loaded segments of the optimal solution:

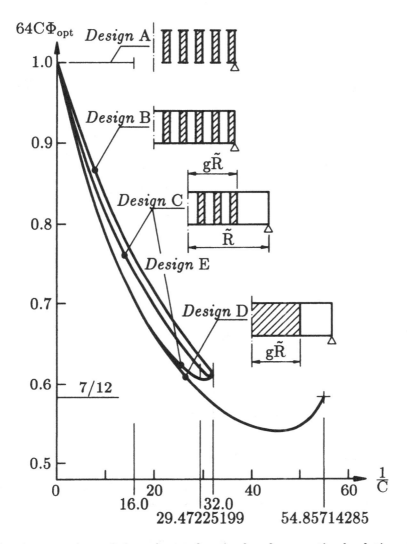

Fig. 109 A comparison of the volume of optimal and non-optimal solutions for elastic perforated plates.

(i) unperforated regions (R_0);

(ii) regions consisting of radial ribs only (R_r).

(H) On the basis of the foregoing findings, optimal solutions were derived for simply supported and clamped circular plates with uniformly distributed full and partial loading as well as a central point load and for simply supported plates with a uniform radial moment applied along their edge.

(I) The above results were confirmed by optimizing a number of intuitively selected

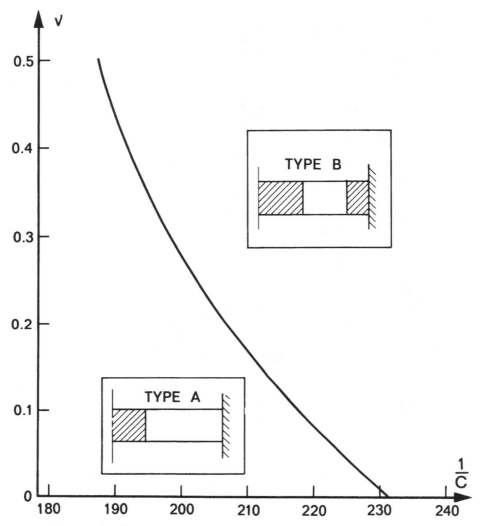

Fig. 110   Range of validity of various types of solutions; clamped perforated plates with
$\nu \neq 0$.

designs with respect to their free geometrical parameters and also by independent
numerical solutions by W. Booz (Essen University). In Fig. 109, for example, the
volume of various partially optimized intuitive designs is compared as a function
of the reciprocal compliance $1/C$. As predicted by the optimality criteria method,
Design D is optimal for all $1/C$ values.

(J) More recently (Ong, Rozvany and Szeto, 1988), the above results were extended
to plates with a *non-zero Poisson's ratio* ($\nu \neq 0$). Whereas for $\nu = 0$ the specific
compliance c was given by $c = M_1^2/s_1 + M_2^2/s_2$ where M_1 and M_2 are principal

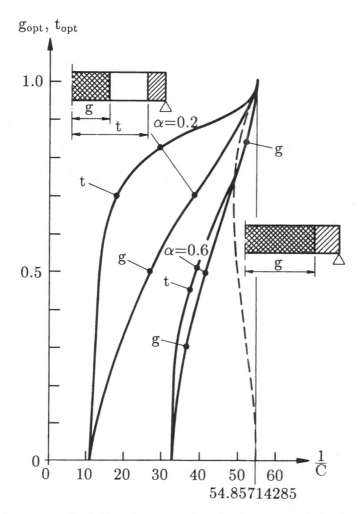

Fig. 111   Various types of solutions for composite, simply supported circular plates.

moments and s_1 and s_2 principal stiffnesses, for $\nu \neq 0$ the specific compliance can be expressed as $c = M_1^2/s_1 + M_2^2/s_2 - \nu M_1 M_2$ where s_1 and s_2 are not the principal stiffnesses any more but have the same relations to the rib densities (d_1, d_2) as for $\nu = 0$:

$$s_1 = d_1, \quad s_2 = d_2/(1 + d_1 d_2 - d_1). \tag{236}$$

Then the specific cost function can be shown to be the one in (235). Optimal solutions have been derived for various axisymmetric loading and boundary conditions. The least-weight solution for clamped circular perforated plates may consist of either one or two unperforated regions and a region with radial ribs. The range of validity of these solutions is shown in Fig. 110.

(K) Further extensions of the above approach concerned *composite plates* in which the perforations are filled with an isotropic material of lower stiffness and lower specific cost. The specific cost function for the stiffness and cost ratios $\alpha = \beta = 0.2$ is shown in Fig. 108b. It has been found that the optimal solution for *axisymmetric* composite plates may consist of:

(i) regions consisting entirely of the stiffer material (cross-hatched in Fig. 111);

(ii) regions consisting entirely of the less stiff material (hatched in Fig. 111); and

(iii) regions with radial ribs made out of the stiffer material and the gaps filled with the less stiff material (unhatched in Fig. 111 which gives the optimal solution for simply supported, uniformly loaded circular plates with $\alpha = \beta = 0.2$).

(L) The above investigation has also been extended to

(i) plates with a prescribed deflection at a given location; and

(ii) allowance for the effect of shear on the compliance.

References — Chapters 1-10

Bendsøe, M.P. 1989: Optimal shape design as a material distribution problem. *Struct. Optim.* **1**, 193-202.

Berke, L.; Khot, N.S. 1987: Structural optimization using optimality criteria. In: Mota Soares, C.A. (Ed.) *Computer aided optimal design: structural and mechanical systems*, pp. 271-312. Springer-Verlag, Berlin.

Berke, L.; Khot, N.S. 1988: Performance characteristics of optimality criteria methods. In: Rozvany, G.I.N.; Karihaloo, B.L. (Eds.) *Structural optimization* (Proc. IUTAM Symposium, Melbourne, 1988), pp. 39-46. Kluwer, Dordrecht.

Cheng, K.-T.; Olhoff, N. 1981: An investigation concerning optimal design of solid elastic plates. *Int. J. Solids Struct.* **17**, 305-323.

Foulkes, J. 1954: The minimum-weight design of structural frames. *Proc. Royal Soc.* **223**, No. 1155, 482-494.

Hemp, W.S. 1973: *Optimum Structures*. Clarendon, Oxford.

Hegemier, G.A.; Prager, W. 1969: On Michell trusses. *Int. J. Mech. Sci.* **11**, 209-215.

Heyman, J. 1959: On the absolute minimum weight design of framed structures. *Quart. J. Mech. Appl. Math.* **12**, 3, 314-324.

Hill, R.H.; Rozvany, G.I.N. 1977: Optimal beam layouts: the free edge paradox. *J. Appl. Mech.* **44**, 696-700.

Hill, R.H.; Rozvany, G.I.N. 1985: Prager's layout theory: a nonnumeric computer method for generating optimal structural configurations and weight-influence surfaces. *Comp. Meth. Appl. Mech. Engrg.* **49**, 1, 131-148.

Kirsch, U. 1989: On the relationship between structural topologies and geometries. *Struct. Optim.* **2**, 39-45.

Kirsch, U.; Rozvany, G.I.N. 1992: Design considerations in the optimization of structural topologies. In: Rozvany, G.I.N. (Ed.) *Proc. NATO/DFG ASI, Optimization of large structural systems* (held 23 September - 4 October 1991, Berchtesgaden). Kluwer, Dordrecht.

Kohn, R.V.; Strang, G. 1986: Optimal design and relaxation of variational problems, I, II and III. *Comm. Pure Appl. Math.* **39**, 113-137, 139-182, 353-377.

Kozlowski, W.; Mróz, Z. 1969: Optimal design of solid plates. *Int. J. Solids Struct.* **5**, 8, 781-794.

Lagache, J.-M. 1981: Developments in Michell theory. In: Atrek, E.; Gallagher, R.H. (Eds.) *Proc. Int. Symp. on Optimum Structural Design* (held in Tucson, Oct. 1981), pp. 4.9-4.16. University of Arizona, Tucson.

Lurie, K.A.; Cherkaev, A.V. 1983: Optimal structural design and relaxed controls. *Opt. Control Appl. Meth.* **4**, 4, 387-392.

Lurie, K.A.; Cherkaev, A.V. 1984a: G-Closure of a set of anisotropically conducting media in the two-dimensional case. *J. Optimiz. Theory Appl.* **42**, 2, 283-304.

Lurie, K.A.; Cherkaev, A.V. 1984b: Exact estimates of conductivity of composites formed by two isotropically conducting media taken in prescribed proportions. *Proc. Roy. Soc. Edinburgh* **99 A**, 71-87.

Lurie, K.A.; Cherkaev, A.V. 1984c: G-closure of some particular sets of admissible material characteristics for the problem of bending of thin elastic plates. *J. Opt. Theory Appl.* **42**, 2, 305-316.

Lurie, K.A.; Cherkaev, A.V.; Fedorov, A.V. 1982: Regularization of optimal design problems for bars and plates. *J. Optimiz. Theory Appl.* **37**, 4, 499-522, 523-543.

Lurie, K.A.; Fedorov, A.V.; Cherkaev, A.V. 1984: On the existence of solutions for some problems of optimal design for bars and plates. *J. Optimiz. Therory Appl.* **42**, 2, 247-281.

Marcal, P.V.; Prager, W. 1964: A method of optimal plastic design. *J. de Mécan.* **3**, 4, 509-530.

Masur, E.F. 1970: Optimum stiffness and strength of elastic structures. *J. Eng. Mech. ASCE* **96**, EM6, 1093-1106.

Masur, E.F. 1975: Optimality in the presence of discreteness and discontinuity. In: Sawczuk, A.; Mróz, Z. (Eds.) *Optimization in structural design* (Proc. IUTAM Symp. held in Warsaw, Aug. 1973), pp. 441-453, Springer-Verlag, Berlin.

Maxwell, J.C. 1872: On reciprocal figures, frames and diagrams of force. *Trans. Roy. Soc. Edinburgh* **26**, 1. Also in: *Scientific Papers* **2** [Niven, W.D. (Ed.), 1890] University Press, Cambridge, 174-177.

Michell, A.G.M. 1904: The limits of economy of material in frame-structures. *Phil. Mag.* **8**, 47, 589-597.

Morley, C.T. 1966: The minimum reinforcement of concrete slabs. *Int. J. Mech. Sci.* **8**, 305-319.

Mróz, Z. 1963: Limit analysis of plastic structures subject to boundary variations. *Arch. Mech. Stos.* **15**, 1, 63-76.

Murat, F.; Tartar, L. 1985: Calcul des variations et homogénéisation. In: *Les méthodes de l'homogénéisation: théorie et applications en physique.* Coll. del la Dir. des Etudes et recherches de Elec. de France, Eyrolles, Paris, pp. 319-370.

Nagetegaal, J.C.; Prager, W. 1973: Optimal layout of a truss for alternative loads. *Int. J. Mech. Sci* **15**, 7, 583-592.

Olhoff, N.; Rozvany, G.I.N. 1982: Optimal grillage layout for given Natural frequency. *J. Engrg. Mech. ASCE* **108**, EM5, 971-975.

Ong. T.G. 1987: *Structural optimization via static-kinematic optimality criteria.* Ph. D Thesis, Monash Univ., Melbourne, Australia.

Ong T.G.; Rozvany, G.I.N.; Szeto, W.T. 1988: Least weight design for perforated elastic plates for given compliance: non-zero Poisson's ratio. *Comp. Meth. Appl. Mech. Engrg.* **66**, 301-322.

Prager, W. 1974: *Introduction to structural optimization.* (Course held Int. Centre for Mech. Sci. Udine. CSIM **212**) Springer-Verlag, Vienna.

Prager, W.; Rozvany, G.I.N. 1977a: Optimization of structural geometry. In: Bednarek, A.R.; Cesari, L. (Eds.) *Dynamical systems.* Academic Press, New York.

Prager, W,: Shield, R.T. 1967: A general theory of optimal plastic design. *J. Appl. Mech.* **34**, 1, 184-186.

Prager, W.; Taylor, J. 1968: *Problems of optimal structural design. J. Appl. Mech.* **35**, 102-106.

Prager, W.; Rozvany, G.I.N. 1977b: Optimal layout of grillages. *J. Struct. Mech.* **5**, 1, 1-18.

Rozvany, G.I.N. 1974: Analytical treatment of some extended problems in structural optimization, Part II. *J. Struct. Mech.* **3**, 4, 387-402.

Rozvany, G.I.N. 1976: *Optimal design of flexural systems.* Pergamon Press, Oxford. Russian translation: Stroiizdat, Moscow, 1980.

Rozvany, G.I.N. 1977: New trends in structural optimization. *Proc. 6th Australian Conf. Mech. Struct. Mater.* (held in Christchurch, New Zealand), Univ. Canterbury, pp. 391-398.

Rozvany, G.I.N. 1979: Optimal beam layouts: allowance for cost of shear. *Comp. Meth. Appl. Mech. Engrg.* **19**, 1, 49-58.

Rozvany, G.I.N. 1981a: Variational methods and optimality criteria. In: Haug, E.J.; Cea, J. (Eds.) *Optimization of distributed parameter structures* (Proc. NATO ASI held in Iowa City), pp. 82-111. Sijthof and Noordhof, Alphen aan der Rijn, The Netherlands.

Rozvany, G.I.N. 1981b: Optimal criteria for grids, shells and arches. In: Haug, E.J.; Cea, J. (Eds.) *Optimization of distributed parameter structures* (Proc. NATO ASI held in Iowa City), pp. 112-151. Sijthof and Noordhof, Alphen aan der Rijn, The Netherlands.

Rozvany, G.I.N. 1984: Structural layout theory: the present state of knowledge. In: Atrek, E.; Gallagher, R.H.; Ragsdell, K.M.; Zienkiewicz, O.C. (Eds.) *New directions in optimum structural design,* pp. 167-195. Wiley & Sons, Chichester, England.

Rozvany, G.I.N. 1989: *Structural design via optimality criteria (the Prager approach to structural optimization).* Kluwer, Dordrecht.

Rozvany, G.I.N.; Booz, W.; Ong, T.G. 1987: Optimal layout theory: multiconstraint elastic design. In: Teo, K.L.; Paul, H.; Chew, C.L.; Wang, C.M. 1987: *Proc. Int. Conf. on Opti-*

mization: Techniques and Applications (held in Singapore, April 1987), pp. 138-151. Nat. Univ. Singapore.

Rozvany, G.I.N.; Gollub, W. 1990: Michell layouts for various combinations of line supports, Part I. *Int. J. Mech. Sci.* **32**, 1021-1043.

Rozvany, G.I.N.; Gollub, W.; Zhou, M. 1989a: Optimal design of large discretized systems by iterative otimality criteria methods. In: Topping, B.H.V. (Ed.) *Proc. NATO ASI, Optimization and decision support systems in civil engineering* (held 25 June - 7 July 1989, Edinburgh). Kluwer, Dordrecht.

Rozvany, G.I.N.; Gollub, W.; Zhou, M. 1989b: Layout optimization in structural design. In: Topping, B.H.V. (Ed.) *Proc. NATO ASI, Optimization and decision support systems in civil engineering* (held 25 June - 7 July 1989, Edinburgh). Kluwer, Dordrecht.

Rozvany, G.I.N.; Gollub, W.; Zhou, M. 1992: Michell layouts for various combinations of line supports, Part II (to be submitted to *Int. J. Mech. Sci.*).

Rozvany, G.I.N.; Hill, R.H. 1976: General theory of optimal force transmission by flexure. *Advances in Appl. Mech.* **16**, 184-308.

Rozvany, G.I.N.; Hill, R.H. 1978a: Optimal plastic design: superposition principles and bounds on the minimum cost. *Comp. Meth. Appl. Mech. Engrg.* **13**, 2, 151-173.

Rozvany, G.I.N.; Hill, R.H. 1978b: A computer algorithm for deriving analytically and plotting optimal structural layout. In: Noor, A.K.; McComb, H.G. (Eds.) *Trends in computerized analysis and synthesis* (Proc. NASA/ASCE Symp. held in Washington D.C., Oct. 1978), pp. 295-300. Wiley, New York. Aslo: *Comp. and Struct.* **10**, 1, 295-300.

Rozvany, G.I.N.; Nakamura, H.; Kuhnell, B.T. 1980: Optimal archgrids: allowance for self-weight. *Comp. Meth. Appl. Mech. Engrg.* **24**, 3, 287-304.

Rozvany, G.I.N.; Olhoff, N.; Cheng, K.-T.; Taylor, J.E. 1982: On the solid plate paradox in structural optimization. *DCAMM Report* **212** June 1981 and *J. Struct. Mech.* **10**, 1, 1-32.

Rozvany, G.I.N.; Ong, T.G. 1986a: A general theory of optimal layouts for elastic structures. *J. Engrg. Mech. Div. ASCE* **112**, 8, 851-857.

Rozvany, G.I.N.; Ong, T.G. 1986b: Optimal plastic design of plates, shells and shellgrids. In: Bevilacqua, L.; Feijóo, R.; Valid, R. (Eds.) *Inelastic behaviour of plates and shells* (Proc. IUTAM Symp. held in Rio de Janeiro, August 1985), p. 357-384, Springer-Verlag, Berlin.

Rozvany, G.I.N.; Ong, T.G. 1986c: Update to "Analytical methods in structural optimization". In: Steele, C.R.; Springer, G.S. (Eds.) *Applied Mechanics Update*, pp. 289-302, ASME, New York.

Rozvany, G.I.N.; Ong, T.G. 1987: Minimum-weight plate design via Prager's layout theory (Prager Memorial Lecture). In: Mota Soares (Ed.) *Computer aided optimal design: structural and mechanical systems* (Proc. NATO ASI held in Troia, Portugal, 1986), pp. 165-179, Springer-Verlag, Berlin.

Rozvany, G.I.N.; Ong, T.G.; Olhoff, N.; Bendsøe, M.P.; Szeto, W.T.; Sandler, R.: Least-weight design of perforated elastic plates I and II. *Int. J. Solids Struct.* **23**, 4, 521-536, 537-550.

Rozvany, G.I.N.; Ong, T.G.; Sandler, R.; Szeto, W.T.; Olhoff, N.; Bendsøe, M.P. 1987a:

Least-weight design of perforated elastic plates I. *Int. J. of Solids Struct.* **23**, 4, 521-536.

Rozvany, G.I.N.; Ong, T.G.; Sandler, R.; Szeto, W.T.; Olhoff, N.; Bendsøe, M.P. 1987b: Least-weight design of perforated elastic plates II. *Int. J. of Solids Struct.* **23**, 4, 537-550.

Rozvany, G.I.N; Prager, W. 1976: Optimal design of partially discretized grillages. *J. Mech. Phys. Solids* **24**, 2/3, 125-136.

Rozvany, G.I.N.; Prager, W. 1979: A new class of structural optimization problems: optimal archgrids. *Comp. Meth. Appl. Mech. Engrg.* **19**, 1, 127-150.

Rozvany, G.I.N.; Rotthaus, M.; Spengemann, F.; Gollub, W.; Zhou, M. 1990: The Masur paradox. *Mech. Struct. Mach.* **18**, 21-42.

Rozvany, G.I.N.; Wang, C.M. 1983a: Constrained optimal layouts through Prager-Shield criteria. *J. Engrg. Mech. Div. ASCE* **109**, 2, 648-653.

Rozvany, G.I.N.; Wang, C.M. 1983b: On plane Prager-structures (I). *Int. J. Mech. Sci.* **25**, 7, 519-527.

Rozvany, G.I.N.; Wang, C.M. 1984: Optimal layout theory: allowance for selfweight. *J. Engrg. Mech. Div. ASCE* **110**, EM1, 66-83.

Rozvany, G.I.N.; Wang, C.M.; Dow, M. 1982: Prager structures: archgrids and cable networks of optimal layout. *Comp. Meth. Appl. Mech. Engrg.* **31**, 1, 91-113.

Rozvany, G.I.N.; Yep, K.M.; Sandler, R. 1984: Optimal design of long-span truss-grids. In: Nooshin, H. (Ed.) *Proc. 3rd Int. Conf. on Space Structures* (held at the University of Surrey, Guildford, Sept. 1984), pp. 689-694, Elsevier Appl. Sci. Publ., London.

Rozvany, G.I.N.; Yep, K.M.; Sandler, R. 1986: Optimal layout of long-span truss-grids I. *Int. J. of Solids Struct.* **22**, 2, 209-223.

Rozvany, G.I.N.; Zhou, M.; Gollub, W. 1990: Continuum-type optimality criteria methods for large finite element systems with a displacement constraint. Part II. *Struct. Optim.* **2**, 77-104.

Rozvany, G.I.N.; Zhou, M.; Gollub, W. 1991: Layout optimization in structural design. In: B.H.V. Topping (Ed.) *Proc. NATO ASI, Optimization and decision support systems in civil engineering* (held 25 June - 7 July 1989, Edinburgh). Kluwer, Dordrecht.

Rozvany, G.I.N.; Zhou, M. 1991: A new direction in cross-section and layout optimization: the COC algorithm. In: Hernandez, S.; Brebbia, C.A. (Eds.) *Proc. OPTI 91, Optimization of structural systems and industrial applications.* Comp. Mech. Publ., Southampton.

Rozvany, G.I.N.; Zhou, M. 1992: Continuum-based optimality criteria (COC) methods. In: Rozvany, G.I.N. (Ed.) *Proc. NATO/DFG ASI, Optimization of large structural systems* (held 23 September - 4 October 1991, Berchtesgaden). Kluwer, Dordrecht.

Save, M. 1972: A unified formulation of the theory of optimal design with convex cost function. *J. Struct. Mech.* **1**, 2, 267-276.

Shield, R.T. 1960: Plate design for minimum weight. *Quart. Appl. Math.* **18**, 2, 131-144.

Strang, G.; Kohn, R.V. 1986: Optimal design in elasticity and plasticity. *Int. J. Num. Math. Eng.* **22**, 183-188.

Spillers, W.R.; Lev. O. 1971: Design for two loading conditions. *Int. J. Solids Struct.* **7**, 1261-1267.

Wang, C.M. 1987: Optimization of multispan plane Prager-structures with variable support locations. *J. Eng. Struct.* **9**, 157-161.

Wang, C.M.; Rozvany, G.I.N. 1983: On plane Prager-structures (II) — Non-parallel external loads and allowance for selfweight. *Int. J. Mech. Sci.* **25**, 7, 529-541.

Wang, C.M.; Rozvany, G.I.N.; Olhoff, N. 1984: Optimal plastic design of axisymmetric solid plates with a maximum thickness constraint. *Comp. and Struct.* **18**, 4, 653-665.

Zhou; M. 1990: *Iterative continuum-type optimality criteria methods in structural optimization*. Res. Report, Essen University.

Chapter 11

CAD-INTEGRATED STRUCTURAL
TOPOLOGY AND DESIGN OPTIMIZATION

N. Olhoff
University of Aalborg, Aalborg, Denmark

M.P. Bendsøe
Technical University Denmark, Lyngby, Denmark

J. Rasmussen
University of Aalborg, Aalborg, Denmark

1. INTRODUCTION

Structural optimization [1,2] can be essentially conceived as a rational search for the optimal spatial distribution of material within a prescribed admissible structural domain, assuming the loading and boundary conditions to be given. In the general case, this problem consists in determining both the optimal topology and the optimal design of the structure. Here the label "optimal design" covers the optimal shape or sizing of the design.

Recent years have witnessed a definite tendency towards development and augmentation of large, practise−oriented optimization systems, se e.g. [3–15]. However, despite many attractive features, these systems are as of yet only capable of conducting shape optimization within the framework of a given structural topology. Thus, until recently, methods for topology optimization were only available for truss−like structures composed of slender members, see e.g. [16], so only systems that include sizing optimization of such structures may, up to now, have been endowed with some sort of tool for topology optimization. This picture is being rapidly changed, however, as substantial current research efforts are devoted to integration of methods of topology optimization [17,18] and optimal design, see [19–27].

Along these lines, these lecture notes deal with the development of complementary methods for optimization of structural topologies and designs, and as a special feature we describe their actual integration and implementation in an interactive CAD−based structural optimization system [13–14]. The presentation is restricted to optimization problems involving linearly elastic, two−dimensional structures and components. In Section 2 we present a method of optimization of structural topology, and then discuss in linearized strains. H is the set of kinematically admissible displacements. The problem is defined on a fixed reference domain Ω and the components of the tensor of elasticity E_{ijkl} depend on the design variables used. For a so–called second rank layering constructed as in fig. 1, we have a relation

$$E_{ijkl} \equiv E_{ijkl}(\mu,\gamma,\theta) \tag{4}$$

where μ and γ denote the densities of the layering and θ is the rotation angle of the layering. The relation (4) can be computed analytically [18] and for the volume we have

$$Volume = \int_{\Omega}(\mu +\gamma - \mu\gamma)d\Omega \tag{5}$$

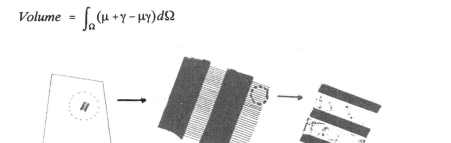

MACRO-SCALE

MICRO-SCALE 1

MICRO-SCALE 2

Fig. 1. *Construction of a layering of second rank.*

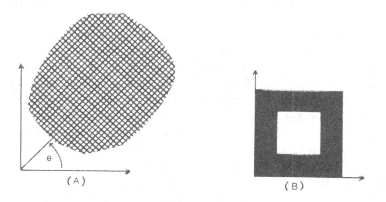

(A)

(B)

Fig. 2. *(A): a periodic microstructure with square holes rotated the angle θ.*
(B): a square cell with a square hole.

Layered materials (used for topology generation in the example in Section 6.1) is just one possible choice of microstructure that can be applied. The important feature is to choose a microstructure that allows density of material to cover the complete range of values from zero (void) to one (solid), and that this microstructure is periodic so that effective properties can be computed (numerically) through homogenization (theory of cells). This excludes circular holes in square cells, while square or rectangular holes in square cells, see fig. 2, are suitable choices of simple microstructures. For the case of a rectangular hole in a square cell, the volume is also given by eq.(5), with μ and γ denoting the amount of material used in the directions of axes of the cell and hole; for this microstructure the angle of rotation of the unit cell becomes a design variable. (This type of microstructure is used for topology generation in the examples of Section 6.2 og 6.3).

The optimization problem can now be solved either by optimality criteria methods [18] or by duality methods [28], where advantage is taken of the fact that the problem has just one constraint. The angle θ of layer or cell rotation is controlled via the results on optimal rotation of orthotropic materials as presented in [29,30].

As stated above, the optimum topology is determined from the condition of minimum compliance subject to a bound on the total structural volume. Shape- and sizing optimization systems, on the other hand, must be able to handle a much larger variety of formulations of which stress and volume minimizations are the most frequent. However, in spite of the incompatible formulations, compliance optimized topologies tend to perform well also from a stress minimization point of view. This is due to the fact that a relatively high amount of energy is stored in possible areas of stress concentration which then become undesirable also in the compliance minimization. The initial topology optimization can therefore in many cases lead to substantial improvements of the final result even though the actual aim is to perform an optimization of a different type.

The topology optimization thus results in a prediction of the structural type and overall lay-out, and gives a rough description of the shape of outer as well as inner boundaries of the structure. This motivates an integration of topology and design optimization [19–27]. In order to gain the full advantage of these design tools, it is necessary that they be integrated and implemented in a flexible, user–friendly, interactive CAD environment with extensive computer graphics facilities. This type of integration will be discussed in Section 3.

Depending on the amount of material available, the generated topology will basically either define the rough shape of a two–dimensional structural domain, possibly with macroscropic interior holes (which shape optimization procedures cannot create), or the skeleton of a truss- or beam–like structure with slender members. The main idea is that the optimal topology can be used as a basis for procedures of refined shape optimization by means of boundary variations techniques (Section 4), or optimization of member sizes and positions of connections by means of sizing optimization techniques (Section 5). Certain cases may even require methods of lay-out optimization [31–33].

3. INTEGRATION OF TOPOLOGY AND DESIGN OPTIMIZATION IN A CAD ENVIRONMENT

The actual topology preprocessor HOMOPT [18] is integrated with the structural shape optimization system CAOS (Computer Aided Optimization of Shapes) [13,14]. CAOS is based on the concept [34] of integration of software modules of finite element analysis, sensitivity analysis, and optimization by mathematical programming, and the system is developed in the spirit of refs. [6,7,35,36]. CAOS was originally based solely on the boundary variation method, i.e., the optimized geometry is a result of changing the shapes of the original boundaries of the structure [37,38]. The initial purpose of CAOS was to act as a framework for experiments with various structural optimization techniques, and, above all, to examine the problem of integrating shape optimization into a traditional Computer Aided Design (CAD) environment. The widely used commercial CAD system AutoCAD is used as the basis for CAOS, but the system concept is independent of the AutoCAD data structure and the techniques used in CAOS can therefore be applied in connection with most other CAD systems as well.

The geometric versatility of CAOS is achieved by an adaptation of a so-called "design element technique", see, e.g. [6], which allows the user to describe the continuous shape of the structure by a small number of variables, and solves the problem of automatic updating of the finite element model as the shape is changed. Fig. 3 shows a structure divided into design elements (for more examples, see [14]). The shape of each boundary of the design elements is determined by movements of control points along the directions of the arrows. The magnitude of these movements are treated as the design variables of the problem. This technique in connection with the user interface of the CAD system allows the designer to construct a shape optimization model of an initial structure very quickly. Any two-dimensional initial shape fit for analysis by membrane finite elements can be handled by the system.

The CAD integration of CAOS implies that all definitions of the optimization problem take place in the CAD system using the interactive facilities otherwise available for drawing and modification of geometric entities. Move directions, for instance, are inserted directly into the drawing with the pointing device of the workstation and they are subsequently treated by the CAD system exactly like any other geometric entity. The division of the structure into design elements, specification of objective- and constraint functions and all other definitions necessary to define the problem take place in the same way. Thus, modifications are very easy to perform by simply moving, erasing or otherwise modifying the specifications with the editing facilities already available in the CAD system.

As mentioned above, CAOS was originally based on boundary variation. However, via integration with the HOMOPT system the models created in CAOS can now be subjected either to topology or shape optimization.

The CAOS-HOMOPT connection was easily established because HOMOPT works directly on a finite element model of the design space in question. A system for monitoring the present result of the topology optimization has been developed taking

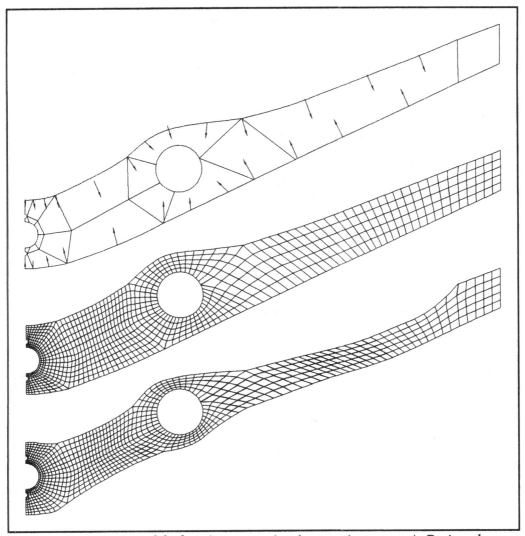

Fig. 3. *Optimization model of engine suspension (symmetric structure). Design elements and move-directions, initial and final FE-models.*

advantage of the graphics capabilities of the CAD system. This enables the user to follow the process from within the CAD system and to transfer the optimized topology directly into the CAD model.

The actual shape optimization model is generated very conveniently by simply drawing the initial structure on top of the generated topology. It is very easy for the designer to approximate the initial position and shape of holes and boundaries while taking the necessary practical constraints, stemming from function, fabrication etc., into consideration. This way, complicated image processing is unnecessary (and probably undesirable) because the designer's choice is often guided by non–geometric information like, eg., the manufacturing capabilities of the company or the relation of the component in question with the rest of the construction of which it is a part.

4. SHAPE OPTIMIZATION

There is a large number of possible formulations of shape optimization problems. One may choose to minimize weight, stress, compliance, displacement or any other property that can be derived from the geometrical model or the output from an analysis program which is usually a finite element module. The same set of possibilities should be available for specification of constraints. This requires a mathematical versatility which is achieved in CAOS with the use of the so–called "bound formulation" [39]. The present version of CAOS handles the following criteria which could act either as the objective or constraint functions:

1. Weight
2. Elastic displacement of a given material point
3. Maximum elastic displacement of any point in the structure
4. Stress (several types) at a given material point
5. Maximum stress (several types) at any point in the structure
6. Compliance

Mathematically, different entries in this list lead to very different optimization problems. Entries 2 and 4 are ordinary scalar functions which can be derived directly from the output from the finite element analysis. Entries 1 and 5 are of integral type and require some postprocessing of the results to be evaluated. Entries 3 and 5 lead to min/max–problems with non–differentiable objective functions.

The usual mathematical program of structural optimization is the following:

$$\text{Minimize} \qquad f(a_i), \qquad i=1..n-1 \qquad (6)$$

$$\text{Subject to} \qquad g_j(a_i) \leq G_j, \qquad i=1..n-1, \ j=1..m \qquad (7)$$

$$\underline{a}_i \leq a_i \leq \bar{a}_i, \qquad i=1..n-1 \qquad (8)$$

where n–1 is the number of design variables, a_i, and m is the number of constraints. Eqs. (8) are side constraints, i.e., upper and lower limits for the design variables. The functions f and g_j are specified by the user as a part of the optimization specification by interactively picking them from the list above.

The bound formulation enables the CAOS system to handle the optimization problem in a uniform way regardless of the blend of scalar–, integral– and min/max–criteria defined by the user. Given the min/max objective function $f = \max(f_j)$, $j=1..p_o$, and a number of constraints, $g_k = \max(g_{kj}) \leq G_k$, $k=1..m$, $j=1..p_k$, we get the following bound formulation of the problem:

Minimize $\quad\quad\quad\quad\quad\quad\quad\quad$ β $\quad\quad\quad\quad\quad\quad\quad\quad\quad\quad\quad$ (9)
a_i, β

Subject to $\quad\quad\quad$ $f_j(a_i) - \beta \leq 0,$ $\quad\quad\quad$ $j=1..p_o$, $i=1..n-1$ $\quad\quad\quad$ (10)

$\quad\quad\quad\quad\quad\quad\quad$ $w_k g_{kj}(a_i) - \beta \leq 0,$ $\quad\quad$ $k=1..m$, $j=1..p_k$, $i=1..n-1$ $\quad\quad$ (11)

$\quad\quad\quad\quad\quad\quad\quad\quad$ $\underline{a}_i \leq a_i \leq \bar{a}_i,$ $\quad\quad\quad\quad$ $i=1..n-1$ $\quad\quad\quad\quad\quad\quad$ (12)

An extra design variable, β, has been introduced, rendering the total number of variables to n. The variable m is still the original number of constraints regardless of whether these are scalar–, integral– or min/max–functions. Each min/max–criterion gives rise to several actual constraints represented by nodal values, the number of which is designated by p_k for min/max–condition k. If condition k is scalar or integral, p_k is obviously 1. The weight factors w_k are imposed on the constraints to allow them to be limited by the same β–value as f. We evaluate w_k prior to the call of the optimizer from the relation:

$$G_k w_k = \beta \rightarrow w_k = \beta/G_k \quad\quad\quad\quad\quad\quad\quad\quad (13)$$

and it is subsequently treated as a constant in the problem.

The tableau (9) through (12) is valid regardless of the blend of functions f and g_k, and the mathematical operations performed are therefore identical for any problem that the user could possibly define. The problem (9)–(12) is solved by sequential programming using either a SIMPLEX algorithm, the CONLIN optimizer [28] or the Method of Moving Asymptotes [40], all of which require the derivatives of the objective and constraint functions to be calculated. For this purpose, a sensitivity analysis [41,42] is performed for each design variable at each iteration. CAOS uses a semi–analytical method based on a differentiation of the finite element equation of equilibrium:

$$[K]\{u\} = \{f\} \quad\quad\quad\quad\quad\quad\quad\quad\quad\quad (14)$$

where [K] is the global stiffness matrix of the structure, $\{u\}$ is the vector of unknown nodal displacements, and $\{f\}$ is the vector of external loads, see, e.g.[43]. Differentiation with respect to a design variable, say a_i, gives:

$$\frac{\partial[K]}{\partial a_i} \{u\} + [K] \frac{\partial\{u\}}{\partial a_i} = \frac{\partial\{f\}}{\partial a_i} \qquad (15)$$

$$\Rightarrow [K] \frac{\partial\{u\}}{\partial a_i} = \frac{\partial\{f\}}{\partial a_i} - \frac{\partial[K]}{\partial a_i} \{u\}$$

$$\equiv \{P_{ps}\} \qquad (16)$$

The right hand side, $\{P_{ps}\}$, of (16) is often termed "Pseudo Load" because it plays the role of an extra load case in the sensitivity analysis. With $\{P_{ps}\}$ known, (16) can be solved using the factorization performed in connection with the initial analysis (14) thus returning the sensitivities of the nodal displacement vector $\{u\}$ with respect to a_i.

We normally assume $\partial\{f\}/\partial a_i = \{0\}$. Furthermore, we shall approximate $\partial[K]/\partial a_i$ by forward finite differences:

$$\frac{\partial[K]}{\partial a_i} \approx \frac{[K(a+\Delta a_i)] - [K(a)]}{\Delta a_i} \qquad (17)$$

where Δa_i is a small perturbation of the i'th component of **a**.
$\{P_{ps}\}$ is thus calculated by:

$$\{P_{ps}\} = - \frac{[K(a+\Delta a_i)] - [K(a)]}{\Delta a_i} \{u\} \qquad (18)$$

The sensitivities of the nodal stresses are easily found from $\partial\{u\}/\partial a_i$. This calculation is also based on finite differences. For each element, the vector of nodal stresses is given by

$$\{\sigma\} = [C][B(x,y)]\{u^e\} \qquad (19)$$

where $[C]$ is the constitutive matrix connecting nodal strains with nodal stresses. The matrix $[C]$ depends only on the material characteristics and remains unchanged by a perturbation of the design variables. The matrix $[B]$ is the geometrical condition connecting nodal displacements with strains. This matrix is a function of the element node coordinates (x,y). The vector $\{u^e\}$ is the part of $\{u\}$ concerning the element in question.

Knowing the derivatives of the element nodal displacements, we can estimate nodal displacements of the perturbed geometry:

$$\{u^{e^*}\} \approx \{u^e\} + \frac{\partial\{u^e\}}{\partial a_i} \Delta a_i \qquad (20)$$

Using these displacements, we can calculate the stresses of the perturbed geometry directly:

$$\{\sigma^*\} = [C][B(x^*,y^*)]\{u^{e^*}\} \qquad (21)$$

where (x^*,y^*) are the perturbed node coordinates. The stress derivative is now approximated by finite differences:

$$\frac{\partial\{\sigma\}}{\partial a_i} \approx \frac{\{\sigma^*\} - \{\sigma\}}{\Delta a_i} \qquad (22)$$

For most problems, the calculation of stress derivatives is surprisingly stable considering the approximations involved. However, for geometries involving large rigid body rotations, the evaluation of displacement sensitivities becomes inaccurate. This, in turn, means that the stress sensitivities loose their reliability and absolute convergence of the problem is unattainable.

5. SIZING OPTIMIZATION

Our structural optimization system possesses capabilities for optimization of 2–D and 3–D truss structures under multiple loading conditions, using cross–sectional areas of bars and positions of joints as design variables. The system is called SCOTS (Sizing and Configuration Optimization of Truss Structures). The development is inspired by [44–46]. In the current setting, weight minimization is the design objective, and constraints include stresses, displacements, and elastic as well as plastic buckling of bars in compression.

The mathematical programming formulation is:

$$\underset{A_i, x_k}{\textit{Minimize}} \quad \sum_{i=1}^{n} \rho_i A_i \ell_i \qquad (23)$$

$$\textit{Subject to} \quad \frac{\sigma_i}{\sigma_i^{yt}} \leq 1, \quad \frac{\sigma_i}{\sigma_i^c} \leq 1, \quad i = 1..n \qquad (24a,b)$$

$$\frac{|d_j|}{\bar{d}_j} \le 1 \qquad j = 1..J \tag{25}$$

$$\underline{A}_i \le A_i \le \bar{A}_i \quad, \quad i = 1..n \quad; \quad \underline{x}_k \le x_k \le \bar{x}_k \quad, \quad 1..k \tag{26a,b}$$

where

$$\sigma_i^c = - \frac{\pi^2 \alpha_i^2 E_i A_i}{\ell_i^2 S_i} \quad for \quad \ell_i \ge \pi\alpha_i \sqrt{\frac{2 E_i A_i}{\sigma_i^{yc} S_i}} \tag{27a}$$

$$\sigma_i^c = - \sigma_i^{yc} + \frac{(\sigma_i^{yc})^2 S_i \ell_i^2}{4\pi^2 E_i \alpha_i^2 A_i} \quad for \quad \ell_1 \le \pi\alpha_i \sqrt{\frac{2 E_i A_i}{\sigma_i^{yc} S_i}} \tag{27b}$$

Here, A_i, ℓ_i, ρ_i and E_i denote cross–sectional area, length, specific weight and Young's modulus for the i–th bar of the truss. The design variables of the problem, see Eq. (23), are A_i and x_k, where the latter symbol represents an element of the total set of variable components of position vectors of joints in the structure. Eqs. (26a,b) are side constraints for the design variables. For simplicity in notation, the remaining constraints are written for a single loading case. Eqs. (25) are constraints that may be specified for any displacement component for the joints of the truss.

Bar stresses are denoted by σ_i and eqs. (24a,b) express constraints for tensile and compressive stresses, respectively. In (24a), σ_i^{yt} represents the yield stress or some other specified upper stress limit. When buckling constraints are considered, the lower limit – σ_i^c for compressive stresses is design dependent and given by eq. (27a) or (27b), respectively, where the former represents dimensioning against Euler Buckling, and the latter plastic buckling on the basis of Ostenfeld's formula. In (27a,b), we have $\alpha_i^2 = r_i^2/A_i$, where r_i is the radius of inertia of the bar cross–section, S_i the factor of safety against buckling, and σ_i^{yc} (>0) the compressive yield stress of the i–th bar. In the literature on optimization of trusses, it is often the case that yielding rather than buckling constraints are considered for compressive bars. This simplification is obtained, if we set $\sigma_i^c = -\sigma_i^{yc}$.

The analyses associated with the mathematical program (23–27) are based on a finite element formulation of the type (14), and the determination of displacement sensitivities follows eqs. (15) and (16) with the exceptions that the design derivatives of the stiffness matrix [K] are derived analytically, and that selfweight loading can be taken into account, whereby $\partial\{f\}/\partial a_i \neq 0$.

6. EXAMPLES

In this section we present examples where topology optimization as described in Section 2 has been used as a preprocessor for subsequent refined sizing or shape optimization depending on the type of structure predicted. In Section 6.1 an example of a truss–like structure will be discussed, and in Sections 6.2 and 6.3 we shall present some detailed examples of postprocessing topology results by refined shape optimization, taking advantage of the CAD–integrated boundary variations techniques of CAOS [13,14].

6.1. Truss–like structure

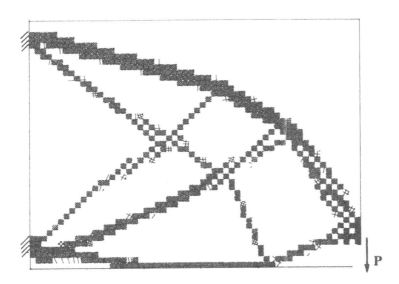

Fig. 4. Solution of topology optimization problem.

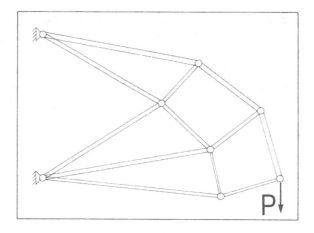

Fig. 5. *Truss interpretation of result in Fig. 4 with bar areas determined by sizing for fixed positions of joints.*

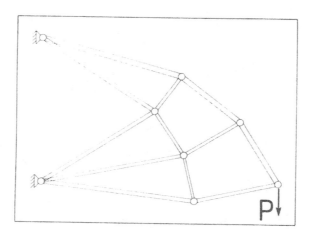

Fig. 6. *The result when both bar cross–sectional areas and positions of unrestrained joints are used as design variables.*

Fig. 4 presents the result of optimizing the topology of a structure within the rectangular design domain shown in the figure. The structure is required to carry a vertical load P, see fig. 4, and is offered support (displacement constraint) along the two hatched parts of the left hand side of the design domain. The domain is subdivided into 2610 finite elements, and the solid volume fraction is 16%.

The result is interpreted as a truss with the number and positions of joints shown in fig. 5. This figure also indicates the individual areas of the bars after a sizing optimization (weight minimization) subject to fixed positions of the nodal points and with the value of the compliance constrained to be equal to that obtained by the topology optimization.

Fig. 6 shows the result of minimizing the weight at the same compliance value, but using both the bar cross–sectional areas and joint positions as design variables (only the horisontal movement of the load carrying joint and the downward movement of the lower–most joint are restrained). We now obtain a slightly different configuration and distribution of bar areas relative to the result in fig. 6, but the optimum weight/compliance ratio is very close to that obtained by the initial topology optimization (fig. 4).

The present example is also considered in [23–25], where the truss solution has been compared with several competitive truss topologies, and proven its superiority. We refer to Refs. [19–25] for several interesting examples of topology optimization of similar type.

6.2 Example: Bearing Pedestal

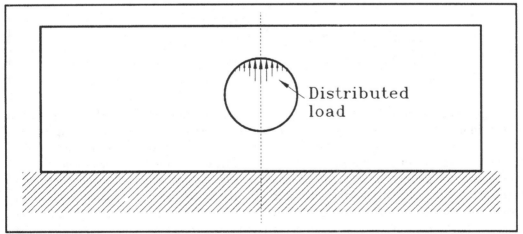

Fig. 7. Initial geometry for pedestal bearing. This geometry completely fills the available space.

In this example [14], we seek to minimize the maximum stress in the bearing pedestal of fig. 7. The initial geometry completely occupies the available space. We shall use 20% of this area as the upper limit for the volume, i.e., the topology optimization starts with an evenly distributed density of 20% in the entire structure with the exception of the rim of the hole which is required to have 100% density in order to properly position the bearing. Due to the symmetry of the geometry and loading, only the left half of the structure is considered.

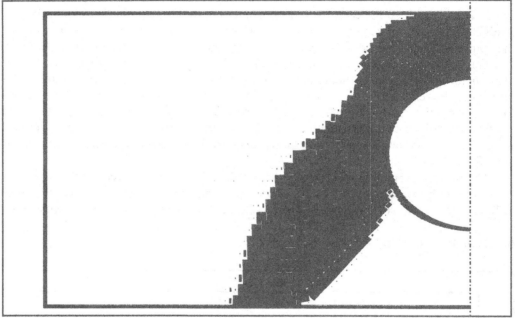

Fig. 8. Optimized topology illustrated by filling the elements by lumps of material corresponding to their final density.

Fig. 8 is an illustration of the optimized topology. Graphically, each element is filled with a lump of material corresponding to its final density. This creates the impression that, in some regions, isolated lumps of material remain outside the solid part of the structure. This is not necessarily the case. The lumps are merely a convenient way of illustrating the porous material. When using CAOS for the actual shape optimization based on an optimized topology, the user has to decide upon the actual position of interfaces between material and void based on the filling of elements by lumps of material. In the present case, the position of interfaces is relatively clear and leads to the shape optimization design element configuration shown in fig. 9. The initial shape has been slightly modified in comparison with the optimized geometry of fig. 8. This is out of practical considerations. The additional material provides a basis for the possible attachment of the pedastal to the underlying surface by, e.g., a bolt joint. The shape optimization model is defined to ensure a minimum thickness of this region. Furthermore, as in the topology model, a minimum thickness of the material surrounding the hole is required.

Due to limited analysis facilities, CAOS is incapable of handling contact problems. Thus, the analysis model presumes that the joint, regardless of its type, provides full contact with the underlying surface in all cases.

We now perform the actual shape optimization based on the optimized topology. We seek to minimize the volume subject to a bound on the maximum von Mises stress of 100 units and a bound on the vertical displacement of the loaded surface of $150 \cdot 10^{-4}$ units. Fig. 10 shows The final finite element model.

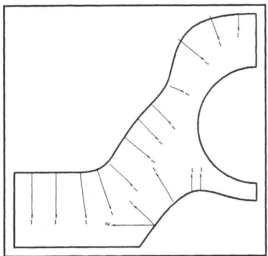

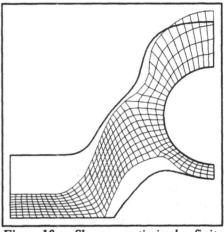

Fig. 10. Shape optimized finite element model.

Fig. 9. Shape design model of optimized topology.

The optimized geometry can be transferred back into the CAD system where the final geometrical adjustments are easily performed by the designer, yielding for instance the geometry of fig. 11. Compared to real–life structures, the geometry of Fig. 11 seems somewhat fragile. This frequently happens when the optimization is performed with respect to a single well–defined load case. Real–life structures are usually designed to handle multiple load cases, and for practical use, the shape optimization system must meet this demand. CAOS presently handles only single load case problems but the design of the system allows for an expansion to multiple load case problems.

The problems of fabrication are often mentioned as a serious drawback of shape optimized structures. The CAD integration of CAOS to a large extent solves this problem. Several excellent CAM interfaces are available that will provide automatic numerical machining based on the CAD model of fig. 11.

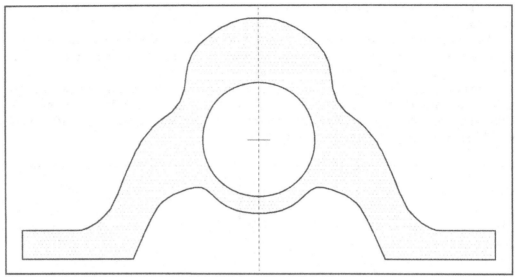

Fig. 11. The final geometry could look something like this.

6.3 Example: MBB–Beam

This section describes the optimization of a support beam from a civil aircraft produced by Messerschmitt–Bölkow–Blohm GmbH, München, BRD. The structure (see fig. 12) has the function of carrying the floor in the fuselage of an Airbus passenger carrier and must meet the following requirements:

1. The upper and lower surfaces must be planar and the distance between them cannot be changed.

2. The maximum deflection of the beam must not exceed 9.4 mm under the given load.

3. The maximum von Mises stress should not exceed 385 N/mm².

4. There must be a number of holes in the structure to allow for wires, pipes etc. to pass through.

The purpose of the optimization is to find the shapes of the holes that minimize the weight of the beam while not violating any of the requirements mentioned above. Because of the symmetry of the structure, we shall analyze only the right hand half of the beam with boundary conditions as indicated in fig. 13 that also shows the finite element mesh

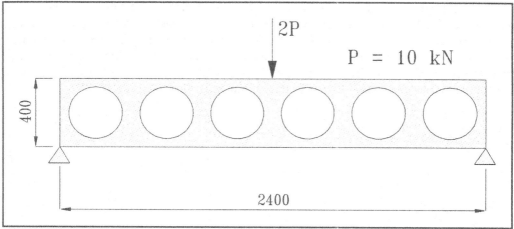

Fig. 12. Initial geometry with loads and boundary conditions.

used for analysis of the initial structure. The data for the initial structure are found to be:

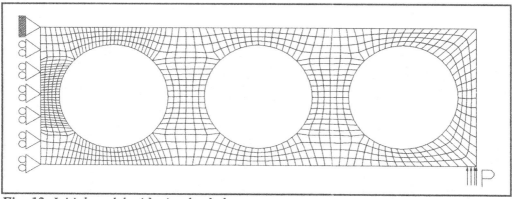

Fig. 13. Initial model with circular holes.

Volume = $1.07 \cdot 10^6$ mm^3
Deflection = 10.1 mm
Max. Stress = 292 N/mm^2

The prescribed upper limit on deflections in the vertical direction is 9.4 mm, so the initial design is infeasible by at least 7.4% because the displacement based finite element method overestimates the stiffness of the structure.

To perform optimization via variation of the boundaries of the holes, we represent these boundaries by b–splines, and introduce a number of master nodes in order to give the system sufficiently many design parameters for the optimization. We shall require symmetry about the horizontal mid–axis of the geometry and utilize link facilities implemented in CAOS to link the movements of master nodes above this line to the corresponding master nodes below. The design model is illustrated in fig. 14.

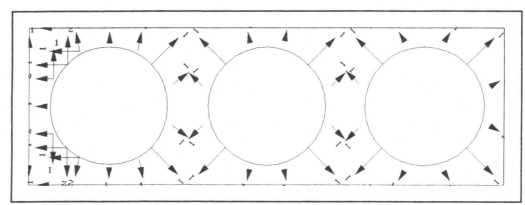

Fig. 14. Design model with b–splines as hole boundaries.

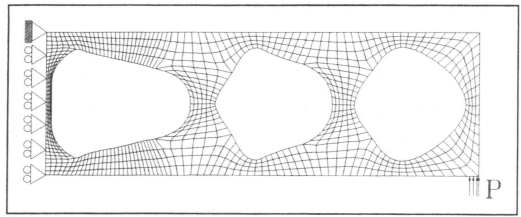

Fig. 15. Final finite element model.

During the optimization of this problem, the stress constraint never becomes active. It is therefore interesting to notice that the final design (fig. 15) is one of very smooth shapes. This is unexpected because there is no smoothness requirement imposed on the transitions between the individual b-splines that make up the holes. This, in combination with the absence of active stress constraints, would in most cases lead to the generation of sharp vertices.

The final design has the following data:

$$\text{Volume} = 1.02 \cdot 10^6 \text{ mm}^3$$
$$\text{Deflection} = 9.4 \text{ mm}$$
$$\text{Max. Stress} = 372 \text{ N/mm}^2$$

i.e., with this model, we manage to create a feasible design and save 5.2% of the volume.

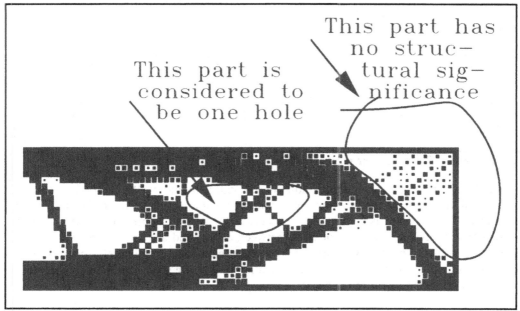

Fig. 16. Result of topology optimization.

The fact that the possible volume reduction even with a b-spline model is rather modest leads to the suspicion that the three-hole topology is not well suited for a structure of this type. It is therefore tempting to start the redesign procedure by a topology optimization.

As discussed in Section 2, the topology optimization requires a volume constraint to be defined. The topology optimization system will then distribute the available volume in the available domain such that the stiffness is maximized. The system enables the user to specify regions or boundaries which are required to be solid, that is, of density 1. We shall use this facility in the present example because the function of the structure requires that the outer contour, except for the left vertical symmetry boundary, remains unchanged.

The original geometry with three circular holes of radius 150 mm has a volume of $1.07 \cdot 10^6$ mm^3. A full beam has a volume of $1.92 \cdot 10^6$ mm^3, i.e., the volume of the initial geometry is 56% of that full beam.

A topology optimization with a bound on the volume corresponding to the volume of the initial structure with three circular holes and no additional requirements was initially attempted. In addition to the volume constraint, we require the rim of the structure to remain solid. The resulting topology is shown in fig. 16. It is evident that a number of holes allowing for the necessary passage of wires, pipes etc. have emerged.

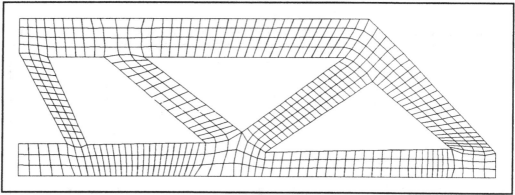

Fig. 17. Initial finite element mesh of optimized topology.

We shall now attempt a shape optimization based on this topology. We therefore return to the original definition of the problem, i.e. minimize volume with a bound on displacement and stress. The problem is a difficult one because the overall impression of the type of the structure is that it is on the interface between a disk and a frame or truss structure. Thus, the division of the geometry into design elements is relatively complicated. Furthermore, while creating the shape optimization model, we shall have to take a number of practical considerations into account:

1.   Due to the cost of manufacture, the complexity of the geometry should be kept at a minimum, i.e., there is a limit to the number of holes that are practical for a structure like this.

2.   The sizes of the individual holes should be comparable to the holes of the initial structure in order to allow for the passage of the same components.

The upper right corner of the frame has been removed. This part of the geometry has a function, but it is structurally insignificant, and can therefore be excluded from the shape optimization and added to the modified structure afterwards. This simplification greatly facilitates the generation of the design model. Fig. 16 illustrates the modifications that have been imposed on the optimized topology and the resulting initial finite element model is shown in fig. 17. This structure has the data:

Volume = $1.10 \cdot 10^6$ mm³
Deflection = 6.0 mm
Max. Stress = 227 N/mm²

The volume of this geometry is slightly larger than the volume of the initial geometry with three circular holes. However, due to the topology optimization, this geometry has significantly larger stiffness. The maximum stress has also been reduced, but this value is unreliable because of the vertices of this geometry. Mathematically, the stress is infinite at sharp concave vertices, but the stress functions of the finite element model in question are unable to model such a state correctly. From a physical point of view, neither vertices of infinite curvature nor infinite stresses exist.

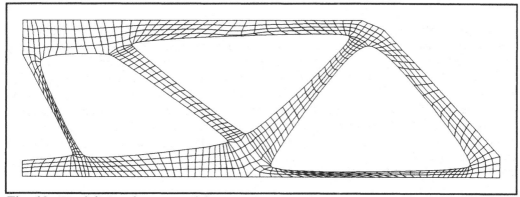

Fig. 18. Final finite element model.

CAOS proceeds by reducing the volume significantly. Because of the structure's resemblance with a truss structure, small relocations of the master nodes may lead to large distortions of the finite element mesh. It has therefore been necessary to perform a redefinition of the finite element mesh topology on the half way between the initial and the final designs. The final design is illustrated in fig 18. It has the data:

Volume = $0.624 \cdot 10^6$ mm³
Deflection = 9.4 mm
Max. Stress = 305 N/mm²

The volume is reduced by 42% in comparison with the initial design with circular holes. The final geometry is a frame–like structure. The stress constraint is not active because, for practical reasons, a minimum thickness is specified for the members of the resulting geometry.

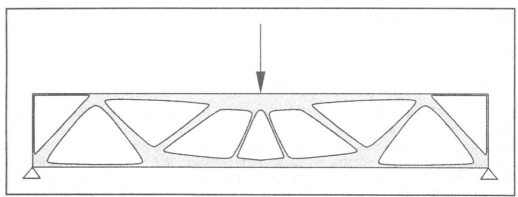

Fig 19. Example of final geometry slightly modified by the designer. The upper right corner has been added again.

Unfortunately, the geometry of the solution introduces the problem of stability which is not covered by CAOS. It is also a problem that the generation of a suitable finite element mesh for a very thin–webbed structure like this is difficult and often requires the topology of the mesh to be redefined. However, the use of membrane elements for a problem like this creates a result that is rich in the sense that it has details that could never have been found by the use of a dedicated system for truss or frame optimization.

Based on the final design of fig. 18 the designer can update his geometrical model and perform the final adjustments, e.g. add the structurally insignificant upper right corner that was removed in order to facilitate the generation of an analysis model, and thereby yield the final design of fig. 19.

CONCLUDING REMARKS

This paper substantiates that initial topology optimization allows shape or sizing optimization to arrive at much better final results. Particularly for problems where there are large possibilities for geometrical variations, the topology optimization is an invaluable tool in the design process. It is the experience from the last example in this paper that topology optimization should be used in the early stages of the development in order to inspire the designer and lead him/her in a beneficial direction. The result of the topology optimization is merely a crude guess and can therefore safely be modified by the designer to meet practical requirements, before the more detailed shape optimization is performed.

It has been the aim of our paper to demonstrate the importance of setting a toolbox of various design facilities at the disposal of the designer. Structural optimization enforces rather than removes the creative aspect of designing, and the final result is therefore very difficult to predict. The collection of structural optimization facilities must be versatile enough to allow the designer to continue work no matter what type of structure emerges. The final design must be a product of creativity rather than availability or lack of analysis facilities.

ACKNOWLEDGEMENTS – The authors are indebted to MSc.s Jan Thomsen and Oluf Krogh, Institute of Mechanical Engineering, Aalborg University, for providing examples for Section 6.1. The work presented in these notes received support from the Danish Technical Research Council (Programme of Research on Computer Aided Design).

REFERENCES

[1]   L.A. Schmit, Structural Synthesis – Its Genesis and Development (AIAA Journal 20, pp.992–1000, 1982)

[2]   N. Olhoff and J.E. Taylor, On Structural Optimization (J. Appl. Mech. 50, pp.1134–1151, 1983)

[3]   H.A.   Eschenauer,   P.U.   Post   and   M.   Bremicker,   Einsatz   der Optimierungsprozedur SAPOP zur Auslegung von Bauteilkomponenten (Bauingenieur 28, pp.2–12, 1988)

[4]   P. Bartholomew and A.J. Morris, STARS: A Software Package for Structural Optimization (in: Proc. Int. Symp. Optimum Structural Design, Univ. of Arizona, USA, 1981)

[5]   J. Sobieszanski–Sobieski and J.L. Rogers, A Programming System for Research and Applications in Structural Optimization (in: E. Atrek et. al.: New Directions in Optimum Structural Design, pp.563–585, Wiley, Chichester, 1984)

[6]   C. Fleury and V. Braibant, Application of Structural Synthesis Techniques (in: C.A. Mota Soares: Preprints NATO/NASA/NSF/USAF Conf. Computer Aided Optimal Design, Troia, Portugal, 1986, Vol.2, pp.29–53, Techn. Univ. Lisbon, 1986)

[7]   B. Esping and D. Holm, Structural Shape Optimization Using OASIS (in: G.I.N. Rozvany and B.L. Karihaloo: Structural Optimization, Proc. IUTAM Symp., Melbourne, Australia, 1988, pp.93–101, Kluwer, Dordrecht, 1988)

[8]   R.T. Haftka and B. Prasad, Programs for Analysis and Resizing of Complex Structures (Computers and Structures 10, pp.323–330, 1979)

[9]   G. Kneppe, W, Hartzheim and G. Zimmermann, Development and Application of an Optimization Procedure for Space and Aircraft Structures (in: H.A. Eschenauer and G. Thierauf: Discretization methods and structural Optimization

– Procedures and Applications, Proc. of a GAMM–Seminar, Siegen, FRG, 1988, pp.194–201, Springer–Verlag, Berlin, 1989)

[10]   L.X. Qian, Structural Optimization Research in China (in: Proc. Int. Conf. Finite Element Methods, Shanghai, China, pp.16–24, 1982)

[11]   G. Lecina and C. Petiau, Advances in Optimal Design with Composite Materials (in: loc.cit.[6], Vol.3, pp.279–289]

[12]   J.S. Arora, Interactive Design Optimization of Structural Systems (in: loc. cit. [9], pp.10–16)

[13]   J. Rasmussen, The Structural Optimization System CAOS (Structural Optimization, 2, pp. 109–115, 1990)

[14]   J. Rasmussen, Collection of Examples, CAOS Optimization System, 2nd Edition (Special Report No. 1c, Institute of Mechanical Engineering, Aalborg University, Denmark, 1990)

[15]   S. Kibsgaard, N. Olhoff and J. Rasmussen, Concept of an Optimization System (in: C.A. Brebbia and S. Hernandez: Computer Aided Optimum Design of Structures: Applications, pp.79–88, Springer–Verlag, Berlin, 1989)

[16]   U. Ringertz, A. Branch and Bound Algoritm for Topology Optimization of Truss Structures (Engineering Optimization, 10, pp.111–124, 1986)

[17]   M.P. Bendsøe and N. Kikuchi, Generating Optimal Topologies in Structural Design Using a Homogenization Method (Comp. Meths. Appl. Mechs. Engrg. 71, pp.197–224, 1988)

[18]   M.P. Bendsøe, Optimal Shape Design as a Material Distribution Problem (Structural Optimization, 1, pp.193–202, 1989)

[19]   M.P. Bendsøe and H.C. Rodrigues, Integrated Topology and Boundary Shape Optimization of 2–D Solids, (Rept. No. 14, 31 pp. Mathematical Institute, Technical Univ. Denmark, 1989)

[20]   K. Suzuki and N. Kikuchi, A Homogenization Method for Shape and Topology Optimization (Comp. Meths. Appl. Mechs. Engrg., submitted, 1989)

[21]   K. Suzuki and N. Kikuchi, Generalized Layout Optimization of Shape and Topology in Three–Dimensional Shell Structures (Rept. No. 90–05, Dept. Mech.

Engrg. and Appl. Mech., Comp. Mech. Lab., University of Michigan, USA, 1990)

[22]   N. Kikuchi and K. Suzuki, Mathematical Theory of a Relaxed Design Problem in Structural Optimization (Paper for 3rd Air Force/NASA Symp. Recent Advances in Multidisciplinary Analysis and Optimization, San Francisco, USA, Sept. 1990)

[23]   P.Y. Papalambros and M. Chirehdast, An Integrated Environment for Structural Configuration Design (J. Engrg. Design, 1, pp. 73–96, 1990)

[24]   M. Bremicker, M. Chirehdast, N. Kikuchi and P.Y. Papalambros, Integrated Topology and Shape Optimization in Structural Design (Techn. Rept. UM–MEAM–DL–90–01, Design Laboratory, College of Engrg., Univ. of Michigan, USA, 1990)

[25]   M. Bremicker, Ein Konzept zur Integrierten Topologie – und Gestaltoptimierung von Bauteilen (in: H.H. Müller–Slany: Beiträge zur Maschinentechnik, pp.13–39, Festschrift für Prof. H. Eschenauer, Research Laboratory for Applied Structural Optimization, University of Siegen, FRG, 1990)

[26]   J.M. Guedes and N. Kikuchi, Pre and Postprocessings for Materials based on the Homogenization Method with Adaptive Finite Element Methods (Rept., Dept. Mech. Engrg. and Appl. Mech., University of Michigan, USA, 1989)

[27]   J.M. Guedes and N. Kikuchi, Computational Aspects of Mechanics of Nonlinear Composite Materials (Rept., Dept. Mech. Engrg. and Appl. Mech., University of Michigan, USA, 1989)

[28]   C. Fleury and V. Braibant, Structural Optimization: A New Dual Method Using Mixed Variables (Int. J. Num. Meth. Engrg. 23, pp.409–428, 1986)

[29]   P. Pedersen, On Optimal Orientation of Orthotropic Materials (Structural Optimization, 1, pp. 101–106, 1989)

[30]   P. Pedersen, Bounds on Elastic Energy in Solids of Orthotropic Materials (Structural Optimization, 2, pp. 55–63, 1990)

[31]   G.I.N. Rozvany, Structural Layout Theory – the Present State of Knowledge (in: loc.cit. [5], Chapter 7)

[32]   G.I.N. Rozvany, Structural Design via Optimality Criteria (Kluwer Academic

Publishers, Dordrecht, The Netherlands, 1989)

[33]   G.I.N. Rozvany and M. Zhou, Applications of the COC Algorithm in Layout
        Optimization (Paper for Proc. Int. Conf. Engineering Optimization in Design
        Processes, Karlsruhe, FRG, 3–4 September 1990)

[34]   P.Pedersen, A Unified Approach to Optimal Design (in: H. Eschenauer and N.
        Olhoff: Optimization Methods in Structural Design, Proc. Euromech –
        Colloquium 164, Univ. Siegen, FRG, 1982, pp.182–187, Bibliographishes
        Institut, Mannheim, FRG, 1983)

[35]   V. Braibant and C. Fleury, Shape Optimal Design Using B–splines (Comp.
        Meths. Appl. Mech. Engrg. 44, pp. 247–267, 1984)

[36]   J.A. Bennett and M.E. Botkin, Structural Shape Optimization with Geometric
        Description and Adaptive Mesh Refinement (AIAA Journal 23, pp. 458–464,
        1985)

[37]   R.T. Haftka and R.V. Gandhi, Structural Shape Optimization – A Survey
        (Comp. Meths. Appl. Mech. Engrg. 57, pp.91–106, 1986)

[38]   Y. Ding, Shape Optimization of Structures: A Literature Survey (Computers and
        Structures 24, pp.985–1004, 1986)

[39]   N. Olhoff, Multicriterion Structural Optimization via Bound Formulation and
        Mathematical Programming (Structural Optimization 1, pp.11–17, 1989)

[40]   K. Svanberg, The method of Moving Asymptotes – A new Method for
        Structural Optimization (Int. J. Num. Meth. Engrg. 24, pp. 359–373, 1987)

[41]   E.J. Haug, K.K. Choi and V. Komkov, Design Sensitivity of Structural Systems
        (Academic Press, New York, 1986)

[42]   R.T. Haftka and H.M. Adelmann, Recent Developments in Structural Sensitivity
        Analysis (Structural Optimization, 1, pp. 137–151, 1989)

[43]   G. Cheng and L. Yingwei, A New Computation Scheme for Sensitivity Analysis
        (Eng. Opt. 12, pp. 219–234, 1987)

[44]   P. Pedersen, On the Minimum Mass Layout of Trusses (Advisory Group for
        Aerospace Research and Development, Conf. Proc. No. 36, Symposium on
        Structural Optimization, Istanbul, Turkey, AGARD–CP–36–70, 1970)

[45]   P. Pedersen, On the Optimal Layout of Multi–Purpose Trusses (Computers and Structures, 2, pp. 695–712, 1972)

[46]   P. Pedersen, Optimal Joint Positions for Space Trusses (Journal of The Structural Division, ASCE 99, pp. 2459–2476, 1973)

Chapter 12

STRUCTURAL OPTIMIZATION OF LINEARLY ELASTIC STRUCTURES USING THE HOMOGENIZATION METHOD

N. Kikuchi and K. Suzuki
University of Michigan, Ann Arbor, Michigan, USA

1. Introduction

There are three major structural optimization problems of a linearly elastic structure ; namely, 1) sizing, 2) shape, and 3) layout(topology) optimization problems.   The characteristics of these problems can be summarized as follows :

Sizing Problem   A typical setting of the problem is to find the optimal thickness distribution of a linearly elastic plate that is supported on its boundary and is subject to a given loading condition on the plate or its boundary.   The optimal thickness of the plate is obtained so as to, e.g., minimize (or maximize) a certain physical quantity such as the mean compliance, while the state (i.e. equilibrium) equation of the structure is maintained as well as various constraints on the state and design variables.   Here, the state variable can be the deflection of the plate, and the thickness of the plate becomes the design variable.   If a beam is subject to axial torsional and bending forces, the sizes of the cross section of the beam such as the radius, height, and width would be optimized to construct the strongest beam structure within a given design restriction.   That is, the design variable of the sizing problem is a physical dimension of the structure.

The main feature of the sizing problem is that the domain of the design and state variables is *a priori* known and is fixed in the optimization process.   For example, the optimum thickness distribution is considered over the whole plate, the domain of the thickness and deflection functions is the whole middle surface of the plate that is not altered by design optimization.

Shape Problem   On the other hand, the shape problem is defined on a domain which is unknown *a priori*. For certain restricted problems, the shape of the domain may be described parametrically using a finite number of parameters with appropriate basis functions such as Bezier splines, but the state equation is defined on a "variable" domain. The optimum shape of the domain is obtained so as to minimize e.g. the mean compliance of the structure, while the equilibrium (i.e. state) equation is satisfied on such a domain to be determined. Thus, if the finite element method is applied to solve the equilibrium equation, its discrete model must be developed at each iteration of optimization process according to the change of the domain.

Layout (Topology) Problem   Ambiguity of the layout (topology) problem is substantial in its mathematical description. In this case, only the known quantities are applied loads, desired support conditions, various design restrictions, and possibly the volume of a structure constructed. The purpose of optimization is to find the optimal layout of a structure so that given applied loads are transmitted to desired supports in a specified region using a given amount of material while equilibrium and design constraints are satisfied. In this problem, not only the physical size but also the shape of the structure are unknown, and further we cannot define these geometrical quantities with appropriate parametric representation ! In order to define a form of layout, topology of the domain (structure) must be specified, but it is not known *a priori*.

A formal description of the structural optimization problem may be defined by minimizing (or maximizing) a functional with respect to the design variables consisting of {size, shape, topology} subject to {the state equation, constraints on the state and design variables}. A typical mathematical setting is given by

$$
\begin{aligned}
&\underset{d}{\text{Minimize}} \quad f(d,u) \\
&\textit{subject to} \\
&u \in V : a(u,v) = L(v), \ \forall v \in V \\
&g_u(u) \le g_{u\,\text{max}} \\
&g_d(d) \le g_{d\,\text{max}}
\end{aligned}
\tag{1}
$$

Here, $f(d,u)$ is the "cost" function, d and u are the design and state variables, respectively, "$a(u,v)=L(v) \; \forall v$ " is a generalized (weak) form of the state equation, and g_u and g_d represent the constraints on the state and design variables, respectively.

As a concrete example, we shall define a shape problem for a linearly elastic plane structure to determine the optimal shape $h(x)$ which is parametrically represented by a set of basis functions $\{ h_1(x), h_2(x), \ldots , h_n(x) \}$, i.e.,

$$h(x) = \sum_{i=1}^{n} d_i h_i(x)$$

In this case the design variable is defined by $d = \{ d_1, \ldots, d_n \}$.

If the constitutive relation of the material of the structure is given by

$$\sigma = \begin{Bmatrix} \sigma_1 \\ \sigma_2 \\ \sigma_{12} \end{Bmatrix} = \begin{bmatrix} D_{11} & D_{12} & D_{13} \\ D_{12} & D_{22} & D_{23} \\ D_{13} & D_{23} & D_{33} \end{bmatrix} \left(\begin{Bmatrix} \varepsilon_1 \\ \varepsilon_2 \\ \gamma_{12} \end{Bmatrix} - \begin{Bmatrix} \varepsilon_{01} \\ \varepsilon_{02} \\ \gamma_{012} \end{Bmatrix} \right) = D(\varepsilon - \varepsilon_0) \tag{2}$$

where σ and ε are the stress and strain vectors in the contracted notation, respectively, and ε_0 is a specified initial strain, the generalized form of the state (equilibrium) equation is given by the principle of virtual displacement :

$$u \in V : \qquad a(u,v) = L(v) \qquad\qquad \forall v \in V \tag{3}$$

Here

$$a(u,v) = \int_{\Omega} \varepsilon(v)^T D\varepsilon(u)d\Omega + \int_{\Gamma} v^T kud\Gamma \tag{4}$$

$$L(v) = \int_{\Omega} \varepsilon(v)^T D\varepsilon_0 d\Omega + \int_{\Omega} v^T bd\Omega + \int_{\Gamma} v^T (kg + t)d\Gamma \tag{5}$$

$$V = \left\{ v = \{v_1, v_2\} : v_i \in H^1(\Omega) \right\} \tag{6}$$

$$\varepsilon(v)^T = \{\varepsilon_1(v) \quad \varepsilon_2(v) \quad \gamma_{12}(v)\} = \left\{ \frac{\partial v_1}{\partial x_1} \quad \frac{\partial v_2}{\partial x_2} \quad \frac{\partial v_1}{\partial x_1} + \frac{\partial v_2}{\partial x_1} \right\} \tag{7}$$

b is the applied body force in the domain Ω, $H^1(\Omega)$ is the Sobolev space defined on Ω, and Γ is the boundary of the domain Ω. Here the generalized boundary condition

$$\begin{bmatrix} n_1 & 0 & n_2 \\ 0 & n_2 & n_1 \end{bmatrix} \begin{Bmatrix} \sigma_1(u) \\ \sigma_2(u) \\ \sigma_{12}(u) \end{Bmatrix} = -\begin{bmatrix} k_1 & k_{12} \\ k_{12} & k_2 \end{bmatrix} \left(\begin{Bmatrix} u_1 \\ u_2 \end{Bmatrix} - \begin{Bmatrix} g_1 \\ g_2 \end{Bmatrix} \right) + \begin{Bmatrix} t_1 \\ t_2 \end{Bmatrix}$$

i.e.

$$N\sigma(u) = -k(u - g) + t \tag{8}$$

is assumed on the boundary Γ, where $n = \{ n_1, n_2 \}$ is the unit vector outward normal to the boundary, k is the elastic tensor of spring distributed, g is a given specified displacement vector, and t is the applied traction on the boundary. If the "cost" function f is defined as the mean compliance of the elastic structure, this is given by

$$f(d,u) = L(u) \tag{9}$$

If the shape optimization is restricted by the upper bound Ω_0 of the volume of a structure, the constraint on the design variables d is specified so as to $\Omega \le \Omega_0$. However, the explicit representation of this volume constraint is a rather difficult task in the parametric definition of the shape of a structure. Furthermore, the domain Ω and its boundary Γ are functions of the design variables but they are written in very indirect manner. Thus, finding the derivative of the cost function and the constraints with respect to d is not obvious. In other words, existence of the optimum shape design may be shown by using compactness argument, since only a finite fixed number of design variables, but computing the optimum shape could be very difficult. In addition, a different choice of the basis functions $h_i(x)$ may yield a significantly different optimum shape. There is no assurance that a unique optimum shape can be obtained by different choice of the basis functions.

2.   Difficulties in Structural Optimization

If the structural optimization problem can be defined using a finite number of parameters together with well-behaved appropriately chosen basis functions, we do not have much difficulties in showing existence of a solution by standard compactness argument. Furthermore, if the cost function and the constraints are explicit functions of such a finite number of parameters, i.e., the design variables, computation of their sensitivity is straightforward, and then application of appropriate optimization algorithm yields the optimum without much difficulty in computation. However, there are so many optimum design problems outside of this category.

For example, let a typical sizing problem be considered to determine the optimum thickness of a linearly elastic so as to minimize the mean compliance, or that maximizes the first eigenvalue of the free vibration problem of a plate. If the thickness is represented by a fixed finite number of C^∞- basis functions, the optimum can be easily obtained, see, e.g., Banichuk[1]. However, as Cheng and Olhoff[2] shown, a different selection of a finite number of basis functions yields the different optimum, and the true optimum does contains many discrete ribs in various size. We also see many discrete rib reinforcement in plates and shells in engineering practice. It is clear that infinitely many basis functions are required to represent such discrete ribs in various size and location. Thus, it is unrealistic to define the thickness optimization problem using a fixed finite number of parameters and basis functions.

From the examination of the principle of virtual displacement that is a weak form of the equilibrium equation, it can be also determined that convergence of parametric representation of the thickness $h^n = \text{span}\{h_1, ..., h_n\}$ is expected at most in $L^\infty(\Omega)$ as n goes to ∞. This is implied by the fact that the coefficient tensor of the differential operator need be merely in $L^\infty(\Omega)$ in the weak form, and then the assumption of the isotropic plate could be insufficient, since convergence of the thickness of a plate

$$h^n(x,y) = \begin{cases} h_a & \text{if} \quad \dfrac{2(i-1)}{2n} \leq x \leq \dfrac{2i-1}{2n} \quad \text{and} \quad \dfrac{2(j-1)}{2n} \leq y \leq \dfrac{2j-1}{2n} \\[3mm] h_b & \text{if} \quad \dfrac{2i-1}{2n} < x < \dfrac{2i}{2n} \quad \text{and} \quad \dfrac{2j-1}{2n} < y < \dfrac{2j}{2n} \end{cases}$$

where $i,j = 1,......, n$, as n goes to ∞, yields an orthotropic plate. That is, we have to state the equilibrium equation in a "relaxed" form using the assumption of an orthotropic plate that possibly possesses variety of microstructures instead of restricting our attention only for too well-behaved isotropic plate that assumes smooth variation of thickness as well as Young's modulus and Poisson's ratio without having many microstructure.   Lurie, Fedorov, and Cherkaev[3] studied this nature mathematically using the theory of G-closure, and provided a clear answer how this problem should be stated and be solved, see also Bendsøe[4].   In other words, even in the sizing problem, if the design variable is distributed and is expected in L^∞ space, the state equation must be extended to the relaxed form that can allow existence of microstructure.

In the shape problem difficulty we encounter is not only for the convergence of parametric representation of the shape of a structure, but also for representation itself. Furthermore, there is no assurance that a singly connected domain is optimal.   In fact, we know there should be holes in a structure for a certain case from our experience.   But, we do not know the number of holes and their location as well as shapes *a priori*.   Can we explicitly represent this situation parametrically with a finite number of preassigned basis functions ?   The answer is negative.   Thus, the shape problem can only be solved with many additional restriction such as a specified number of holes, i.e., topology.   In this case only sub-optimal solutions can be obtained, and we do not have any guarantee that change of the number of holes would not yield significant difference on the optimum shape obtained. Even with this restriction to a fixed topology, there still exists considerable difficulty in the shape problem when it is solved by using the finite element method.   Because of a finite number of parametric representation of the shape, the sensitivity analysis for defining a sequential linearized problem is rather straightforward, but solving the state equation using the finite element method requires development of a discrete finite element model of the structure.   If the initial shape of the domain is considerably different from the "current" one, development of a discrete model of the structure at each optimization step becomes a significant task, since it must be done automatically without any manual help nor terminating computation for optimization.   Since the design variable is defined on a fixed domain, the sizing problem is free from this modeling difficulty.   An automatic finite element mesh generation must be introduced to solve this into an optimization system.

It is needless to say that what kind of difficulties we can find in the layout (topology) problem, since it is a complex combination of the sizing and shape problems if the topology of a structure is fixed with a finite number of parameters.   Again it is not known whether or not the optimal topology can be represented by a finite number of parameters.   Only available approach is to set up an initial set of truss joints and to generate additional ones within the limited total volume of the structure using, e.g. dynamic programing methods, see Palmer[5].   In this case, the type of structure must be predetermined, for example, as a truss structure that cannot transmit the bending moment at joints. But, this choice need not be the best again. Furthermore, we can deal with only a fixed finite number of joints. There is, again, no assurance that a convergent optimal structure can be obtained as the number of joints is increased.

3.   A New Approach Based on the Homogenization

As shown in above, there are many difficulties in structural optimization. A common difficulty in the sizing, shape, and layout problems is convergence of a finite number of parametric representation.   This means that once a method is introduced to solve the convergence problem for one of these problems, it may be applicable for the others, and that there is a possibility of existence of a method to solve these three problems at once.   To solve the shape problem, a common approach is to define the shape using certain functions and to vary it to achieve the optimality. Variation of the shape generates substantial difficulty in computation. Thus, we should abandon this approach. That is, the shape should not be defined by a set of functions, and the problem must be solved in a fixed domain as for the sizing problem so that finite element(difference) model need not be renewed during optimization.   Furthermore, since the topology is unknown *a priori*, it should not be specified. From the very beginning, an "infinite" number of holes should be prepared at "everywhere" to solve both the shape and topology problems.

Does there exist such a method ? To answer to this, we shall recall that convergence in the sizing problem is solved by applying the concept of the homogenization, i.e., G-closure by Lurie, Fedorov, and Cherkaev[3], and their method is applied to solve the thickness optimization problem by e.g. Bendsøe[4].   Now let us also recall the paper by Murat and Tartar[6] presented in 1983 on optimality conditions and homogenization. In this paper, Murat and Tartar stated "Where the variable is a domain, the computing of variations was done as early as 1950 by Hadamard, by pushing the boundary along the normal and then computing the induced variation of the functional. Turning this idea into theorems is not easy, and moreover already one can see another defect of this method : a given domain is compared only with domains of the same shape ; it is impossible to make a hole inside the domain or to add a few small pieces far away by means of this technique.   ..... The real difficulty lies in the fact that the set of domains, i.e., characteristic functions, does not possesses natural paths from one domain to another : there is no manifold structure that enables us to use classical derivatives.   ..... In the problem where the variable appears as a domain and some partial differential equation is involved, there is another phenomenon that we discovered ten years ago (it was later called homogenization) : generalized domains appear which are the analogue of a mixture of two different materials and the effective properties of these mixtures have to be understood (they are not obtained by averaging certain quantities in more than one dimension)."

The very similar but more directly related idea for structural optimization is given by Kohn and Strang[7] : "The need for relaxation is reflected in the design problem by the possibility that there may be no optimal design. Though initially surprising, this phenomenon is easily understood : it is sometimes advantageous to perforated parts of Ω by many fine holes with a suitably chosen geometry. If the optimal characteristics can be realized only in the limit as the scale of the perforation tends to zero, then the design problem has no geometrical solution - rather, a solution exists only in a suitably generalized class of designs, which allows the use of composites obtained by perforation at some points of Ω. ..... We first extend the design problem, allowing for the use of composite materials. Then we choose the best composite at each point."

It should be now clear that there is a method to solve quite wide range structural optimization problems using the notion of homogenization after extending the design problem by allowing possibly perforated structures. Since perforation is expected in the microstructure, and since the degree of perforation as well as location is not specified *a priori*, arbitrary shape and topology of the structure may be well represented. It is certain that we may end up a perforated structure as the optimum because of the relaxation (i.e. extension) of the original design problem that is restricted to the use of only a solid nonperforated material. Because of the convergence property in the sense of homogenization, a precise mathematical theory should be able to furnish to the relaxed i.e. extended problem of structural optimization. In short, application of the theory of homogenization is an answer to the optimal design problem in structural analysis.

4. Relaxed Optimal Design Problem

We shall describe a relaxed design problem using microscale rectangular holes to perforate a structure, see Bendsøe and Kikuchi[1]. Suppose that the total volume of microscale holes is specified in a given design domain Ω, that is, the volume Ω_s of "solid" material distributed in the design domain is specified. For simplicity, the design domain is plane so that plane stress analysis is sufficient to compute displacements and stresses, while the shape of microscale voids is assumed to be rectangular as shown in Figure 1.

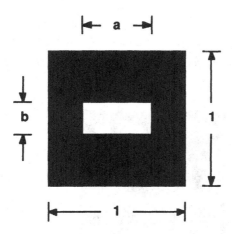

Figure 1.   A Unit Cell Describing the Microstructure with a
Rectangular Hole

Rectangular holes are chosen because they can realize the complete void ($a=b=1$) and solid ($a=b=0$) as well as generalized perforated medium ($0<a<1$, $0<b<1$). If circular holes are assumed, they cannot reach to the complete void, and then are not appropriate to our purpose. It is noted that there are other choices to represent microscale holes such as using a generalize ellipse defined by $(y_1/a)^n+(y_2/b)^n = 1$, where a and b are the principal radii of the ellipse and n is the power defining the shape. However, in order to develop the complete void in the unit cell both a and b must be 1 as well as n goes to infinity for the generalize ellipse, while only a and b need be 1 if the rectangular hole is applied.   Thus, using rectangular holes is simpler. Since holes are rectangular in the unit cell that characterize the microstructure of a perforated medium for the design problem, their orientation is important in the macroscopic problem for stress analysis.   Indeed *the anisotropic elasticity tensor in the macroscopic problem strongly depends on the orientation of microscale holes.* Thus, *the sizes a and b and the orientation θ of microscale rectangular holes must be the design variables of the relaxed problem.*

Suppose that a, b, and θ are functions of the position x of an arbitrary point of a macroscale domain of a linearly elastic perforated structure Ω in the two-dimensional Euclidean space $\mathbb{R}^2$: $a=a(x)$, $b=b(x)$, and $\theta=\theta(x)$.   Functions a, b, and θ may not be so smooth, but we shall assume they are sufficiently smooth, for example, $a,b,\theta \in H^1(\Omega)$. Assuming that a periodic microstructure characterized by $a(x)$, $b(x)$, and $\theta(x)$ exists in a small neighborhood of an arbitrary point x in Ω, and assuming that such microstructure at x

need not be the same with the one at a different point x^*, see Figure 2 which shows a schematic setting of varying microstructures, a homogenized elasticity tensor $\mathbb{E}^H(x)$ is computed in order to solve a macroscopic stress analysis problem of a perforated structure.

Figure 2.   Assumption of "Continuous" Change of Microstructures

The homogenized elasticity tensor is computed by solving the problem defined in the unit cell in which a rectangular hole is placed : Find the characteristic deformations

$\chi^{(kl)}=\{\chi_1^{(kl)},\chi_2^{(kl)}\}\in V_Y$ satisfying

$$\sum_{i,j,m,n=1}^{2}\int_Y E_{ijmn}\frac{\partial \chi_m^{(kl)}}{\partial y_n}\frac{\partial v_i}{\partial y_j}dY = \sum_{i,j=1}^{2}\int_Y E_{ijkl}\frac{\partial v_i}{\partial y_j}dY \quad ,\forall v \in V_Y \tag{10}$$

where $V_Y = \{v=\{v_1,v_2\} : v_i \in H^1(Y) , v_i$ is Y-periodic in the unit cell $Y, i=1,2\}$, and the unit cell is defined by $Y=(-1/2,1/2)\times(-1/2,1/2)$. The elasticity tensor $\mathbb{E}$ is chosen either for the plane stress or for plane strain problem depending on the structure to be designed.   The elasticity tensor $\mathbb{E}$ is zero if y is located in the hole, and coincides with the one of the "solid" material that is utilized to form a structure if y is outside of the hole.   It is noted that Young's modulus does not affect to the optimum perforation of the relaxed design problem while Poisson's ratio may imply change to the optimum.   It is also noted that the material of the "solid" portion need not be isotropic. For example, if the layout of a fiber-reinforced laminate is considered, the elasticity tensor of the "solid" portion must be the one for the laminate, and is anisotropic. After obtaining the characteristic deformations $\chi^{(kl)}$, the homogenized elasticity tensor $\mathbb{E}^H$ is computed by

$$E_{ijkl}^{H} = \sum_{m,n=1}^{2} \int_{Y} \left(E_{ijkl} - E_{ijmn} \frac{\partial \chi_{m}^{(kl)}}{\partial y_{n}} \right) dY \tag{11}$$

Since the sizes $\{a,b\}$ of rectangular holes are functions of the position x, the homogenized elasticity tensor $\mathbb{E}^H$ varies in Ω. This means that the characteristic deformations must be obtained everywhere in the design domain Ω. Solving the unit cell problem (10) at everywhere is unrealistic. Thus, we shall solve (10) for several sampling points $\{a_i, b_j : i,j=1,....,n\}$ of the sizes $\{a,b\}$ of rectangular holes, where $0 \le a_i \le 1$ and $0 \le b_j \le 1$, and we shall form a function $\mathbb{E}^H = \mathbb{E}^H(a,b)$ by an appropriate interpolation. The last step of obtaining the elasticity tensor for stress analysis of the macroscopic perforated structure is rotation of $\mathbb{E}^H$ by the angle θ. Defining the rotation matrix R by

$$R(\theta) = \begin{bmatrix} \cos\theta & -\sin\theta \\ \sin\theta & \cos\theta \end{bmatrix}$$

the elasticity tensor $\mathbb{E}^G$ for stress analysis is computed by

$$E_{ijkl}^{G} = \sum_{I,J,K,L=1}^{2} E_{IJKL}^{H}(a,b) R_{iI}(\theta) R_{jJ}(\theta) R_{kK}(\theta) R_{lL}(\theta) \tag{12}$$

for $i,j,k,l = 1$, and 2, at arbitrary point x in Ω. It is clear that $\mathbb{E}^G$ is a function of the sizes $\{a,b\}$ and the rotation θ of microscale rectangular holes, i.e. the design variables. Now these $\mathbb{E}^G$ define the D matrix in the principle of virtual work (4) for stress analysis of the perforated structure.

For the optimization problem, $d=\{a,b,\theta\}$, and the upper bound Ω_s is specified that is the total volume of "solid" material forming the perforated structure, and it defines a constraint on the design variable :

$$\int_{\Omega}(1-ab)d\Omega \leq \Omega_s < \Omega \tag{13}$$

where Ω is the total volume of the design domain in which an optimum perforated structure is placed, that is, the layout of a structure is given.

The main feature of this setting of the optimum layout of a structure is that size, shape, and topology are represented by the distributed design variables $d=\{a,b,\theta\}$, and that the problem is defined on a fixed design domain Ω. In other words, the layout problem is writen as a sizing optimization problem. Thus, a geometric model of the finite element analysis defined at the beginning need not be modified during the optimization process, and if the design domain is sufficiently simple, its discrete model by finite elements can be developed easily without introducing any special spline functions to define the shape of a structure designed in the design domain. Only the change in optimization is the design variables $d=\{a,b,\theta\}$, that is, the D matrix in (4), and the domain Ω is independent of the design variables. Thus, sensitivity of the cost function and the constraints, and then the optimality condition of the minimization problem (1) can be obtained without any difficulty. Once sensitivity and the optimality condition are obtained, introduction of an optimization method is straightforward, despite that the number of discrete design variables obtained from $d=\{a,b,\theta\}$ would be fairly large. For example, if the design domain Ω is decomposed into N finite elements, and if $d=\{a,b,\theta\}$ are represented by constants in each finite element, we generate 3N discrete design variables.

Details of computational procedure including an optimization method to solve the design problem (1) can be found in Suzuki and Kikuchi[9], and will not be discussed here. Since the design variables a, b, and θ are distributed functions on the design domain Ω, they must be approximated by the discrete ones to solve the optimization problem (1). To this end, these are approximated by constant functions in each finite elements of the discrete model of the design domain. This yields that the microstructure is assumed to be constant in an element, and then the homogenized elasticity tensor is also constant in each finite element. From the experience we have so far, the approximation by piecewise constant functions can provide satisfactory results. Because of this choice, the optimality condition of the discrete approximation of the constrained minimization problem (1) can be obtained easily, and then an iteration scheme based on the optimality criteria method can be derived from this.

In the following examples, we shall assume that Young's modulus and Poisson's ratio of the isotropic "solid" material in the unit cell are given as E=100GPa and v=0.3, respectively. The homogenized elasticity tensor $\mathbb{E}^H$ is computed for 6 x 6 sampling points in the design variables a and b, respectively, and is interpolated by the Legendre polynomials. The homogenization problem (10) is solved using 12 x 12 finite elements in the unit cell as shown in Figure 3, where a rectangular hole with $a=1/3$ and $b=2/3$ is assumed. The components E^H_{1111}, E^H_{1122}, and E^H_{1212} of the homogenized elasticity tensor $\mathbb{E}^H$ are computed as shown in Figures 4, 5, and 6, respectively.

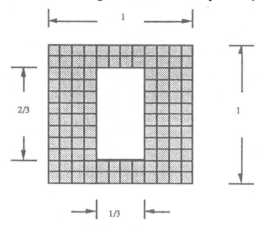

Figure 3   A finite element model of the unit cell with a rectangular hole

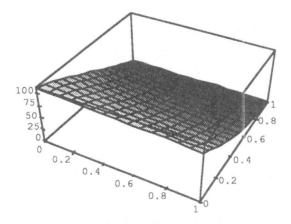

Figure 4   Homogenized elasticity constant E^H_{1111} with respect to the size a and b

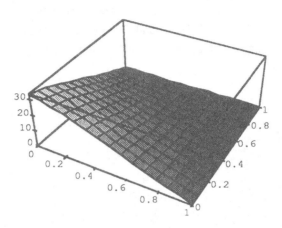

Figure 5   Homogenized elasticity constant E^H_{1122} with respect to the size a and b

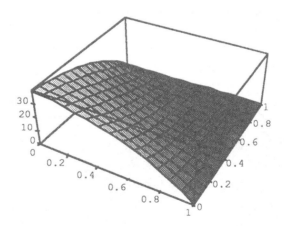

Figure 6   Homogenized elasticity constant E^H_{1212} with respect to the size a and b

5.   A Verification of the Relaxed Design Problem

We shall examine the present method to solve a simple structural design problem, the exact solution of which can be obtained analytically using a simple structural model such as a truss or a beam.   To this end, a two-bar framed structure shown in Figure 7 is considered.

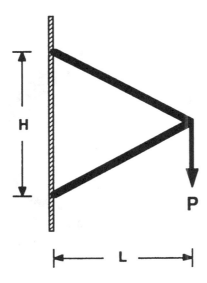

Figure 7.   A Two-Bar Truss

For the fixed values of the applied load P and the horizontal length L of the frame, the optimal height H is determined by minimizing the mean compliance. If the cross section of the truss-bars is unchanged and is rectangular with the unit width, the optimal height H is obtained as H=2L under the constant volume constraint.

We shall now obtain the optimum layout by minimizing the mean compliance with a single load P applied at the center of the right edge of the design domain, shown in Figure 8, that is larger than the size Lx2L, where L=10cm, so that it can contain the optimum layout obtained by a two bar truss structure inside of the design domain. The relaxed design problem (1) is solved for the discrete finite element model of the design domain consisting of 40 x 96 uniform rectangular finite elements.   As shown in Figure 9, the present homogenization method forms a two-bar framed structure if the amount of the "solid" material is specified to be rather small without assuming any type of structures such as trusses, beams, and frames.   It automatically identifies the optimum structure and its configuration. Furthermore, if details are examined, it is clear that the size of the bar is also determined.   As the point is close to the end portion where a load P is applied, the height of the cross section of the bar becomes is small.

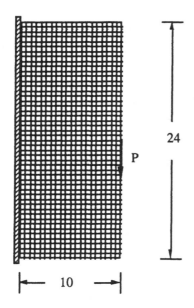

24

P

|← 10 →|

Figure 8   Design Domain and the
          Support/Loading Condition

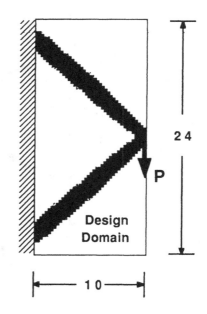

2 4

P

Design
Domain

|← 1 0 →|

Figure 9.   Optimal Relaxed Design
          (Volume = 40 cm^2)

Figure 10   The "optimum" configuration with constraints in the design variables

Now we shall solve the same problem by assuming that the design variables { a, b, θ } are restricted to be

$$a = b \qquad \text{and} \qquad \theta = 0 ,$$

that is, the microscale hole is square and is not rotated during the optimization. Figure 10 shows the configuration obtained under this restriction. Its configuration is similar to the one shown in Figure 9, but it cannot reproduce the two-bar framed structure, and its compliance is also much higher than the case of no constraints on the design variables. The main reason of this difference is caused by the restriction on the angle θ of the microscale hole. Since holes cannot be oriented to the direction of the principal stresses if the angle θ is restricted, the mean compliance is not minimized. This suggests that the simplistic idea of removing finite elements whose stress is low does not work well to obtain the optimum layout of a structure, because the angle θ is not involved in the design problem.

6. Layout of a Beam Structure

We shall solve a layout problem of a beam structure using the above formulation and make a comparison to the result of the beam height optimization that can be found in standard textbooks and monographs of structural optimization. If the problem is regarded as a sizing problem that finds the optimum thickness of the rectangular cross section of a beam with a fixed width, see Figure 11, this can be solved easily by applying a standard technique of structural optimization.

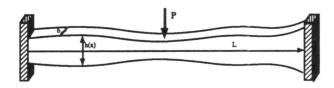

Figure 11 Thickness optimization of a clamped beam subject to a point load P

Minimizing mean compliance subject to equilibrium equation and volume constraint is equivalent to

$$\max_{h} \min_{w} \left\{ \int_0^L \frac{1}{2} Ebh^3 \left(\frac{d^2w}{dx^2} \right)^2 dx - Pw \big|_{x=L/2} \right\}$$

(14)

subject to

$$\int_0^L bh\,dx - V \le 0$$

(15)

Here w is vertical displacement, h is height of beam that is to be designed, and b is width of beam that is fixed, E is Young's modulus, P is applied point load, and V is upper bound on volume. Using the Lagrange multiplier method, necessary condition for optimum can be derived as

$$h^2 \left(\frac{d^2w}{dx^2} \right)^2 = \text{constant} \quad (0 \le x \le L)$$

(16)

It follows from the boundary conditions that the optimal thickness is given by

$$h(x) = \frac{3\bar{V}}{2bL} \sqrt{\left| 1 - \frac{x}{(L/4)} \right|} \quad (0 \le x \le L/2)$$

(17)

Symmetry is applied for the rest of the beam : $L/2 < x < L$, and the optimum thickness is shown in Figure 12.

Figure 12. Analytical solution of beam thickness design

The optimal thickness involves two hinges at x=L/4 and 3L/4, i.e., h = 0 at x=L/4 and 3L/4, while the maximum thickness is obtained at x=0, L/2, and L. The gradient h' of the thickness is ∞ at a=L/4 and 3L/4.

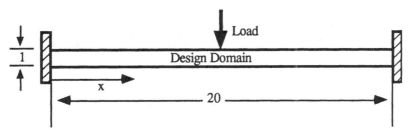

Figure 13. A Design Problem of a Beam Structure

Now, if this problem is solved by the present method by specifying different volume of the solid material, see Figure 13, while the design domain is fixed, the optimal layout of a structure is obtained as shown in Figure 14. It is clear that hinges appear at x=5 and 15, and is the same to the case of the standard thickness optimization. Since it is possible to generate internal holes inside the structure, holes are "naturally" formed to increase the stiffness against bending moment. i.e. forms sandwich beam. In most of thickness optimization published so far, internal holes are not presented. Thus, the method introduced in this paper can provide far more sophisticated optimal structures. If the volume of solid material is reduced, very truss like structures are formed. Because of the restriction on the design domain, reinforcement can be placed only inside the rectangular domain. Large reinforcement is observed in the vicinity of the fixed end points as well as the center at which the point force is applied. It is clear that the present method can provide the shape and topology of the overall structure as well as the optimal sizes of members of a "truss" structure, and that there have been no other methods which have comparable capability to the present one.

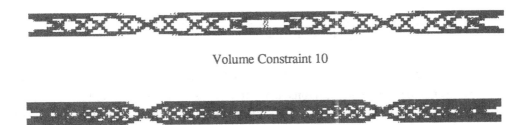

Volume Constraint 10

Volume Constraint 15

Volume Constraint 18

Figure 14.   Optimal Layout of a Beam like Structure

7.   Convergence of the Finite Element Approximation

Next issue to be discussed is whether the shape and topology, i.e., the layout of the structure obtained as the optimum in the relaxed design problem, converges to the unique design as finite element meshes are uniformly refined, while other conditions are fixed. Importance of this test can be recognized by the observation in Cheng and Olhoff that shows if the relaxed approach is not applied, the optimal rib distribution is highly mesh dependent. That is, refinement of the discrete model may lead completely different optimal solutions.

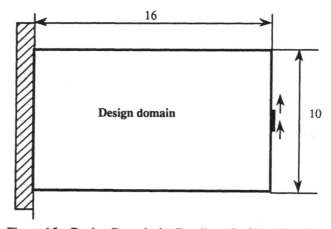

Figure 15.   Design Domain for Bending of a Short Cantilever

To check convergence property, let us solve the relaxed optimization problem (1) for a short cantilever subject to the vertical force at the free end, see Figure 15. For the volume 60 of the "solid" material, (1) is solved by using 32x20, 48x30, 64x40, and 80x50 equal size uniformly divided rectangular finite elements covering a rectangular design domain Ω. Applying the same homogenized elasticity tensor to the first example, the optimal configurations are obtained as shown in Figure 16. It is clear that the optimal configurations are convergent as the size of finite elements is reduced.   Even every coarse meshes can provide sufficient idea of the topology and shape of the optimal structure. A very truss-like framed structure is built if the amount of the solid material is sufficiently smaller than that of the design domain. Each member is even straight. If the amount of the solid material becomes large, the optimal design may not be a truss-like structure. Curved frames can be generated, and more continuum-like shapes are formed. In this problem, despite of the relaxation that allows perforated composite at everywhere of the design domain, the optimal structure is not perforated at all. In other words, the relaxed problem provides the "classical" optimal solution to the design problem using only solid structural members.

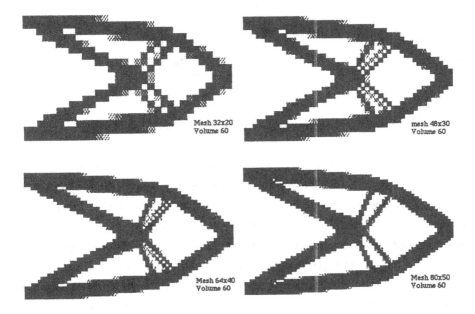

Figure 16. Convergence of the Optimal Configuration ($\Omega_S = 60$)

Figure 17 presents convergence of the mean compliance as the size of finite element goes to zero. Monotonic decreasing of the mean compliance is clearly observed.

We shall now examine the stress distribution of the optimum layout of a structure obtained by minimizing the mean compliance. Since the mean compliance is a global quantity related to the strain energy of a structure, it cannot control the stress locally. In other words, the stress cannot be bounded by a given value at *every* point of a structure.

Figure 18 shows the optimum layout and the distribution of the Mises stress for the three different volume of the "solid" material. Here the design domain and the loading/support

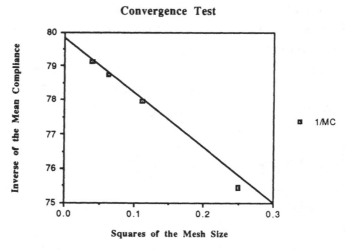

Figure 17. Convergence of the Mean Compliance

condition is the same to the one in Figure 15. It is clear that the fully stressed condition is achieved in most of the identified structure except the neighborhood of the support/loading points, especially if the volume of the "solid" material becomes small. In this sense, minimizing the mean compliance tends to yield a fully stressed structure in the present homogenization approach because not only the shape and size but also the topology of a structure can be arbitrarily changed in the present formulation. However, if the optimum layout is not a "structure" such as a truss, beam, and frame, then the stress distribution is not unidirectional, and is constant. Thus, the fully stressed condition cannot be related to the mean compliance.

8. Layout of a Triangular Membrane

So far only rectangular finite elements are applied to solve the optimal design problem (1) and have obtained rather discrete truss-like structures. Now let us examine a case that non-rectangular finite elements must be applied to form a discrete model, and that the present approach can also provide smooth shape of a structure if the volume of the "solid" portion is large enough in the design domain. To do this, let a "triangular" design domain is considered as shown in Figure 19.

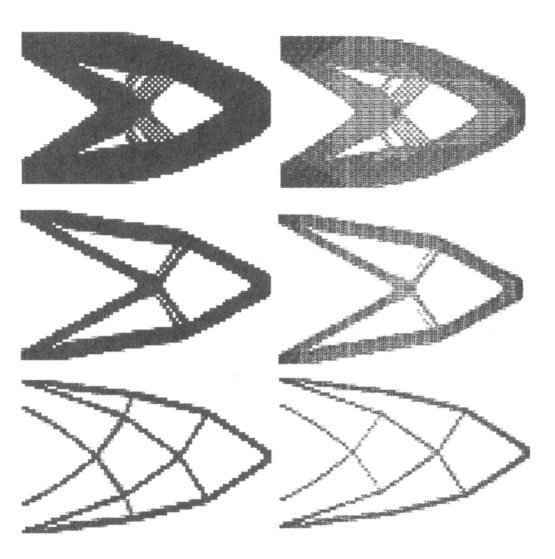

Figure 18   Optimum layout and Mises stress distribution of a structure
for the three different volume of the "solid'" material

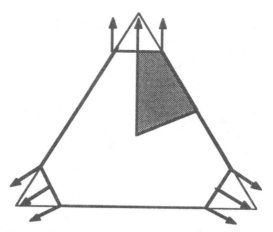

Figure 19   A Triangular Membrane Subject to Tension Forces

Applying symmetry condition, only 1/6 part of the design domain is modeled by the finite element method.   The boundary condition and physical dimensions are specified in Figure 20.   The domain is discretized by 40 x 40 meshes which are not orthogonal.   Using this example, we shall examine transition from the shape to the topology optimization.   The volume of the "solid" material is varied in a wide range from 50 cm^2 to 180 cm^2, while the area of the 1/6 of the original domain is fixed to be 317.5 cm^2.

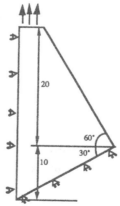

Figure 20   Design Domain for the Optimization(1/6 portion of the original domain)

As shown in Figure 21, if there are sufficient amount of "solid" material, the optimal structure is simply connected without having holes inside of the domain. The obtained shape is also very smooth that is almost same quality to the one obtained by the boundary variation method which is traditionally applied to solve the shape optimization problem. When the amount of "solid" is reduced, holes starts appearing interior of the domain. It is also noted that the loading portion becomes porous when the amount of "solid" material is really small in order to diverse the applied traction uniformly in all the directions equally, while the three compressed link bars in 30,150, and 270 degree directions, are also porous. Thus, if these porous bars are replaced by solid bars maintaining the "solid" material volume, the size of bars becomes very small. Thin solid bars which are in tension, are allocated in 90, 210, and 330 degree directions in the optimal structure. The outer frames are completely solid. Transition from the simply connected domain to the multi-connected domain occurs at Ω_s=130 cm^2 - 140 cm^2.

It is also noted that from our experience the perforated composite appears only in the vicinity of the portion where distributed traction or body forces are applied. If a finite number of discrete "point" forces is applied, the optimal structure by the relaxed problem tends to generate a very discrete solid structure rather than a perforated composite structure. In the present example, a distributed load is applied on the three end surfaces, the optimal structure is perforated in the vicinity of distributed loads while most of the rest of the part of the structure forms a discrete solid structure, i.e. a truss like structure without any perforation. Instead of a distributed load applied along a portion of the boundary, if the displacement is specified over there, perforation does not also occur in the vicinity of such boundary. In this sense, despite of possibility of perforation and extensive composite type microstructure, the optimum structure is rather discrete, and it tends to avoid perforation.

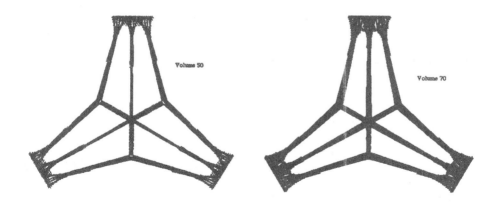

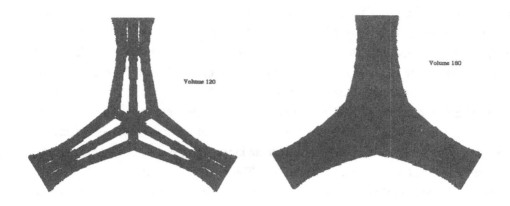

Figure 21 Optimal Configurations for Various Volumes of "solid" Material

9. Optimal Layout of a Structure Subject to Multi-Loads

In practice, a structure must often be designed under the multi-loading condition. That is, a structure must be safe for a set of various loadings. How can we extend the present homogenization method to solve such multi-loaging optimization problem ? Here we shall introduce a "non-standard" way to define the problem, although most of multi-loading problems are defined by introducing an appropriate linear combination of the cost functions for each loading. To this end, if the mean compliance is regarded as the cost function, we have the following relation :

$$L(u) = a(u,u)$$
$$(18)$$

Thus, the optimization problem (1) for a single load can be written as

$$\underset{d}{\text{Minimize}} \quad a(u,u)$$

$$\textit{subject to}$$
$$u \in V : a(u,v) = L(v), \ \forall v \in V$$
$$\int_{\Omega} (1-ab) d\Omega \leq \Omega_0$$
$$(19)$$

Now, let the principle of virtual work be represented by

$$u_i \in V : a(u_i, v) = L_i(v), \forall v \in V \qquad , \quad i = 1, \ldots, m \tag{20}$$

for the i-th loading case

$$L_i(v) = \int_\Omega \varepsilon(v)^T D\varepsilon_{0i} d\Omega + \int_\Omega v^T b_i d\Omega + \int_\Gamma v^T (kg_i + t_i) d\Gamma \tag{21}$$

where $i = 1, \ldots, m$, and then let a special functional $a^m(u, u)$ be defined by

$$a^m(u,u) = \int_\Omega \underset{i=1,\ldots,m}{Max} \left\{ \varepsilon(u_i)^T D\varepsilon(u_i) \right\} d\Omega + \int_\Gamma \underset{i=1,\ldots,m}{Max} \left\{ u_i^T k u_i \right\} d\Gamma \tag{22}$$

Using this special functional, we shall define the optimization problem

$$\underset{d}{Minimize} \qquad a^m(u,u)$$

subject to
$$u_i \in V : a(u_i, v) = L(v), \forall v \in V, i = 1, \ldots, m$$
$$\int_\Omega (1 - ab) d\Omega \le \Omega_0 \tag{23}$$

This formulation yields a "conservative" optimum design in the sense that the resulted structure is safe for all the loads applied independently, and yields a sub-optimum solution of the standard multi-lodaing optimization problem :

$$\underset{d}{Minimize} \qquad \underset{i=1,\ldots,m}{Max} \; a(u_i, u_i)$$

subject to
$$u_i \in V : a(u_i, v) = L(v), \forall v \in V, i = 1, \ldots, m$$
$$\int_\Omega (1 - ab) d\Omega \le \Omega_0 \tag{24}$$

Despite of disadvantage of the possibility that only a sub-optimum solution may be obtained from formulation (23), this problem can be solved very similarly to the single load case with minor modification, while solving the standard multi-loading optimization problem (24) is rather difficult.

As an example of multi-loading, we shall solve the design problem shown in Figure 22 in which a structure pin-supported at two different size circular holes is subject to three loadings, tension P1, bending downward P2, and bending upward P3. Furthermore, we shall restrict design to the case that parts of the constructed structure cannot pass through the center portion of the design domain.

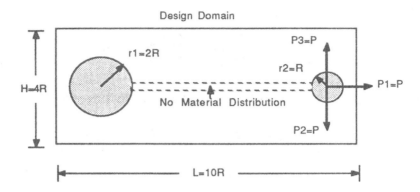

Figure 22. Design Domain, Restriction on Design, and Multiple Loadings

The relaxed design problem for the multi-loading (23) is solved using the similar homogenization method for the single load case (1), and the optimum layout is obtained for two different volume of the solid material, as shown in Figure 23. Figure 24 shows the optimum layout for the case that the design restriction is ignored, i.e., that structure can be placed along the center horizontal line connecting two pins. Significant design

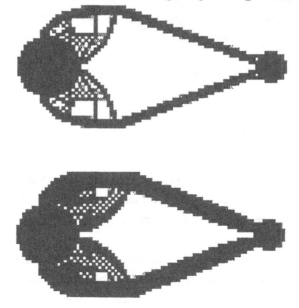

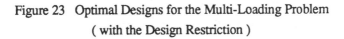

Figure 23   Optimal Designs for the Multi-Loading Problem
(with the Design Restriction)

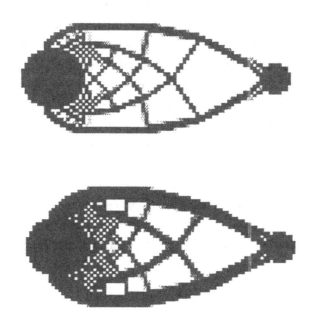

Figure 24   Optimal Design of the Multi-Loading Problem
(without the Design Restriction)

change is observed in this problem with/without design restriction.   If design restriction is
abandoned, the optimum structure contains elements of Michell truss, i.e., is formed by
nearly orthogonal truss/beam network.   Design restriction yields rather stiff two bars in most
of the portion, and they are supported by two small scale Michell truss-like substructures.   It
is natural that the "mean compliance" of the case without design restriction is lower than that
with design restriction.

10.   Extension to Plates Subject to Bending

Since the homogenized D matrix for plane stress has been obtained in above, this can
be applied to formulate plate bending problems by replacing the D matrix for isotropic plate.
However, it must be realized that this simple replacement may not yield an orthotropic plate
obtained by the homogenization process by assuming microstructure due to infinitely refined

ribs as studied by Cheng and Olhoff, and Bendsφe[4]. The plate model obtained by simple replacement of the *D* matrix computed for plane stress problems is an intermediate model of an isotropic plate and an orthotropic plate with microscale ribs, and it yields rather discrete "finite" size ribs in the optimum layout of a plate-like structure as shown in the following example. Since microscale ribs might be only interested in theory, the present formulation due to simple replacement may have much practical meaning for design of plate structures.

As an example of the optimum layout of a plate subject to transverse loads, let a square design domain be considered subject to a concentrated force at the center and a uniformly distributed pressure force. Suppose that a plate is simply supported along the boundary of the square design domain, and suppose that there are no other design constraints except the total volume of the "solid" material. The optimum layouts are obtained as shown in Figure 25. Majority of the solid portion forms a cross-like structure, and is reinforced by thin beams along the diagonals.in both loading cases. It is also noted that hinges are formed at the quarter points of the diagonals where the slip lines are passing through in the limit analysis of a plate. As shown in Figure 25, the optimum structures are not perforated in most of the design domain.

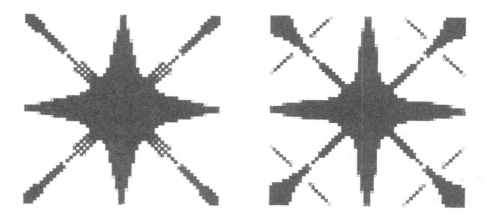

Figure 25   Optimum Layout of Plate-Like Structure Subject to
a Transverse Point Force at the Center and a Uniformly Distributed Pressure

11.   Extension to a Shell-like Structure

Extension of the present method can be also made for shell-like three-dimensional structures similarly to plate-like structures. As an example, we shall consider a layout of reinforcement of a thin shell supported by three hinges and is subject to a uniformly distributed pressure. As shown in Figure 26, three edges are straight, while a curved edge is defined by a sin function. Interior portion of the shell is interpolated linearly.

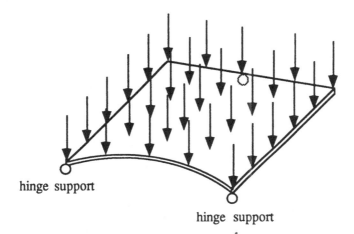

hinge support

hinge support

Figure 26   Shell Structure Reinforced

The design problem is to find the optimal reinforcement by adding material on the already buildup thin shell structure, say 1mm thick shell. As shown in Figure 27, the optimum reinforcement is very discrete, i.e., there are little portion where perforated composite is assigned to minimize the mean compliance. Very clear rib reinforcement is obtained. Basically two main ribs are generated along the lines connecting the three support points, while two rather narrow ribs are formed along the flat boundary edges to increase rigidity. If the shell is curved, the applied force is decomposed into the membrane and bending ones, and then the curved edge can resist to transverse loads more than the flat edge. Thus reinforcement is required along the flat edges of the shell considered as shown in Figure 27.

12.   Extension to Three-Dimensional Solid Structures
The same concept is applicable to the layout problem of three-dimensional solid structures. In this case, we shall make perforation by microscale rectangular parallelepipeds with rotation about the three coordinate axes. Thus the design variables become six at an

arbitrary point x : d = { $a, b, c, \theta_1, \theta_2, \theta_3$ }, i.e. three sizes and three rotations of rectangular parallelepipeds at each point of a design domain. The rest of the part is exactly the same to the case of plane problems. As an example, let us solve an optimum layout

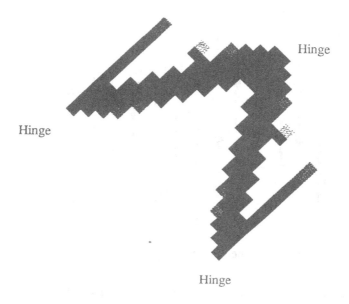

Figure 27   Top View of the Rib Reinforcement

problem of a structure subject to multi-loads, $P_x=P_y=3P_z=1$, at the free end of a rectangular parallelepiped design domain as shown in Figure 28. Figure 29 shows the optimum layout obtained by the present method. It is clear that there are many difficulty to find such a layout by applying the standard boundary variation method for the shape optimization of three-dimensional structures by specifying the shape using appropriate spline surface representation. Even in this simple design domain and applied multi-loads, the optimum layout is rather complicated in the sense that it requires many patches of spline surfaces to develop its solid modeling, and then many control points must be introduced for shape optimization to vary the shape of a structure in optimization process. Although the number of finite elements and discrete design variables becomes fairly large in the homogenization method for three-dimensional solid structures, we are free from defining

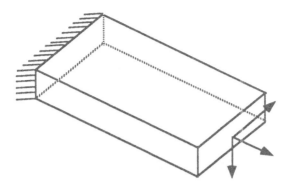

Figure 28   Design Domain of a Three-Dimensional Solid Structure

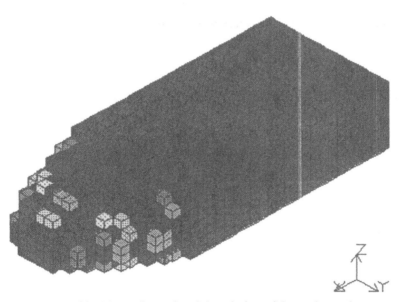

Figure 29   Three-dimensional description of the optimum layout
of a three-dimensional solid under the multi-loading condition

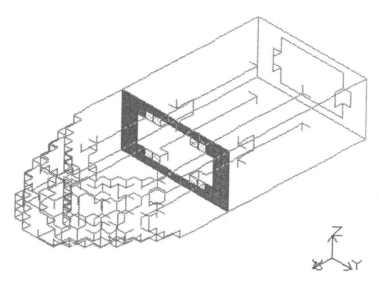

Figure 30   Optimum Layout in a cross-section perpendicular to the x axis

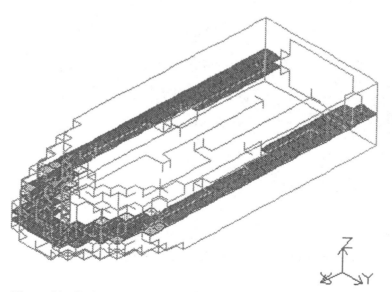

Figure 31   Optimum layout in a cross-section perpendicular to the z axis

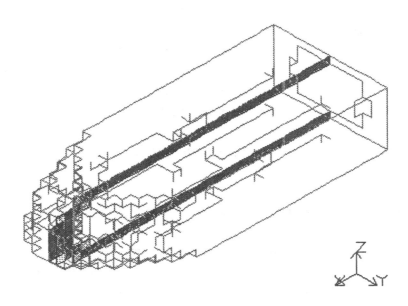

Figure 32   Optimum layout in a cross-section perpendicular to the y axis

the solid model of a structure and automatic decomposition into finite elements.   Topology and shape are obtained as the result of optimization by accumulating "pixels" for the solid portion.   The result in Figure 29 is obtained by a 34x14x8 uniform mesh, and then it involves 22,848 discrete design variables and 14,175 degrees of freedom in the finite element model for stress analysis.   Figures 30,31, and 32 shows the layouts of the structure in its three orthogonal cross sections.   It is clear that a hole is generated inside of the structure.

13.   Construction of an Integrated System

As shown in above the homogenization method can provide the optimum layout of an elastic structure subject to a single load as well as multiloads, and is a fixed domain method that makes it possible to solve the problem using a fixed finite element model during optimization.   As the result obtained, the topology and shape of a structure are identified by the distribution of the microscale holes of perforation.   Since these are represented by "gray scale of pixels" of a finite element model of the design domain, we have to recognize them by defining appropriate spline surfaces based on such gray scale map if function representation of the shape must be obtained. Furthermore, it should note that the homogenization method introduced here is defined by minimizing the mean compliance only with volume constraint. However, practice of design optimization of elastic structures requires much more : the "cost" function minimized can be the maximum Mises stress with various other constraints on displacement, strain, stress, and requirement to be stable structures as well as constraints due to manufacturability and others. Thus, the method presented here cannot answer to all of the requirement in practice of design optimization. In order to fulfill such requirement, we may construct an integrated system of a topology-generating into the design optimization process with the following general scheme:

Phase I: Generate information about the optimum topology for the
            structure.
Phase II: Process and interpret the topology information.
Phase III: Set up a model for detailed design optimization and optimize the
                design applying standard structural optimization techniques.

This scheme is the basis for an integrated structural optimization system which is rudimentarily drafted in Figure 33.

Phase I is accomplished by the homogenization method describe in above. As demonstrated, the method requires only loading and boundary conditions, a definition of the design domain (i.e., the domain where the designer wants material to be distributed in an optimal way), the objective function (i.e. the mean compliance) and the volume constraint. What we achieve here is an *analytically derived approximate configuration*, that corresponds to results of a "concept evolution" stage of the design process, based on satisfying the primary design requirement (based on the stiffness of a structure) from the structural viewpoint. For a practical design tool, the ability to refine the design and account for other requirements is necessary. The method produces an optimum material distribution described by density, $1 - ab$, data which can be represented as images. The density information is generally "noisy" and has to be processed and smoothed with computer vision tools, producing higher level representations that are easier to be interpreted than the original ones. Based on these smoothed pictures the designer can now generate an "initial design", choosing geometric elements (like straight lines, circles, and appropriate splines) that meet the manufacturing requirements. Any commercial CAD program featuring standard data

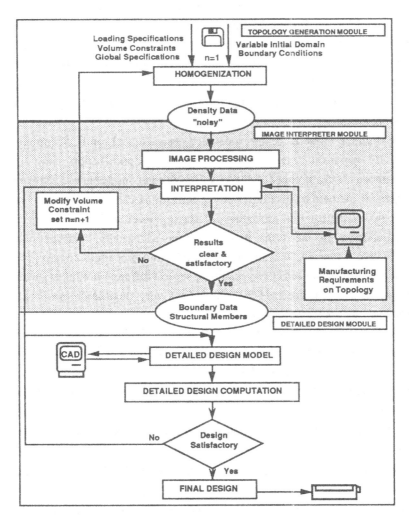

Figure 33   Basic flow chart for the three-Phase
integrated structural optimization system

interfaces may be used for this step. For example, if a component is intended for automated assembly, certain symmetries will substantially reduce the assembly cost, such as adding a non-functional hole to achieve easy part orientation during assembly. If the topological change is sufficiently drastic, a new topology optimization may be necessary with a modified initial domain specification for the homogenization. At the same time a mechanical interpretation has to be done, i.e., structural elements (like trusses, beams or solids) have to be recognized. The steps "image processing" and "interpretation" constitute Phase II of the design process and are represented in the center part of the diagram shown in Figure 33. The final result of Phase II is an initial design represented by boundary data and information about structural members.

Starting from the Phase II design, a detailed design model can be set up in order to perform Phase III operation : standard sizing and shape optimization using structural optimization methods. As the number of design variables and constraints is usually moderate in this step, general mathematical programming algorithms can be utilized to locate the solution, which makes the solution procedure adjustable concerning versatile objective and constraint specifications. This means that the designer may add specific model definitions that have been neglected in Phase I. The structural optimization procedure consists of a finite element program for stress and sensitivity analyses, several different optimization algorithm and standard pre- and post-processors that allow for a very flexible definition of design variables, objective functions and constraints.

Since the topology of a structure is fixed to the one obtained by the homogenization method (Phase I), and since its geometry is represented by a set of appropriate spline functions, Phase III involves only standard optimization procedures for sizing and shape optimization.   As stated in Introduction, if the topology of a structure is fixed, the shape optimization problem can be solved by introducing an automatic mesh generation scheme to the system of optimization consisting of finite element analysis, design sensitivity analysis, and optimization methods, that is, one of the key features of the integrated system is the implementation of an automatic mesh generation method inside of the optimization algorithm. For the shape optimization problem, the domain discretized into finite elements may be largely deformed during the optimization process despite that the topology is fixed.   If the initial finite element connectivity and the total numbers of nodes and elements are kept throughout the optimization, no matter how sophisticated mesh moving schemes are introduced, many finite elements would be excessively distorted.   It is natural that this implies unnecessary approximation error resulting in non-reliable stress analysis for the optimization.   In order to avoid this difficulty, implementation of an automatic mesh generation method to regenerate finite element discretization at each optimization step, is indispensable as pointed out by Bennett and Botkin [10]

Various algorithms of automatic mesh generation have been introduced in the last two decades after Fukuda and Suhara [11] first derived an algorithm for two-dimensional domain using triangular elements. This work is extended by Cavendish [12], and then the concept of automatic mesh generation becomes widely recognized among engineers as well as researchers. A significant conceptual development is made by Lo [13]. After this, automatic mesh generation becomes a commonly applied tool at least for two-dimensional and three-dimensional shell-like structures. A program for automatic mesh generation can be developed relatively easily, if the speed of computation is ignored. It is also noted that algorithms applicable both for triangular and quadrilateral elements are developed based on the QUADTREE concept by Shephard [14], and extensive success of their application in practical fields is reported. This is extended even to three-dimensional solid structures. In the present work, an algorithm developed by Tezuka [15] which is a variation of Lo's method, is applied.

We shall briefly describe the algorithm of automatic mesh generation applied in the present study :

(1) A given domain is divided into several subdomains in order to introduce different mesh "density" , i.e., refinement. If a uniform discretization is required, only one single domain needs to be introduced.

(2) Nodal points are placed on the boundary of each subdomain in order to define their geometry. Boundary segments of subdomains are represented by Besizer splines with given control points. If the total number of nodes on each segment is specified, nodes are automatically or manually generated on the segment. Some of segments become the design boundary of the shape optimization.

(3) In each subdomain, compute the average mesh size from the nodes on the boundary, and then span a rectangular grid over the subdomain.

(4) Identify grid points which are interior of the subdomain. Note that grid points being too close to the boundary should be excluded to generate finite elements which are not excessively distorted.

(5)   Apply a triangulation algorithm from the nodes on the boundary using the interior nodes identified in the previous step. Guideline for triangulation is that a triangle having the maximum of the minimum inner angle should be chosen.

(6)   After triangulation, apply a smoothing scheme to have smooth gradual refinement as well as almost regular shape triangular elements.

Step 4 and Step 5 could be time consuming, since search algorithms are extensively used. Especially, if a large number of nodes is introduced, these search algorithms are executed over "all" of the nodes.   Sophistication must be introduced to reduce the necessary computing time for these searches.

The mesh generator described in above is included into the pre-processor of a model of finite element stress analysis, and the coordinates of control points of the design boundary is modified by an optimization algorithm based on the result of sensitivity analysis. Only the fixed information is the control points of the design boundary segments during the optimization, while nodal points on such segments are changed at each design step, sensitivity analysis must be performed for the coordinate of the control points. Since the total number of the control points of the design boundary is rather small, say at most $10 \sim 20$ for plane problems, and since its explicit relation between to the nodal coordinates of the finite element model generated can be expressed by the coordinate transformation of all the elements, sensitivity may be computed by the finite difference approximation.   If semi-analytical or analytical method is applied to compute sensitivity, we have to evaluate derivatives of the stiffness matrix for all the nodes in the model either approximately or analytically, and this is unrealistic.   The best way in this setting is application of the finite difference method to evaluate design sensitivity with respect to the location of the control points.   It is also noted that representation of design constraints on geometry such as the total volume of a structure and geometric restriction to define its configuration is rather difficult if spline representation is applied for the design boundary.

Figure 34 shows a bracket design problem in plane.   Phase I, the homogenization method for the optimum layout results the configuration shown in Figure 35, and its result is processed through image processing and interpretation of the geometry (Phase II), and implies the finite element model generated by the automatic mesh generation method. Applying SAPOP system, [16,17] the optimum shape is obtained as shown in Figure 36.

Details of the integrated system and an example described here can be found in Bremicker, Chirehdast, Kikuchi, and Papalambros[18].

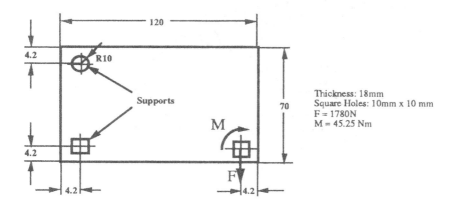

Thickness: 18mm
Square Holes: 10mm x 10 mm
F = 1780N
M = 45.25 Nm

Figure 34   Boundary specifications and the initial design domain

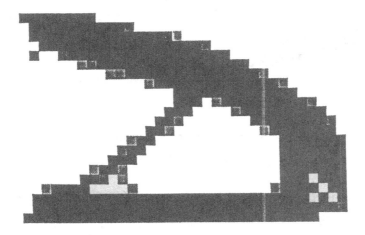

Figure 35   Optimum layout by the homogenization method (Phase I)

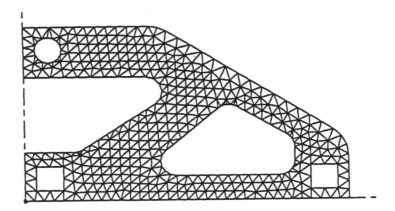

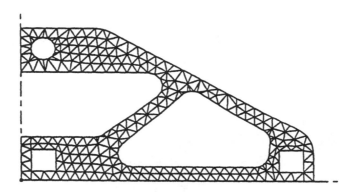

Figure 36 The initial model and the optimum shape of the structure (Phase III)
Initial design model and its optimum

Acknowledgement During the work, the authors are supported by NASA Lewis Research Center, NAG 3-1160, Office of Naval Research, ONR N-00014-88-K-0637, DHHS-PHS-G-2-R01-AR34399-04, and RTB Corporation.,Ann Arbor. The authors also express their sincere appreciation to Professors Martin Bendsøe, Technical University of Denmark, Alejandro Diaz, Michigan State University, and Panos Papalambros, The University of Michigan, for their valuable discussions on this subject, and to Dr. Mike Bremicker and Mr. Y.K. Park for their help to complete some of examples described in this article.

References

[1]   Banichuk, N.V., *Problems and methods of optimal structural design*, Plenum
        Press, New York (1983)

[2]   Cheng, K.T., and Olhoff, N., An investigation concerning optimal design of solid
        elastic plates, *Int. J. Solids and Structures* 17 (1981) 305-323

[3]   Lurie, K.A., Fedorov, A.V., and Cherkaev, A.V., Regularization of optimal design
        problems for bars and plates, Parts I and II, *J. Optim. Theory Appl.* 37(4) (1982)
        4999-521, 523-543

[4]   Bendsøe, M.P., Generalized plate models and optimal design, in J.L. Eriksen, D.
        Kinderlehrer, R. Kohn and J.L. Lions, eds., *Homogenization and effect moduli of
        materials and media*, The IMA Volumes in Mathematics and Its Applications,
        Spriger-Verlag, Berlin, 1986, 1-26

[5]   Palmer, A.C., Dynamic Programing and Structural Optimization, in R.H.
        Gallagher and O.C. Zienkiewicz, eds., *Optimum Structural Design*, John Wiley &
        Sons, Chichester (1973) 179-200

[6]   Murat, F., and Tartar, L., Optimality conditions and homogenization, in A.
        Marino, L. Modica, S. Spagnolo, and M. Degiovanni, eds., *Nonlinear variational
        problems*, Pitman Advanced Publishing Program, Boston, 1985, 1-8

[7]   Kohn, R., and Strang, G., Optimal Design and relaxation of variational problems,
        Parts I, II, and III, *Communications on Pure and Applied Mathematics*, XXXIX
        (1986) 113-137, 139-182, 353-378

[8]   Bendsøe, M.P., and Kikuchi, N., Generating optimal topologies in structural
        design using a homogenization method, *Comput. Mechs. Appl. Mech. Engrg.*, 71
        (1988) 197-224

[9]   Suzuki, K., and Kikuchi, N., Shape and topology optimization using the
        homogenization method, *Comput. Mechs. Appl. Mech. Engrg.* too appear (1991)

[10]   Bennett, J. A. and Botkin, M.E., Structural shape optimization with geometric
        problem description and adaptive mesh refinement, *AIAA J.*, 23(3), 458-464
        (1985)

[11]   Fukuda, J., and Suhara, J., Automatic Mesh Generation for Finite Element
        Analysis, in Advance, in: *Computational Methods in Structural Mechanics and
        Design*, (Ed. Oden, J.T., and Yamamoto, Y.), UAH Press, Huntsville, Alabama,
        U.S. (1972)

[12]   Cavendish, J.C., Automatic Triangulation of Arbitrary Planar Domains for the
        Finite Element Method, *International Journal for Numerical Methods in
        Engineering*, 8, 679-696 (1974)

[13]   Lo, S.H., A New Mesh Generation Scheme for Arbitrary Plannar Domains,
        International Journal for Numerical Methods in Engineering, 21, 1403-1426

(1985)

[14]   Shephard, M.S., and Yerry, M.A., An approach to automatic finite element mesh generation, in: *Computers in Engineering 1982*, 3., edited by Hulbert, L.E., The American Society of Mechanical Engineers, New York, 21-28 (1982)

[15]   Tezuka, A, *A Development of Automatic Mesh Generator with Arbitrary Geometry-Based Input Description*, MS Thesis, Department of Mechanical Engineering and Applied Mechanics, The University of Michigan, Ann Arbor, MI, U.S., (1988)

[16]   Eschenauer, H., Post, P.U. and Bremicker. M., Einsatz der Optimierungsprozedur SAPOP zur Auslegung von Bauteilkomponenten. *Bauingenieur 63*, 515-526 (1988)

[17]   Bremicker,M., Eschenauer, H., Post, P., Optimization Procedure SAPOP - A General Tool for Multicriteria Structural Design. In: Eschenauer, H., Koski, J., Osyczka, A., Multicriteria Design Optimization. Berlin, Springer Verlag (to appear May 1990)

[18]   Bremicker, M., Chirehdast, M., Kikuchi, N., and Papalambros, P.Y., Integrated Topology and Shape Optimization in Structural Design, Journal of Mechanics of Structures and Machines, (in Review) 1989

Chapter 13

INTRODUCTION TO SHAPE SENSITIVITY
THREE-DIMENSIONAL AND SURFACE SYSTEMS

B. Rousselet
University of Nice, Nice, France

1 Introduction

In these lectures we present some basic material for the shape optimization of structures. We emphasise the so - called continuous approach with few results on numerical approximation with finite elements or boundary integrals; this approach is traditional in mathematics and theoretical mechanics, whereas in mechanical engineering the tendency is to first approximate the behaviour of the structure with finite elements and afterwards to tackle optimization.

The choice of one of these approaches depends on the habits of thought; in many cases, discretisation in the first or second step yields the same results; this has been proved when one uses <u>conformal</u> finite elements (S. Moriano 1988 , M. Masmoudi 1987). If one is interested in deriving necessary optimality conditions and finding explicit solutions, then the continuous approach is necessary; this is the route followed by the Pragerian school.

However in connection with finite elements, the continuous approach is quite versatile: it enables the addition of design sensitivity to a commercial finite element code (Melao Barros & Mota soares 1987, Chenais & Knopf-Lenoir 1988); but it also enables the inclusion of design sensitivity in an open finite element library such as Modulef (1985) and makes good use of existing software (Mehrez-Palma-Rousselet 1991 to appear).

Moreover, formulae obtained with the continuous approach can be implemented with boundary elements (Masmoudi (1987), Mota soares, Rodrigues Choi (1984)).

It should also be pointed out that these techniques may be used and are used in other fields of application; for example in acoustic (Masmoudi 1987) and in fluid mechanics (Pironneau 1984).

However what is shape optimization ? It is an optimal design problem where the design variable is the shape of the domain Ω occupied by the physical system; the best shape of a fillet in a tension bar will provide a classsical engineering example (Haug,Choi-Komkov 1986):

we want to find the best shape of Γ_0 to minimize volume with constraints on Von-Mises

yield stress.

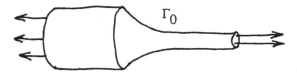

One of the first publications seems to originate with Hadamard (1908) but the pioneers of research oriented toward the use of computers seems to be Céa,Gioan, Michel (1974). Since that date many papers have been devoted to this topic; for example Chenais (1977), Murat, Simon (1976), Rousselet (1976,1977,1982), Dems,Mroz (1984), Pironeau (1984). INRIA schools devoted to shape optimization have been organized by Pironneau (1982) and Céa, Rousselet (1983).

2 shape optimization and continuum mechanics

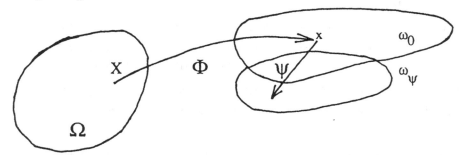

As in conventional optimal design, the clue of the approach is to obtain first - order estimates of the variation of a functional of the state of the system; but for shape optimization one soon realizes that the set of possible domains has no standard vector space structure, so that it seems that classical differential calculus and calculus of variations cannot apply here.

Indeed, these techniques can be used if one realizes that for a given topology and regularity of the boundary, it is natural to look for domains as mappings of a given domain Ω; we shall denote

$$\omega_{\Psi} = (\Phi + \tilde{\Psi})(\Omega) = \{ x \in E / x = \Phi (X) + \tilde{\Psi} (X) \; \forall \; X \in \Omega \}$$

where E is the usual Euclidean space (in one, two or three dimensions); Ψ is an element of a vector space of functions; it will enable to define variations of $\omega = \Phi(\Omega)$.

Anyone familiar with the foundations of continuum mechanics should realise that this is what we are doing when we are using a Lagrangian representation of the movement of a continuous medium; this is the usual representation in solid mechanics. For the implementation of the analysis of large deformations of solids, it is usual to use an updated Lagrangian formulation: this amounts to linearizing the behaviour of the solid around a configuration obtained with a fraction of the load.

Here we are going to linearize around the given domain ω, but we should keep in mind that in the overall process of optimization we shall update the domain ω around which we linearize the cost functional and the constraints.

Basic tools for this linearization are well - known in continuum mechanics, but were derived independently for shape optimization by several authors including Dervieux-Palmerio (1975), Murat-Simon (1976), Rousselet (1976). These tools are recalled in the next two sections.

3 Differential calculus and linearization around a given domain

To join domain sensitivity and surface sensitivity, we recall some basic notations of curvilinear coordinates; in fact the mapping $\Phi: \Omega \longrightarrow \omega$,

$X \longmapsto x = \Phi(X)$ defines <u>curvilinear coordinates</u> in ω ; we denote by

$$g_i(x) = \frac{\partial \Phi}{\partial X_i}$$ the <u>local basis</u>; generally it is not orthonormal so that it is

convenient to use the dual basis g^i

defined by $g^i \cdot g_j = \delta^i_j$

With these notations the matrix of $\dfrac{\partial \Phi}{\partial X}$ is $(g_1\, g_2\, g_3)$

Let $f : \omega \longrightarrow R$ be a scalar funtion; if we set $f_{,i} = \dfrac{\partial f(\Phi(X))}{\partial X_i}$ the chain

rule yields $\dfrac{\partial f}{\partial x} = f_{,i}\, g^i$; it is usual to set $\text{grad } f = g^{ij} f_{,i}\, g_j$

where $g^{ij} = g^i \cdot g^j$

For future reference we recall that

$$\int_\omega f(x)\, dx = \int_\Omega f(\Phi(X))\, |D\Phi|\, dX$$

with $|D\Phi| = \det(g_1\, g_2\, g_3) = \sqrt{g}$

where $g = \det(g_{ij})$ with $g_{ij} = g_i \cdot g_j$

<u>Note</u>. <u>*In the following repeated latin indeces are summed from 1 to 3 and greek indeces*</u>
<u>*from 1 to 2.*</u>

<u>For a vector field v defined in Ω_Φ , the chain rule also yields :</u>

$$\frac{\partial v}{\partial x} = v_{,i}\, g^i$$ where $v_{,i} = \dfrac{\partial v(\Phi(X))}{\partial X_i}$

but if we express v in the local basis g_i :

and wish to express $\dfrac{\partial v}{\partial x}$ with these components it is classsical to introduce Christoffel

symbols $\quad \Gamma^i_{kj} = g^i \cdot g_{k,j} \quad$ so that

$$\dfrac{\partial v}{\partial x} = v^i{}_{|j}\, g_i \otimes g^j \quad \text{where} \quad v^i{}_{|j} = v^i{}_{,j} + \Gamma^i_{jk}\, v^k \quad \text{and} \quad g_i \otimes g^j \quad \text{is}$$

the linear maping defined by $\quad (g_i \otimes g^j)\, (h) = g_i\, h^j$

The divergence operator is well-known in continuum mechanics; we recall here some formulae which have similar features when applied to surfaces. We first consider as a <u>definition</u> the following equality which should hold for any continuously differentiable function f with compact support in ω :

$$\int_\omega f\, \text{div } v\, dx = - \int_\omega \dfrac{\partial f}{\partial x}\, v\, dx$$

where component-wise $\quad \dfrac{\partial f}{\partial x}\, v = f_{,i}\, (g^i\, v) = f_{,i}\, v^i$

to obtain an expression of $\quad$ div $v \quad$ in local basis it is usefull to state

<u>Lemma 3.1</u>

(i) $\quad \dfrac{\partial g}{\partial g_{i\,j}} = g\, g^{i\,j}$

(ii) $\quad \dfrac{\partial g}{\partial x_i} = 2\, g\, \Gamma^j_{i\,j} \quad$ or $\quad \dfrac{\partial \sqrt{g}}{\partial x_i} = \sqrt{g}\, \Gamma^j_{i\,j}$

<u>Proof.</u>

(i) $\quad$ Comes from $\quad g_{i\,j}\, g^{j\,k} = \delta^k_i$

(ii) $\quad$ The proof uses (i) and some manipulations.

<u>Proposition 3.2</u> $\quad$ Let $\quad v = v^i\, g_i$

(i) The following expressions hold

$$\text{div } v = \frac{1}{\sqrt{g}} (v^i \sqrt{g})_{,i} = v^i{}_{|i} = g^i \cdot v_{,i}$$

(ii) The following identity holds

$$\text{div } (f\, v) = f \text{ div } v + \frac{\partial f}{\partial x} v$$

<u>Proof</u>.

(i)   The first identity comes from the definition (3,11) and (3,5), (3,6), (3,7).

(ii)   Is straightforward in components.

The formula which provides **the first-order change of an integral over a domain** ω **with respect to changes of its shape** is well-known in continuum mechanics (see for example Germain 1979) and is now widely used in shape optimal design (see e.g. Céa (1975, 1986), Masmoudi (1987)). Here we try to provide a presentation which is introductory to the more complex case of surface variation.

We recall from § 2 that $\omega_\Psi = (\Phi + \tilde\Psi)(\Omega) = (I + \Psi) (\omega)$ with $\omega = \Phi(\Omega)$

and $\tilde\Psi = \Psi \circ \Phi$

To make precise the first variation of a function f_Ψ defined on a variable domain ω_Ψ we set the

<u>Definition 3.3.</u>

Let f_Ψ be a function defined on ω_Ψ ; this function may depend explicitly on the vector field Ψ and implicitly through the position of the point $x = x + \Psi(x)$ where it is evaluated; we set

$$\tilde f_\Psi(X) = f_\Psi(\Phi(X) + \Psi \circ \Phi(X))   \qquad \text{and}$$

$$\delta f(x) = \lim_{t \to 0} \frac{f_\Psi(x + t\, \Psi(x)) - f(x)}{t} =$$

Remark 1. $$\frac{d}{dt} f_\Psi \circ (\Phi + t\, \tilde\Psi) \big|_{t=0} = \frac{\partial}{\partial t} \tilde f_\Psi (X) \big|_{t=0}$$

We note that δf is a function defined on ω ; it is linear with respect to ψ; if f does not depend explicitly on Ψ, the chain rule yields $\delta f = \frac{\partial f}{\partial x} \Psi$; on the other hand if $\Psi(x) = 0$, $\delta f(x)$ is just the partial derivative with respect to Ψ . Moreover if f does not depend on x, δf is just the directional derivative with respect to Ψ.

Remark 2.

The usual rules for computing derivatives of a sum or of a product of functions hold for the operator δ

Remark 3.

In continuum mechanics, when a flow $t \mid\!\!-\!\!-\!\!-\!\!-\!\!-\!\!-\!\!> \quad x(t)$ is defined on ω,

the <u>material derivative</u> of $f(t, x(t))$ is $\quad \lim_{\delta t \to 0} \dfrac{f(t, x(t + \delta t))- f(t, x(t))}{\delta t}$

we note that δf is a particular case when $x(t) = X + t\Psi(X)$; this simple flow is what is needed to define first order variation of ω . Herein we call <u>material derivative</u> the δ operator

To compute the variation of the integrals we need the following lemma.

<u>Lemma 3.4.</u>

Let $g = \det(g_{ij}) \quad$ then

$\delta\sqrt{g} = \sqrt{g}\,\text{div}\,\Psi$

<u>Proof.</u>

We have $\quad \delta g = \dfrac{\partial g}{\partial g_{ij}}\,\delta\,g_{ij}$; as $g_{ij} = g_i \cdot g_j$ we obtain

$\delta\,g_{ij} = \tilde{\Psi}_i \cdot g_j + g_i \cdot \tilde{\Psi}_j$; then using Lemma 3.1 we obtain

$\delta\,g = 2\,g\,g^j.\tilde{\Psi}_j = 2g\,\text{div}\,\Psi$

<u>Proposition 3.5.</u>

$$\delta \int_\omega u \, d\omega = \int_\omega \delta u \, d\omega + \int_\omega u \, \text{div}\,\Psi \, d\omega$$

<u>Proof.</u>

We obtain from (3.5), (3,6): $\quad \displaystyle\int_\omega u \, d\omega = \int_\Omega \tilde{u}\,\sqrt{g}\,dX$

The result then comes from the definition and Lemma 3.4.

<u>Example.</u>

$\text{vol}(\omega) = \displaystyle\int_\omega d\omega \quad \text{yields}$

$$\delta \, vol(\omega) = \int_\omega div\Psi \, d\omega = \int_{\partial\omega} \Psi . v \, d\sigma$$

as could be expected.

We now state the **variation of a derivative**.

Proposition 3.6.

Let u be a function defined in ω_Ψ ; we have

$$\delta \frac{\partial u}{\partial x} = \frac{\partial}{\partial x} \delta u - \frac{\partial u}{\partial x} \cdot \frac{\partial \Psi}{\partial x}$$

Proof.

We note that ; $\dfrac{\partial u_{t\Psi}}{\partial x} = \dfrac{\partial \tilde{u}_{t\Psi}}{\partial X} \left(\dfrac{\partial (\Phi + t \, \tilde{\Psi})}{\partial X} \right)^{-1}$ We use the definition of δ

$$\delta \frac{\partial u_{t\Psi}}{\partial x} = \frac{\partial}{\partial t}\left(\frac{\partial \tilde{u}_{t\Psi}}{\partial X} \right)\Big|_{t=0} \left(\frac{\partial \Phi}{\partial X} \right)^{-1} + \frac{\partial \tilde{u}_{t\Psi}}{\partial X}\Big|_{t=0} \frac{\partial}{\partial t}\left(\frac{\partial (\Phi + t \, \tilde{\Psi})}{\partial X} \right)^{-1}\Big|_{t=0}$$

$$= \frac{\partial}{\partial X}\left(\frac{\partial \tilde{u}_{t\Psi}}{\partial t} \right)\Big|_{t=0} \left(\frac{\partial \Phi}{\partial X} \right)^{-1} - \frac{\partial \tilde{u}_0}{\partial X} \left(\frac{\partial \Phi}{\partial X} \right)^{-1} \frac{\partial \tilde{\Psi}}{\partial X} \left(\frac{\partial \Phi}{\partial X} \right)^{-1}$$

$$= \frac{\partial}{\partial x} \delta u - \frac{\partial u}{\partial x} \cdot \frac{\partial \Psi}{\partial x}$$

which proves the result.

4. Shape sensitivity for a model system.

We apply the previous results to shape sensitivity of the simplest example: a membrane prestressed with an inplane tension T and submitted to a normal density of force f ; the normal deflection is the solution of:

$$\text{(4,1)} \begin{cases} -T \, \Delta \, u = f & \text{in} \quad \omega \\ u = 0 & \text{on the part } \gamma_1 \quad \text{of the boundary } \partial \omega \text{ where it is fixed.} \\ T\dfrac{\partial u}{\partial n} = 0 & \text{on} \quad \gamma_2 \quad \text{where it is free} \end{cases}$$

We denote by V the space of kinematically admissible displacements; the principle of virtual work states that

$$\text{(4,2)} \qquad \forall \, v \in V \quad a(u, v) = l(v) \qquad\qquad\qquad\qquad \text{where}$$

$$\text{(4,3)} \quad a(u, v) = \int_\omega T \frac{\partial u}{\partial x} \frac{\overline{\partial v}}{\partial x} \, dx \qquad \text{and} \qquad l(v) = \int_\omega f \, v \, dx$$

Note. The overbar denotes the vector associated to a linear form and vice versa: $\dfrac{\overline{\partial v}}{\partial x}$ is the grdient of v.

The variation of the solution is itself the solution of an equation as stated below.

Proposition4.1

Let u be the solution of (4.2), then its first-order variation δu satisfies

$$\text{(4,4)} \qquad \forall \, v \in V \qquad a(\delta u, v) = -(\delta a)(u, v) + (\delta l)(v)$$

where δa and δl are variations of a and l *for fixed u and v* :

$$\text{(4,5)} \qquad (\delta a)(u, v) = \int_\omega T \left(\frac{\partial u}{\partial x} \cdot \frac{\overline{\partial \psi}}{\partial x} \cdot \frac{\overline{\partial v}}{\partial x} \, dx \right) + \frac{\partial v}{\partial x} \cdot \frac{\overline{\partial \psi}}{\partial x} \cdot \frac{\overline{\partial u}}{\partial x} \, dx)$$

$$\qquad\qquad + \int_\omega T \frac{\partial u}{\partial x} \cdot \frac{\overline{\partial v}}{\partial x} \, \text{div} \, \psi \, dx$$

$$\text{(4,6)} \quad (\delta l)(v) = \int_\Omega v \, \delta f \, dx + \int_\Omega f \, v \, \text{div} \, \Psi \, dx$$

The proof is a direct application of Propositions 3.4 and 3.5 .

Shape sensitivity of a functional .

We consider the simplest case

(4,7) $J = \int_\omega \alpha(u)\ dx$

The proposition 3.5 yields

(4,8) $\delta J = \int_\omega \alpha'(u)\ \delta u\ dx\ +\ \int_\omega \alpha(u)\ \text{div}\ \Psi\ dx$

As in conventional design sensitivity this expression is not explicit with respect to Ψ: δu is defined through equation 4.4; but this expression may be transformed.

Proposition 4.2

Let $L(u,v) = J(u) + a(u,v) - l(v)$

and set p the solution of

(4,9) $\forall\ w \in V\ \ \dfrac{\partial L}{\partial u}\ w = 0$ or $a(w,p) = -\dfrac{\partial J}{\partial u}\ w$

then $\delta J = (\delta L)(u,p)$ where the variation of L is computed at u and p fixed; or more precisely:

(4,10) $\delta J = \int_\Omega \alpha(u)\ \text{div}\ \psi\ dx + (\delta a)(u,p) - (\delta l)(p)$

with δa and δl given in the previous Proposition.

5 Surface differential calculus.

We consider now a surface S imbedded in a 3D space E^3, parametrized by a single-valued Φ from a reference open domain Ω of a 2D space E^2. The striking difference with section 3 is that Φ is a mapping from a 2D space to a 3D space. With simplifications all the material presented would be adequate for plane curves, although the use of arc length would simplify some formulae.

To emphasize that Φ stems from a two - dimensional space, we denote by ξ the variable in Ω and greek indeces are implicitly running from 1 to 2 ; repeated indeces mean summation, from 1 to 2.

The local basis is noted

(5,1) $a_\alpha = \dfrac{\partial \Phi}{\partial \xi_\alpha}$; it is a basis of the tangent space to S at the point $m = \Phi(\xi)$.

The <u>dual basis is defined by</u> $a^\alpha . a_\beta = \delta^\alpha_\beta$

where the dot means the usual scalar product of E^3.

So $(a_1\ a_2)$ is the matrix of $\dfrac{\partial\Phi}{\partial\xi} = a_\alpha \otimes e^\alpha$ where $e^\alpha = e_\alpha$ is the standard

basis of E^2 . We note that

$$\begin{pmatrix} {}^t a^1 \\ {}^t a^2 \end{pmatrix} \begin{pmatrix} a_1 & a_2 \end{pmatrix} = I_{R^2}$$

or in tensor notations $(e_\alpha \otimes a^\alpha) . (a_\alpha \otimes e^\alpha) = e_\alpha \otimes e^\alpha$

but $\begin{pmatrix} a_1 & a_2 \end{pmatrix} \begin{pmatrix} {}^t a^1 \\ {}^t a^2 \end{pmatrix} = a_1 \otimes a^1 + a_2 \otimes a^2$

is the matrix of Π , the orthogonal projection onto the tangent plane.

Let now f : S --------> R be a real function defined on S. If we set

(5,2) $\dfrac{\partial f}{\partial m} = \dfrac{\partial \tilde{f}}{\partial \xi_\alpha} a^\alpha$ it is easy to check that this linear mapping from the

tangent plane to R is independent of the parametrization; we also have $\dfrac{\partial f}{\partial m} a\alpha = \dfrac{\partial \tilde{f}}{\partial \xi_\alpha}$

or $\dfrac{\partial f}{\partial m} \dfrac{\partial \Phi}{\partial \xi} = \dfrac{\partial \tilde{f}}{\partial \xi}$

In the following all the notions introduced are independent of the parametrization with the exception of the Christoffel symbols.

The integral over the surface may be written with a parametrization:

(5,3) $\displaystyle\int_S f(m)\ dS = \int_\Omega f(\Phi(\xi))\ \sqrt{a}\ d\xi$

where

(5,4) $a = \det(a_{\alpha\beta})$ with $a_{\alpha\beta} = a_\alpha . a_\beta$ or

$\sqrt{a} = ||\ a_\alpha \times a_\beta\ ||$ (area element).

The differentiation of a vector field is here more intricate; this is intuitively obvious with a circle: let $T(\theta)$ be a unitary tangent vector field. It is clear that when $T(\theta)$ is near $T(\theta_0)$, the first-order change is not tangent but rather orthogonal to the circle; thus we need to introduce the orthogonal projection Π onto the tangent plane; Note that its matrix is

$$a_1 \otimes a^1 + a_2 \otimes a^2 \qquad \text{we set} \quad a_{\alpha,\beta} = \frac{\partial a_\alpha}{\partial \xi^\beta}$$

where $\Pi\, a_{\alpha,\beta}$ is a tangent vector. Its decomposition in the local basis is classically expressed with Christoffel symbols (they do depend on the parametrization!):

$$\Pi\, a_{\alpha,\beta} = \Gamma^\lambda_{\alpha\beta}\, a_\lambda$$

Note that $\qquad \Gamma^\lambda_{\alpha\beta} = \Gamma^\lambda_{\beta\alpha} \qquad$ as $\quad a_{\alpha,\beta} = a_{\beta,\alpha}$

Similarly for a tangent vector field $v_t = v^\alpha a_\alpha$ as $\dfrac{\partial v_t}{\partial \xi^\alpha}$ is not tangent, we

consider

$$(5,6) \quad \Pi \frac{\partial v_t}{\partial m} = v^\alpha_{|\beta}\, a_\alpha \otimes a^\beta \qquad \text{with}$$

$$(5,7) \qquad v^\alpha_{|\beta} = v^\alpha_{,\beta} + \Gamma^\alpha_{\lambda\beta}\, v^\lambda$$

We turn now to the **divergence of a tangent vector field** also defined by an integration by parts.

For any continuously differentiable function f with compact support in S:

$$(5,8) \quad \int_S f\, \mathrm{div}\, v_t\, dS = - \int_S \frac{\partial f}{\partial m}\, v_t\, dS$$

To obtain an expression in the local basis it is convenient to note:

Lemma 5.1

(i) $\dfrac{\partial a}{\partial a_{\alpha\beta}} = a\, a^{\alpha\beta}$

(ii) $\dfrac{\partial a}{\partial \xi^{\alpha}} = 2\, a\, \Gamma^{\lambda}_{\alpha\lambda}$ or $\dfrac{\partial \sqrt{a}}{\partial \xi^{\alpha}} = \sqrt{a}\, \Gamma^{\lambda}_{\alpha\lambda}$

Proof:

It is based on $(a^{\alpha\beta})(a_{\beta\gamma}) = \delta^{\alpha}_{\gamma}$ and

$$a_{\lambda\mu,\alpha} = \Gamma^{\gamma}_{\lambda\alpha}\, a_{\gamma\mu} + \Gamma^{\gamma}_{\mu\gamma}\, a_{\lambda\gamma}$$

which comes directly from the definition of Christoffel symbols.

Proposition 5.2

Let $v_t = v^{\alpha}\, a_{\alpha}$ be a tangent vector field, we then have the following

expressions of the divergence

(5,9) $\operatorname{div} v_t = \dfrac{1}{\sqrt{a}}(\sqrt{a}\ v^{\alpha})_{,\alpha} = v^{\alpha}_{\ |\alpha} = a^{\alpha}\cdot\Pi\,\dfrac{\partial \tilde{v}_t}{\partial \xi^{\alpha}} = a^{\alpha}\cdot\Pi\dfrac{\partial v_t}{\partial m}\, a_{\alpha}$

Proof.

The first equality comes from the definition and (5,3); then we obtain

$$\operatorname{div} v_t = v^{\alpha}_{\ ,\alpha} + v^{\alpha}\,\dfrac{(\sqrt{a})_{,\alpha}}{\sqrt{a}} \qquad\qquad \text{and with Lemma 5.1:}$$

$$\operatorname{div} v_t = v^{\alpha}_{\ ,\alpha} + v^{\alpha}\,\Gamma^{\lambda}_{\alpha\lambda} = v^{\alpha}_{\ |\alpha} \quad ; \ (5,6)\ \text{now gives}$$

$$v^{\alpha}_{\ |\alpha} = a^{\alpha}\cdot\Pi\dfrac{\partial v_t}{\partial m}\, a_{\alpha} = a^{\alpha}\cdot\Pi\,\dfrac{\partial \tilde{v}_t}{\partial \xi^{\alpha}}$$

Because we are interested in variation of S , we shall have to consider vector fields Ψ which are transverse to S ; so now we recall how to compute derivatives of transverse vector fields.

It is usual to introduce a unitary normal vector

$$(5,10) \qquad a_3 = \frac{|| \, a_1 \times a_2 \, ||}{|| \, a_1 \times a_2 \, ||}$$

As $a^3 \cdot a_3 = 1$ we have $\quad \dfrac{\partial \, a_3}{\partial m} \cdot a_3 = 0 \quad$ so that $\quad \dfrac{\partial \, a_3}{\partial m} \quad$ may be

considered as an operator of the tangent plane. Its expression in the local basis is usually written:

$$(5,11) \quad \frac{\partial \, a_3}{\partial m} = - \, b^{\alpha}_{\beta} \, a_{\alpha} \otimes a^{\beta} \qquad \text{so that}$$

$$b^{\alpha}_{\beta} = - \, a^{\alpha} \cdot \frac{\partial \, a_3}{\partial m} \cdot a_{\beta}$$

we also set $\quad b_{\alpha \beta} = - \, a_{\alpha} \cdot \dfrac{\partial \, a_3}{\partial m} \cdot a_{\beta}$

Note that the lowering of indeces is performed systematiclly with the metric tensor $a_{\alpha\beta}$:

$$b_{\alpha \beta} = a_{\alpha \lambda} \, b^{\lambda}_{\beta}.$$

The derivative of a tangent vector a_{α} may be written:

$$(5,12) \qquad a_{\alpha,\beta} = \Gamma^{\lambda}_{\alpha \beta} \, a_{\lambda} + b_{\alpha \beta} \, a_3$$

Now let Ψ be a transversed vector field:

$$(5,13) \qquad \psi = \psi_{\lambda} \, a^{\lambda} + \psi_3 \, a_3 = \psi^{\lambda} \, a_{\lambda} + \psi_3 \, a_3$$

from the previous formula we can obtain:

$$(5,14) \qquad \frac{\partial \tilde{\psi}}{\partial \xi \beta} = (\psi_{\lambda} |_{\beta} - b_{\lambda \beta} \, \psi_3) \, a^{\lambda} + (\psi_{3,\beta} + b^{\lambda}_{\beta} \, \psi_{\lambda}) \, a^3$$

or

$$(5,15) \qquad \frac{\partial \tilde{\psi}}{\partial \xi \beta} = (\psi^{\lambda} |_{\beta} - b^{\lambda}_{\beta} \, \psi_3) \, a_{\lambda} + (\psi_{3,\beta} + b_{\lambda \beta} \, \psi^{\lambda}) \, a_3$$

from which we obtain

(5,16)
$$\frac{\partial \psi}{\partial m} = \psi^{\lambda}_{||\beta} \; a_{\lambda} \otimes a^{\beta} + \psi^{3}_{||\beta} \; a_{3} \otimes a^{\beta}$$
with

$$\psi^{\lambda}_{||\beta} = (\psi^{\lambda}_{|\beta} - b^{\lambda}_{\beta} \psi_{3})$$

$$\psi^{3}_{||\beta} = (\psi_{3,\beta} + b_{\lambda\beta} \psi^{\lambda})$$

An important operator for surface variation is the **tangential divergence of a vector field:**

(5,17)
$$\text{div}_{S} \; \psi = a^{\beta} \cdot \frac{\partial \psi}{\partial m} \; a_{\beta} = (\psi^{\beta}_{|\beta} - b^{\beta}_{\beta} \psi_{3}) = \qquad \psi^{\beta}_{||\beta}$$

We recognize that $\quad \psi^{\beta}_{|\beta} = \text{div} \; \Pi \; \psi \quad$ and it is usual to set $\quad H = - b^{\beta}_{\beta}$

(mean curvature of S), so that we can also write:

(5,18)
$$\text{div}_{S} \; \psi = \text{div} \; \Pi \; \psi + H \; \psi_{3}$$

We note that $\text{div}_{S} \; a^{3} = H$.

Now we are to provide some basic formulas for **surface variation.**
The material derivative operator δ is defined in the same way as for domain variation

(5,19)
$$\delta f(m) = \frac{d}{dt} f_{t\Psi} \circ (\Phi + t \; \tilde{\Psi})(\xi) \; |_{t = 0} = \frac{\partial}{\partial t} \tilde{f}_{t\Psi} (\xi) \; |_{t = 0}$$

We should emphasize some differences:
S is a surface; Ψ is a transverse vector field to S ;

(5,20)
$$S_{t\psi} = \{ \; m \; | \; \forall \; M \in S \quad m = M + t \; \tilde{\psi}(M) \}$$

$f_{t\psi}$ is defined on $S_{t\psi}$

First-order variation of integrals will be obtained with the following lemma.

<u>Lemma.5.3</u>
Let $a = \det (a_{\alpha\beta})$ we have

(5,21) $\delta \sqrt{a} = \sqrt{a} \ \text{div}_S \ \psi$

From which we obtain

Proposition 5.4

(5,22) $\delta \int_S f_S \ dS = \int_S \delta f \ dS + \int_S f_S \ \text{div}_S \ \psi \ dS$

Example.

area $(S) = \int_S dS$ implies $\delta \ \text{area} \ (S) = \int_S \text{div} \Pi \ \psi \ dS + \int_S H \ \psi^3 \ dS$

now a Green's formula yields

$$\int_S \text{div} \Pi \ \psi \ dS = \int_{\partial S} (\Pi \ \psi) \cdot n \ d\sigma$$

n being normal to S in the tangent plane, the interpretation is obvious.

The second term $\int_S H \ \psi^3 \ dS$ means that for a given area the change

depends on H ; this is obvious for the one-dimensional example of the circle: a given area
means $R\alpha = \text{cste}$;

$$\delta \ \text{length} \ \int_S \frac{1}{R} dR \ R \ d\theta = \delta \ R \ \alpha = \delta \ R \ \text{cst} \ / \ R$$

in the limit if R -------> + ∞   (rectilinear segment), the variation of length is zero as it
should be for a normal vector field to a rectilinear segment.

Now we study the **variation of a derivative**. This is more difficult than in the
volumic case. We first state and "prove" a simple but *wrong* result .

(5,23) $\delta \dfrac{\partial u}{\partial m} = \dfrac{\partial \delta u}{\partial m} - \dfrac{\partial u}{\partial m} \Pi \dfrac{\partial \Psi}{\partial m}$

The natural but **wrong** proof is as follows:

$\dfrac{\partial u}{\partial m} = \dfrac{\partial \tilde{u}}{\partial \xi} \left(\dfrac{\partial \Phi}{\partial \xi} \right)^{-1}$ now if $\Phi(\xi) = \Phi_0(\xi) + \tilde{\psi}(\xi)$

$$\frac{\partial \Phi}{\partial \xi} = \frac{\partial \Phi_0}{\partial \xi} + \frac{\partial \tilde\Psi}{\partial \xi} = \frac{\partial \Phi_0}{\partial \xi}\left(I + \left(\frac{\partial \Phi_0}{\partial \xi}\right)^{-1} \frac{\partial \tilde\Psi}{\partial \xi} \right)$$

so that expanding up to first-order

$$\left(\frac{\partial \Phi}{\partial \xi}\right)^{-1} = \left(\frac{\partial \Phi_0}{\partial \xi}\right)^{-1} - \left(\frac{\partial \Phi_0}{\partial \xi}\right)^{-1} \frac{\partial \tilde\Psi}{\partial \xi} \left(\frac{\partial \Phi_0}{\partial \xi}\right)^{-1}$$

as in the volumic case; then expanding $\quad u = u_0 + \delta u + ...$

$$\frac{\partial \tilde u}{\partial \xi}\left(\frac{\partial \Phi_0}{\partial \xi}\right)^{-1} = \frac{\partial u}{\partial m_0} = \frac{\partial u_0}{\partial m_0} + \frac{\partial \delta u}{\partial m_0} + ... \qquad \text{and}$$

$$\frac{\partial \tilde u}{\partial \xi}\left(\frac{\partial \Phi_0}{\partial \xi}\right)^{-1} \frac{\partial \tilde\Psi}{\partial \xi} \left(\frac{\partial \Phi_0}{\partial \xi}\right)^{-1} = \frac{\partial u}{\partial m_0} \Pi \frac{\partial \psi}{\partial m_0} \qquad \text{so that up to first-order}$$

$$\frac{\partial u}{\partial m_0} - \frac{\partial u_0}{\partial m_0} = \frac{\partial \delta u}{\partial m_0} - \frac{\partial u_0}{\partial m_0} \Pi \frac{\partial \psi}{\partial m_0} + ... \quad \text{which is equivalent to (5,23)}$$

What is wrong? The crucial point is that $\dfrac{\partial \Phi_0}{\partial \xi}\left(\dfrac{\partial \Phi_0}{\partial \xi}^{-1}\right)$ cannot be the identity!

$\dfrac{\partial \Phi_0}{\partial \xi}$ is not on-to so it cannot have a right inverse (its image is a 2D tangent plane); it has a

left inverse $B = e_\alpha \otimes a^\alpha$, the matrix of which is $\begin{pmatrix} {}^t a^1 \\ {}^t a^2 \end{pmatrix}$, where a^α means the

components of a^α in an orthonormal basis of E^2. Indeed, we have seen that $\dfrac{\partial \Phi_0}{\partial \xi} B$ is

the orthogonal projection onto the tangent plane; this means

$$I_{E^3} = \frac{\partial \Phi_0}{\partial \xi} B + a^3 \otimes a^3$$

Rather than modifying the previous proof, we are going to use more directly the basis vectors, but first of all the **right** proposition is as following.

Proposition 5.5:

$$\delta\frac{\partial u}{\partial m} = \frac{\partial \delta u}{\partial m} - \frac{\partial u}{\partial m}\Pi\frac{\partial \Psi}{\partial m} + a^3(a_3 . \frac{\partial \Psi}{\partial m} . \frac{\partial u}{\partial m})$$ or component-wise

$$\delta\frac{\partial u}{\partial m} = (\delta \tilde{u})_{,\alpha} a^\alpha - \tilde{u}_{,\alpha}(a^\alpha . \tilde{\psi}_{,\mu}) a^\mu + \tilde{u}_{,\alpha} a^{\alpha\,\mu} (a^3 . \tilde{\psi}_{,\mu}) a^3$$

The proof rests on the following lemma.

Lemma 5.6:

(i) $\delta\, a_\alpha = \dfrac{\partial \tilde{\psi}}{\partial \xi^\alpha}$ or $\delta\dfrac{\partial \Phi}{\partial \xi} = \dfrac{\partial \tilde{\psi}}{\partial \xi}$

(ii) $\delta\, a^3 = - (a^3 . \dfrac{\partial \tilde{\psi}}{\partial \xi^\mu}) a^\mu = - a^3 . \dfrac{\partial \Psi}{\partial m} = - \tilde{\psi}^3_{||\beta} a^\beta$

(iii) $\delta\, a^\alpha = - (a^\alpha . \dfrac{\partial \tilde{\psi}}{\partial \xi^\mu}) a^\mu + a^3 a^\alpha{}^\mu (a^3 . \dfrac{\partial \tilde{\psi}}{\partial \xi^\mu})$

or $\delta\, B = - B\dfrac{\partial \Psi}{\partial m}\Pi + B (a^3 . \overline{\dfrac{\partial \psi}{\partial m}}) a^3$

   (overbar means transposition; see section 4)

Proof:

We set $\Phi = \Phi_0 + \Psi$ we have $a_\alpha = \dfrac{\partial \Phi}{\partial \xi^\alpha}$

(i) so that

$$a_\alpha = a^0_\alpha + \dfrac{\partial \tilde{\psi}}{\partial \xi^\alpha}$$ which gives $\delta\, a_\alpha = \dfrac{\partial \tilde{\psi}}{\partial \xi^\alpha}$

(ii) We use $a^3 . a_\mu = 0$ so that $(\delta\, a^3) . a_\mu = - a^3 . \delta\, a_\mu$

moreover $a^3 . a^3 = 1$ gives $(\delta\, a^3) . a^3 = 0$

then as $\delta\, a^3 = (\delta\, a^3 . a_\mu) a^\mu = - (a^3 . \delta\, a_\mu) a^\mu$

we obtain $\delta\, a^3 = - (a^3 . \dfrac{\partial \tilde{\psi}}{\partial \xi^\mu}) a^\mu$

then as $\quad \dfrac{\partial \psi}{\partial m} = \dfrac{\partial \tilde{\psi}}{\partial \xi^\mu} a^\mu$

we have obtained the second equality; the third one comes just from the notation (5,16)

$$\frac{\partial \psi}{\partial m} = \psi^\lambda_{||\beta} \; a_\lambda \otimes a^\beta + \psi^3_{||\beta} \; a_3 \otimes a^\beta$$

(iii) $\quad a^\alpha \cdot a_3 = 0 \quad$ implies $\quad (\delta\, a^\alpha) \cdot a_3 = - (\, a^\alpha \cdot \delta\, a_3 \,)$

and $\quad a^\alpha \cdot a_\lambda = \delta^\alpha_\lambda \quad$ gives $\quad \delta\, a^\alpha \cdot a_\lambda = a^\alpha \cdot \delta\, a_\lambda$

so that

$$\delta\, a^\alpha = - (a^\alpha \cdot \delta\, a_\lambda\,)\, a^\lambda + (\, a^\alpha \cdot \delta\, a_3\,)\, a^3$$

$$= - (\, a^\alpha \cdot \frac{\partial\tilde{\psi}}{\partial\xi\lambda})\, a^\lambda + a^\alpha \cdot a^\mu (\, a^3 \cdot \frac{\partial\tilde{\psi}}{\partial\xi^\mu})\, a^3$$

On the other hand $\quad B\, h = e_\alpha\, (\, a^\alpha \cdot h\,)$

so that

$$\delta\, B\, h = e_\alpha\, (\, \delta\, a^\alpha \cdot h\,) = - e_\alpha\, (\, a^\alpha \cdot \frac{\partial\tilde{\psi}}{\partial\xi\lambda})\, (a^\lambda \cdot h\,) + e_\alpha\, (\, a^\alpha \cdot a^\mu\, (\, a^3 \cdot \frac{\partial\tilde{\psi}}{\partial\xi^\mu})\,)(a^3 \cdot h\,)$$

Proof of the Proposition5.5:

$$\frac{\partial u}{\partial m} = \tilde{u}_{,\alpha}\, a^\alpha \qquad \text{so that} \qquad \delta\, \frac{\partial u}{\partial m} = (\, \delta\tilde{u}\,)_{,\alpha}\, a^\alpha + \tilde{u}_{,\alpha}\, \delta\, a^\alpha$$

we note that $\tilde{u}$ is computed at a fixed point so that δ and $\dfrac{\partial}{\partial\xi^\alpha}$ commute; secondly we

use (iii) of the previous Lemma:

$$\delta\frac{\partial u}{\partial m} = (\, \delta\tilde{u}\,)_{,\alpha}\, a^\alpha + \tilde{u}_{,\alpha}\, (\, a^\alpha \cdot \frac{\partial\tilde{\psi}}{\partial\xi\lambda})\, a^\lambda + \tilde{u}_{,\alpha}\, a^\alpha \mu\, (\, a^3 \cdot \frac{\partial\tilde{\psi}}{\partial\xi^\mu})\, a^3$$

this is the component-wise formula of the proposition; the intrinsic formula stems from

$$\frac{\partial u}{\partial m} = \tilde{u}_{,\alpha}\, a^\alpha \qquad \text{and} \qquad \frac{\partial\psi}{\partial m} = \tilde{\psi}_{,\lambda}\, a^\lambda$$

6 Sensitivity analysis for surface heat equation.

We still consider a simple surface system; i.e. a stationnary surface heat conduction equation; we set:

   f   surface density of heat source,
   g   line density of heat source,
   q   heat flux vector,
   u   deviation of temperature from the natural state.

We assume Fourier-law for an isotropic homogeneous medium:

$$q = -c \frac{\partial u}{\partial x}$$

the conservation of heat gives: $\text{div } q = f$.

We assume prescribed zero deviation of the temperature on γ_1 ; $u = 0$;

prescibed heat flux on γ_2 : $q . n = -g$;   note that the minus sign is a convention, q is pointing toward the cold subset, g is positive when heat is received and n is outward).

Finally we have

$$\begin{cases} \text{div } q = f & \text{in } S \\ u = 0 & \text{on } \gamma_1 \\ q.n = -g & \text{on } \gamma_2 \\ q = -\frac{\overline{\partial u}}{\partial m} \end{cases}$$

or component-wise

$$\begin{cases} q^\alpha{}_{|\alpha} = f & \text{in } S \\ u^\alpha = 0 & \text{on } \gamma_1 \\ q^\alpha . n_\alpha = -g & \text{in } S \\ q^\alpha = -c \, a^{\alpha\beta} u_{,\beta} \end{cases}$$

In a standard way we consider the variational formulation:

$$\forall v \in V \quad a(u, v) = l(v) \quad \text{where}$$

$$V = \left\{ v \in H^1(S) \mid v|_{\gamma_1} = 0 \right\}$$

$$a(u, v) = \int_S c \frac{\partial u}{\partial m} \cdot \frac{\overline{\partial v}}{\partial m} \, dS$$

$$l(v) = \int_S f v \, dS + \int_{\gamma_2} g v \, d\sigma$$

As in the volumic case δu is the solution of an equation with the same bilinear form a

Proposition 6.1

Let u be the solution of (6.1), then its variation du satisfies

$$\forall \, v \in V \quad a(\, du, v \,) = -(\, \delta a\,)\,(\, u, v \,) + (\, \delta l\,)\,(v)$$

where δa and δl are the variations of a and l *for fixed u and v* :

$$(\,\delta a\,)\,(\,u, v\,) = - \int_S c\{ \frac{\partial u}{\partial m} \Pi \frac{\partial \psi}{\partial m} \cdot \overline{\frac{\partial v}{\partial m}} + \frac{\partial v}{\partial m} \Pi \frac{\partial \psi}{\partial m} \cdot \overline{\frac{\partial u}{\partial m}} \} \, dS$$

$$+ \int_S c \frac{\partial u}{\partial m} \cdot \overline{\frac{\partial v}{\partial m}} \, \mathrm{div}_S \, \psi \, dS$$

$$(\,\delta l\,)\,(v) = \int_S v \, \delta f \, dS + \int_S f \, v \, \mathrm{div}_S \, \psi \, dS + \int_{\gamma_2} v \, \delta g \, d\sigma +$$

$$\int_{\gamma_2} v \, g \, \mathrm{div}_{\gamma_2} \, \psi \, d\sigma$$

As in the volumic case, the <u>proof</u> is simple if one uses the previous results: Propositions 5.4 and 5.5. As a touch of "humour" we note that the "wrong" Proposition 5.5 would give here the same Proposition 6.1. This is because the term

$(\, a_3 \cdot \frac{\partial \Psi}{\partial m} \cdot \frac{\partial u}{\partial m} \,)\, a_3$ has a zero scalar product with $\frac{\partial v}{\partial m}$, which is a tangential

vector.

Surface sensitivity of a functional.

The simplest case is $J(\, u \,) = \int_S \alpha(u) \, dS$

With proposition 5.4 we obtain $\qquad \delta J = \int_S \alpha'(u) \, \delta u \, dS + \int_S \alpha(u) \, \mathrm{div}_S \, \psi \, dS$

To make this expression explicit with respect to Ψ we use the same Proposition 4.3 to obtain

$\delta J = (\delta L)(u, p)$ where $L = J + a(u, p) - l(p)$ and δL is computed at fixed u and p:

$$\delta J = \int_S \alpha(u) \, div_S \, \psi \, dS + \delta a (u, p) - \delta l(p)$$

where δa and δl are given in Proposition 6.1 .

7 Boundary expression of shape sensitivity

We turn here to the volumic case of section 4. It is possible to obtain a different expression of the shape sensitivity of a functional: formula (4,10) may be tranformed to a formula which involves boundary integrals. We need some auxiliary lemmas.

Lemma 7.1

The solution u of (4,1) satisfies

$$(7,1) \quad \forall \, w \in \omega \quad a(\, u, w \,) = l(\, w \,) + l_{\gamma_1}(\, w \,)$$

where W is the space of virtual displacements which do not necessarily satisfy w = 0 on γ_1 and

$$(7,2) \quad \text{and} \quad l_{\gamma_1}(\, w \,) = \int_{\gamma_1} T w \frac{\partial u}{\partial n} \, d\sigma$$

the proof just uses the Stockes formula:

$$\int_{\omega} -T\Delta u \; w \; dx \; = \; \int_{\omega} T \frac{\partial u}{\partial x} \frac{\partial \overline{w}}{\partial x} \; dx \; - \; \int_{\partial\omega} T w \frac{\partial u}{\partial n} \, d\sigma$$

which gives Lemma 7.1 as $T\frac{\partial u}{\partial n} = 0$ on γ_2

Lemma 7.2

$$(i) \quad \frac{\partial u}{\partial x} \frac{\partial \psi}{\partial x} \frac{\partial \overline{p}}{\partial x} \; = \; \frac{\partial}{\partial x}(\frac{\partial u}{\partial x} \, \psi) \, \frac{\partial \overline{p}}{\partial x} \; - \; \overline{\psi} \, \frac{\partial^2 u}{\partial x^2} \, \frac{\partial \overline{p}}{\partial x}$$

$$(ii) \quad \frac{\partial p}{\partial x} \frac{\partial \psi}{\partial x} \frac{\partial \overline{u}}{\partial x} \; = \; \frac{\partial}{\partial x}(\frac{\partial p}{\partial x} \, \psi) \, \frac{\partial \overline{u}}{\partial x} \; - \; \overline{\psi} \, \frac{\partial^2 p}{\partial x^2} \, \frac{\partial \overline{u}}{\partial x}$$

(iii) $\dfrac{\partial u}{\partial x}\dfrac{\partial \psi}{\partial x}\dfrac{\partial \overline{p}}{\partial x}+\dfrac{\partial p}{\partial x}\dfrac{\partial \psi}{\partial x}\dfrac{\partial \overline{u}}{\partial x}$

$$=\dfrac{\partial}{\partial x}(\dfrac{\partial u}{\partial x}\psi)\dfrac{\partial \overline{p}}{\partial x}+\dfrac{\partial}{\partial x}(\dfrac{\partial p}{\partial x}\psi)\dfrac{\partial \overline{u}}{\partial x}-\dfrac{\partial}{\partial x}(\dfrac{\partial u}{\partial x}\dfrac{\partial \overline{p}}{\partial x})\psi$$

(iv) $\dfrac{\partial u}{\partial x}\dfrac{\partial \overline{p}}{\partial x}\operatorname{div}\psi=\operatorname{div}((\dfrac{\partial u}{\partial x}\dfrac{\partial \overline{p}}{\partial x})\psi)-\dfrac{\partial}{\partial x}(\dfrac{\partial u}{\partial x}\dfrac{\partial \overline{p}}{\partial x})\psi$

(v) $-\dfrac{\partial u}{\partial x}\dfrac{\partial \psi}{\partial x}\dfrac{\partial \overline{p}}{\partial x}-\dfrac{\partial p}{\partial x}\dfrac{\partial \psi}{\partial x}\dfrac{\partial \overline{u}}{\partial x}+\dfrac{\partial u}{\partial x}\dfrac{\partial \overline{p}}{\partial x}\operatorname{div}\psi=$

$$-\dfrac{\partial}{\partial x}(\dfrac{\partial u}{\partial x}\psi)\dfrac{\partial \overline{p}}{\partial x}-\dfrac{\partial}{\partial x}(\dfrac{\partial p}{\partial x}\psi)\dfrac{\partial \overline{u}}{\partial x}+\operatorname{div}((\dfrac{\partial u}{\partial x}\dfrac{\partial \overline{p}}{\partial x})\psi)$$

(vi) $p\,\delta f+p\,f\operatorname{div}\psi=\operatorname{div}(p\,f\,\psi)-f\dfrac{\partial p}{\partial x}\psi+p(\,\delta f-\dfrac{\partial f}{\partial x}\psi)$

and if f does not depend explicitly on ω :

$$p\,\delta f+p\,f\operatorname{div}\psi=\operatorname{div}(p\,f\,\psi)-f\dfrac{\partial p}{\partial x}\psi$$

The _proof_ uses the definitions and is left to the reader.

Proposition 7.3

Let u be the solution of the model system (4.1) and $J_\omega(u)$ be the functional (4.7), then its variation given by (4.10) may be also expressed as

(7,3) $\delta J=-T\displaystyle\int_{\gamma_1}\dfrac{\partial u}{\partial n}\dfrac{\partial p}{\partial n}(\,\psi\cdot n\,)\,d\sigma+$

$$\int_{\gamma_2} \alpha(u) \; (\psi. \; n \;) \; d\sigma + \int_{\gamma_2} (\; T \frac{\partial u}{\partial \sigma} \frac{\partial \overline{p}}{\partial \sigma} - pf) \; (\psi \; . \; n) \; d\,\sigma$$

Proof.

Lemma 7.2 (v) enables us to derive from (4.5):

$$(\delta a \,)(u \, , p \,) = -T \int_{\omega} (\; \frac{\partial}{\partial x} (\frac{\partial u}{\partial x} \; \psi) \; \frac{\partial \overline{p}}{\partial x} \; + \; \frac{\partial}{\partial x} (\frac{\partial p}{\partial x} \; \psi) \; \frac{\partial \overline{u}}{\partial x} \;)dx +$$

(7,4) $$T \int_{\omega} div \; ((\frac{\partial u}{\partial x} \; \frac{\partial \overline{p}}{\partial x}) \; \psi \;) \; dx$$

Next we use Lemma 7.2 (vi) with (4,6) and assume for simplicity that f does not depend explicitly on ω : $\delta f = \frac{\partial f}{\partial x} . \psi$

(7.5) $(\delta \, 1 \,) \, (f \,) = \int_{\omega} div (p \; f \; \psi \,) \; dx - \int_{\omega} f \frac{\partial p}{\partial x} \psi \; d x$

Then we note that Lemma 7.1 for the adjoint state (4,9) gives

$$\forall \; w \in W \quad a(p \, , w \,) = -< \frac{\partial J}{\partial u} , w > + \int_{\gamma_1} T \; w \; \frac{\partial p}{\partial n} \; d \; \sigma$$

so that:

(7,6) $-T \int_{\omega} \frac{\partial}{\partial x} (\frac{\partial u}{\partial x} \; \psi) \; \frac{\partial \overline{p}}{\partial x} \; =$

$$< \frac{\partial J}{\partial u}, \frac{\partial u}{\partial x} \, w > - \int_{\gamma_1} T \, (\frac{\partial u}{\partial x} \, \psi) \, \frac{\partial p}{\partial n} \, d\sigma$$

Similarly as the state u of the system sartisfies:

$$\forall \, w \in W \quad a(u , w) = \int_{\omega} f \, w \, dx + \int_{\gamma_1} T \, w \, \frac{\partial u}{\partial n} \, d\sigma$$

we have

$$(7,7) \; -T \int_{\omega} \frac{\partial}{\partial x} (\frac{\partial p}{\partial x} \, \psi) \, \frac{\partial \overline{u}}{\partial x} = \int_{\omega} \frac{\partial p}{\partial x} \, \psi \, dx + \int_{\gamma_1} T \, (\frac{\partial p}{\partial x} \, \psi) \, \frac{\partial u}{\partial n} \, d\sigma$$

Finally we recall from Proposition 4.3 that if

$$(7,8) \; J_{\omega}(u) = \int_{\omega} \alpha \, (u) \, dx$$

$$\delta J = \int_{\omega} \alpha \, (u) \, div(\psi) \, dx + \delta \, a(u , p) - \delta \, l(p)$$

so that (7,4), (7,5), (7,6), (7,7) provide

$$(7,9) \; \delta J = \int_{\omega} \alpha \, (u) \, div(\psi) \, dx + < \frac{\partial J}{\partial u}, \frac{\partial u}{\partial x} \, w > -$$

$$\int_{\gamma_1} T \, (\frac{\partial u}{\partial x} \, \psi) \, \frac{\partial p}{\partial n} \, d\sigma - \int_{\gamma_1} T \, (\frac{\partial p}{\partial x} \, \psi) \, \frac{\partial u}{\partial n} \, d\sigma +$$

REFERENCES

Céa J. , Conception optimale ou identification de domaines: calcul rapide de la derivée directionnelle de la fonction cout, Math. modeling & numerical analysis, vol. 20, 1986, pp 371-402 .

Céa J., Une méthode numérique pour la recherche d'un domaine optimal, Publications IMAN-P2 Nice , 1975.

Céa J, Gioan A., Michel J.,Quelques résultats sur l'identification de domaines, Calcolo, III-IV, 1973.

Céa J. & B. Rousselet B.,Ed., Conception optimale de formes, Ecole INRIA (26-30 septembre 1983, Université de Nice).

Chenais D., Sur une famille de variétés à bord lipschtziennes: une application à un problème d'identification de domaines, Annales de l'Institut fourier (1977), 27, pp 201-231.

Chenais D. Knopf-Lenoir C., Sur la communication entre logiciels éléments finis et optimiseur en controle distribué; Calcul des structures et intelligence artificelle, publié par Fouet J.M. Ladevèze P. Ohayon R., Pluralis 1988.

Germain P., Cours de mécanique, Ecole Polytechnique, Palaiseau, France, 1980.

Grisvard P., Elliptic problems in non smooth domains , Pitman 1985.

Hadamard J., Mémoire sur le problème d'analyse relatif à l'équilibre des plaques elastiques encastrées (1908), Oeuvre de J. Hadamard, CNRS, Paris 1968.

Haug E. 1 Choi K. & Komkov V., Design sensitivity analysis structural systems, Academic Press, New York (1986).

Masmoudi M., Outils pour la conception optimale de formes, thèse d'Etat, Université de Nice, 1987.

Then we note that on γ_1 $u = 0$ and $p = 0$ so that $\dfrac{\partial u}{\partial x} = \dfrac{\partial u}{\partial n} n$ and

$\dfrac{\partial p}{\partial x} = \dfrac{\partial p}{\partial n} n$ and on γ_2 $\dfrac{\partial u}{\partial x} = \dfrac{\partial u}{\partial \sigma^\alpha} a^\alpha$ so that

$$\delta J = - \int_{\gamma_1} T \, \dfrac{\partial u}{\partial n} \, \dfrac{\partial p}{\partial n} \, (\psi . n \,) d \, \sigma + \int_{\gamma_2} \alpha(u) \, \psi . n \, d\sigma +$$

$$+ \int_{\gamma_2} (\, T \dfrac{\partial u}{\partial \sigma} \dfrac{\partial p}{\partial \sigma} - p \, f \,) \, (\, \psi . n \,) d \, \sigma$$

Remark.
        When one uses finite elements to solve (4,1), it has been observed that the boundary expression (7,3) is not very accurate; theoretical support of this fact will be given in section 8.

8 Use of finite elements and boundary integrals

        We consider the model system of section 4 and to make things simpler we assume $u = 0$ is the only boundary condition. We shall give error estimates of δJ when we replace u and p by finite elements approximation; many results of this type may be found in Masmoudi (1987) .

        Firstly we consider the boundary expression (7,3) which in the case of $u = 0$ on $\partial \omega$ is

(8,1) $\delta J^B = -T \displaystyle\int_{\partial \omega} \dfrac{\partial u}{\partial n} \, \dfrac{\partial p}{\partial n} (\, \psi . n \,) \, d\sigma$

        We denote u_h and p_h finite elements approximations of u and p ; h denotes the mesh size. We set

$$(8,2) \quad \delta J_h^B = -T \int_{\partial \omega} \frac{\partial u_h}{\partial n} \frac{\partial p_h}{\partial n} (\psi . n) \, d\sigma$$

For simplification we assume that there is no error in the approximation of the geometry; we need an error estimate of the normal derivative of u and p on $\partial \omega$; an accurate estimation may be based on Rannacher & Scott (1982):

$$(8,3) \quad || u - u_h ||_{1,\infty; \Omega} \le c \, h^k || u ||_{k+1,\infty; \Omega}$$

$$\text{where } || u ||_{k,\infty; \Omega} = || u ||_{L^\infty(\Omega)} + \sum_{l=0}^{l=k} || \frac{\partial^l u}{\partial x^l} ||_{L^\infty(\Omega)}$$

and k is the degree of polynomials used in the finite element approximation. We note that for $k = 1$, the second derivatives of the solution u should be essentially bounded ($\frac{\partial^2 u}{\partial x^2} \in L^\infty(\Omega)$); this is an assumption which, for example, excludes

reintrant corners in $\partial \Omega$ (Grisvard 1985).

Proposition 8.1.

If datas are smooth enough such that (8.3) holds then

$$| \delta J^B - \delta J_h^B | \le c \, h^k || u ||_{k+1,\infty; \Omega} \, || p ||_{k+1,\infty; \Omega} || \psi . n ||_{0,\infty; \Gamma}$$

Proof.

$$| \delta J^B - \delta J_h^B | \le T \int_{\partial \omega} | \frac{\partial u}{\partial n} - \frac{\partial u_h}{\partial n} | \, | \frac{\partial p}{\partial n} | \, |(\psi . n)| \, d\sigma \quad +$$

$$T \quad \int_{\partial \omega} | \left| \frac{\partial u}{\partial n} - \frac{\partial u_h}{\partial n} \right| | \left| \frac{\partial p}{\partial n} - \frac{\partial P_h}{\partial n} \right| | (\psi.n) | d\sigma \quad +$$

$$T \quad \int_{\partial \omega} | \frac{\partial u}{\partial n} | \, | \, | \frac{\partial p}{\partial n} - \frac{\partial P_h}{\partial n} | \, | (\psi. n) \, | d\sigma$$

fom which we obtain:

$$| \delta J^B - \delta J_h^B | \leq T (| \left| \frac{\partial u}{\partial n} - \frac{\partial u_h}{\partial n} \right| |_{0, \infty, \gamma} | \left| \frac{\partial p}{\partial n} \right| |_{0, \infty, \gamma} +$$

$$| \left| \frac{\partial u}{\partial n} - \frac{\partial u_h}{\partial n} \right| |_{0, \infty, \gamma} | \left| \frac{\partial p}{\partial n} - \frac{\partial P_h}{\partial n} \right| |_{0, \infty, \gamma} +$$

$$| \left| \frac{\partial u}{\partial n} \right| |_{0, \infty \, \gamma} | \left| \frac{\partial p}{\partial n} - \frac{\partial P_h}{\partial n} \right| |_{0, \infty; \gamma}) \, | \left| \psi. n \right| |_{0, \infty; \gamma}$$

We note that (8.3) implies:

$$| \left| \frac{\partial u}{\partial n} - \frac{\partial u_h}{\partial n} \right| |_{1, \infty; \gamma} \leq c \, h^k | \left| u \right| |_{k+1, \infty; \omega}$$

and equivalently for p from which we obtain the proposition.

Secondly we consider the domain expression.
We only state the result; we set δJ^D the expression given by (4,10); δJ_h^D means

that u and p are replaced by a finite element approximation u_h and p_h ; k

stands for the degree of polynomial approximation and k' the order of derivatives of ψ which are essentially bounded.

Proposition8.2.

If the datas are smoth enough then

$$| \delta J^D - \delta J_h^D | \leq c(u) \; h^{k+k'} (\| p \|_{K+1;\Omega} +1) \; \| \psi \|_{k'+1,\infty,\Omega}$$

The <u>proof</u> is technical but the result may be understood directly. It means that if the vector field ψ has essentially bounded second derivatives, the error estimate is in h^{k+1} and, moreover, if the second derivatives are small it will be multiplied by a small constant; this error estimate is to be compared with h^k of Proposition 8.1 ; numerical evidence of this result may be found in Rochette (1990).

<u>Thirdly we consider the use of boundary integrals</u> to solve the state equation. For simplification, we consider a membrane with no density of force and prescribed <u>constant displacement</u> on the two pieces of the boundary.

(8,4)
$$\begin{cases} - T \, \Delta u = 0 & \text{in } \omega \\ \\ u = b & \text{on } \gamma_1 \\ \\ u = 0 & \text{on } \gamma_0 \end{cases}$$

The <u>transformation</u> of this boundary value problem into a boundary integral equation is performed by introducing the classical <u>elementaty solution</u> of the Laplacian Δ :

(8,5) $\quad E(x,z) = -\dfrac{1}{2\pi} \log | x - z |$ $\quad$ (in two dimensions);

classically $\quad \Delta_z \, E(x, z) = \delta(x - z)$ $\quad$ holds; and if we consider

(8,6) $\quad v(x) = \displaystyle\int_\gamma E(x, y) \, q(y) \, d\sigma_y$ $\quad$ where γ is the boundary of Ω,

this funtion satisfies

$$(8,7) \quad \begin{cases} \Delta v = 0 \quad\quad \text{outside } \gamma \\ v \quad\quad\quad \text{is continuous in } R^2 \\ \dfrac{\partial v^i}{\partial n} - \dfrac{\partial v^e}{\partial n} = q \quad\text{on } \gamma \end{cases}$$

where $v^i = v\big|_\omega$ and $v^e = v\big|_{R^2-\omega}$

So, the solution of (8,4) is given by $u(x) = \displaystyle\int_\gamma E(x,y)q(y)\, d\sigma_y$

where q is solution of

$$(8,9) \quad \begin{cases} \displaystyle\int_\gamma E(x,y)q(y)\, d\sigma_y = b(x) \quad \forall\, x \in \gamma_1 \\[4mm] \displaystyle\int_\gamma E(x,y)q(y)\, d\sigma_y = 0 \quad \forall\, x \in \gamma_0 \end{cases}$$

Now, we note $u = 0$ outside of γ_0 and $u = b$ inside of γ_1 are the solution of (8,4) with the right boundary conditions, so that using (8,7) we obtain

$$(8,10) \quad \dfrac{\partial u}{\partial n} = q \quad \text{on } \gamma$$

If we consider now the simple functional $J_\omega = \displaystyle\int_\omega |\dfrac{\partial u}{\partial x}|^2 dx$ we note

that $\quad \dfrac{\partial J}{\partial u} \cdot w = \displaystyle\int_\omega \dfrac{\partial u}{\partial x}\dfrac{\partial w}{\partial x}\, dx$

so that the adjoint state is $p = -u$ and

$$\delta J^B = T \int_\gamma |\dfrac{\partial u}{\partial n}|^2 (\psi.n)\, d\sigma = T \int_\gamma q^2\, (\psi . n)\, d\sigma \quad \text{is quite easy to}$$

compute. Moreover it is easy to compute with Green's formula.

$$J_\omega(u) = \int_{\gamma_1} T\, b\, q\, d\sigma$$

The case of an equation with a non zero right-hand side may also be computed with integral equations but we need to use the second Green formula; we skip the details and refer to Masmoudi (1987). Before stating the error estimate of δJ, we should recall that many error estimates for the solution of boundary integral equations have been obtained by Nedelec (1976,1977) and the proof is still in Masmoudi (1987).

Proposition8.3

For smooth enough datas

$$| \delta J^I - \delta J^I_h | \le c\, h^{\min(2k+1,l+1)}\, || q ||_{k+1,\gamma}\, || \Pi ||_{k+1,\gamma}\, || \Psi.n ||$$

where:
- l means the degree of polynomials used to approximate the boundary;
- k the degree of finite elements approximation
- q is the solution of an equation of the type of (8,9) set on γ;
- Π is the solution of a similar equation for the adjoint state.

Remark.   We note that for $k = 2$ we should use $l = 4$ so that the error estimate is in h^5; finally we should note that with a smooth approximation of the boundary the error estimate of δJ is quite good; this is in contrast with what we obtained with finite elements.

B. Rousselet
Laboratoire de Mathematiques
Université de Nice
06034 Nice CEDEX

$$T \int_\omega \text{div} \left(\left(\frac{\partial u}{\partial x} \frac{\partial \overline{p}}{\partial x} \right) \psi \right) dx - \int_\omega \text{div}(p\, f\, \psi)\, dx$$

Then we note $\langle \frac{\partial J}{\partial u}, \frac{\partial u}{\partial x} \psi \rangle = \int_\omega \alpha'(u) \frac{\partial u}{\partial x} \cdot \psi\, dx$, so that the first two terms

of (7,9) provide:

$$(7,10) \quad \int_\omega \alpha\,(u)\, \text{div}(\,\psi\,)\, dx + \int_\omega \alpha'(u) \frac{\partial u}{\partial x} \cdot \psi\, dx = \cdot$$

$$\int_\omega \text{div}\,(\,\alpha\,(u)\,\,\psi\,)dx = \int_{\gamma_2} \alpha(u)\,\psi\cdot n\, d\sigma$$

Using Green's formula in (7,9) we obtain.

$$(7,11) \quad \delta J = \int_{\gamma_2} \alpha(u)\,\psi \cdot n\, d\sigma - \int_{\gamma_1} T \left(\frac{\partial u}{\partial x}\, \psi \right) \frac{\partial p}{\partial n}\, d\sigma$$

$$- \int_{\gamma_1} T \left(\frac{\partial p}{\partial x}\, \psi \right) \frac{\partial u}{\partial n}\, d\sigma + T \int_{\partial\omega} \left(\frac{\partial u}{\partial x} \frac{\partial \overline{p}}{\partial x} \right) (\psi \cdot n)\, d\sigma$$

$$- \int_{\partial\omega} (p\, f\, \psi \cdot n)\, d\sigma$$

Modulef, Une bibliothèque modulaire d'éléments finis, INRIA (1985), Rocquencourt, France.

Mota Soares C.A., Rodrigues H.C. & Choi K.K., Shape optimal structural design using boundary elements and minimum compliance techniques, ASME Journal of mechanisms, transmissions & automation in design, vol 106, pp518-523, 1984.

Murat F. & Simon J., Etude de problèmes d'optimum design, Proceedings of the 7th IFIP Conf., Springer Verlag, Lecture Notes in Computer Sciences, n° 41, 1976, pp 54-62.

Nédelec J.C., Curved finite element methods for a solution of singular integral equation on surfaces in $\mathbf{R}^3$, Comp. meth. applic. mech. eng., 1976.

Nédelec J.C., Approximation des équations intégrales en mécanique et en physique, cours EDF-CEA-INRIA, Centre de mathématiques appliquées, Ecole Polytechnique, Palaiseau, 1977.

Pironneau O., Ed., Optimisation de forme dans les systèmes à paramètres distribués: résolution numérique et applications, Ecole INRIA, Rocquencourt, France (8-10 novembre 1982).

Pironneau O., Optimal shape design for elliptic systems, Springer series in computational physics, Springer Verlag 1984.

Rannacher R. & Scott, Some optimal error estimates for piecewise linear finite element approximation, Math. of comp., vol. 38, pp 437-445.

Rochette M., Conception optimale de formes appliquée aux résistances ajustables, thèse, Université de Nice Sophia-Antipolis, 1990.

Rousselet B., Etude de la régularité desz valeurs propres par rapport à des déformations bilipschitziennes du domaine, CRAS 283, série A, 1976, p 507.

Rousselet B., Problèmes inverses de valeurs propres, pp 77-85, "Optimization techniques", Céa ed., Lect. notes in computer sciences,Springer-Verlag, 1976.

Rousselet B., Shape design sensitivity of a membrane, pp 595-623, JOTA, 1983;

Rousselet B.,   Quelques résultats en optimisation de domaines, thèse d'état, Université de Nice, 1982.

Rousselet B.   Shape design sensitivity, from partial differential equation to implementation. Eng. Opt.,1987,vol 11,pp151-171.

Rousselet B.,   Shape optimization of structures with state constraints, pp 255-264 dans "Control of partial differential equations", A. Bermudez ed; Lect. notes in control and information sciences, vol. 114, Springer Verlag.

Chapter 14

MIXED ELEMENTS IN SHAPE OPTIMAL DESIGN OF STRUCTURES BASED ON GLOBAL CRITERIA

C.A. Mota Soares
CEMUL, Lisbon, Portugal

R.P. Leal
DEM, FCTUC, Coimbra, Portugal

ABSTRACT

In this paper mixed elements are applied to the optimal shape design of two dimensional elastic structures. The theory of shape sensitivity analysis of structures with differentiable objective functions is developed based on mixed elements, and using the Lagrangian approach and the material derivative concept. The mixed finite element model is based on an eight node mixed isoparametric quadratic element, whose degrees of freedom are two displacements and three stresses, per node. The corresponding nonlinear programming problem is solved using the method of sequential convex programming and the modified method of feasible directions, available in the commercial programme ADS (Automated Design Synthesis). The formulation developed is applied to the optimal shape design of two dimensional elastic structures, selecting the compliance as the objective function to be minimized and the initial volume as constraint. The advantages and disadvantages of the mixed elements are discussed with reference to applications.

1 INTRODUCTION

It is well-known that efficient shape optimization requires a good sensitivity analysis. Accurate sensitivity values provide a good relation between shape perturbation and corresponding variations of the objective function and constraints. The regularity of the finite element mesh is important to obtain a numerical solution of the problem with enough accuracy. Poor results in analysis implies bad results in sensitivity analysis. Boundary element methods have been used to overcome this problem, since distortion in the boundary is much smaller than the corresponding distortion of the domain (e.g. Mota Soares *et al.*, 1987).

Sensitivity analysis can be based on analytical or numerical differentiation of finite element equations. These procedures are generally classified as discrete methods and have been considered by Zienkiewicz and Campbell, 1973, Francavilla *et al.*, 1975, and Braibant, 1986, among others.

Alternatively the gradients can be obtained by analytical expressions for the sensitivity of objective and constraint functions or from the explicit expressions of optimality conditions. Banichuk, 1983, Dems and Mroz, 1984, and Haug *et al.*, 1986, have been developing this technique, and call it the continuum or variational method.

In this paper, the theory of shape sensitivity analysis of structures is developed, based on mixed elements and global criteria, and using the Lagrangian approach and the material derivative concept. Analytical formulae obtained by this method are used to calculate sensitivities of objective and constraint functions of the optimization problem. The formulation is simplified for the compliance functional. In the application, the design objective is to minimize the compliance of the structure subjected to an area constraint. In the formulation used, values obtained on the boundary, namely displacements and stresses, are required. Consequently, the employment of displacement finite elements has been shown to be inefficient, since the results obtained on the boundary are not, generally, accurate enough.

Mixed finite elements offer advantages over displacement finite elements, since, in general, the corresponding stresses are more accurate. Thus, it is expected that mixed finite elements are more suitable to optimal design than displacement finite elements.

Only recently has the mixed finite element been applied to the optimal design of structures. The sensitivity analysis of beams and plates with static, dynamic and stability constraints, based on mixed formulations has been developed by Mota Soares and Leal, 1987,1988. The theory has been applied to the minimum weight design of plates, subject to constraints on displacements, stresses, natural frequencies or buckling stresses (Leal and Mota Soares, 1989). For shape optimal design of two-dimensional linear elastic structures, analytical formulae are used to minimize compliance of the structure subjected to an area constraint (Leal and Mota Soares, 1990). Also, Rodrigues, 1988a,b has developed a variational formulation for the shape optimal design of two-dimensional linear elastic structures, using four node isoparametric mixed finite elements based on the functional of Hu-Washizu, to interpolate the stress, strain, and displacement fields; the necessary quality of the mesh was guaranteed by the use of an automatic adaptive mesh generator.

The mixed finite element used in this paper is the isoparametric quadratic element based on the Hellinger-Reissner functional, with 8 nodes and 2 displacements and 3 stresses as degrees of freedom per node. The perturbation field on the boundary is interpolated with linear design elements. The mesh regularity of the finite element discretization is guaranteed by a mesh regenerator.

2 MIXED ELEMENTS IN TWO DIMENSIONAL ELASTICITY

Mixed elements are based on the Hellinger-Reissner principle (e.g. Washizu, 1974), and for two dimensional elasticity this functional is given by

$$
\begin{aligned}
V_R(u_i, \sigma_{ij}) = &\int_\Omega \left[\frac{1}{2} \sigma_{ij}(u_{i,j} + u_{j,i}) - U_0(\sigma_{ij}) - b_i u_i \right] d\Omega \\
&- \int_{\Gamma_t} u_i \hat{t}_i \, d\Gamma - \int_{\Gamma_u} (u_i - \hat{u}_i) t_i \, d\Gamma \qquad i,j = 1,2
\end{aligned}
\tag{1}
$$

where Ω is the domain, Γ_t is the boundary with known forces, Γ_u is the boundary with prescribed displacements and the superscript bar indicates the prescribed forces and displacements on the boundary, x_i are the global coordinates, u_i are displacements, σ_{ij} are stresses, b_i are body forces and t_i are tractions on the boundary. U_0^* is the complementary energy density. Throughout this paper index notation and summation convention is used and a comma denotes differentiation with respect to x, i.e., $u_{j,i} = \partial u_j / \partial x_i$. The stress field σ is defined in the space

$$T(\Omega) = \left\{ \tau(x) = (\tau_{ij}): \tau_{ij} \in L^2(\Omega), \tau_{ij} = \tau_{ji}, i,j \in [1,2] \right\}$$

and the displacement field u is defined in the space of admissible displacements

$$V(\Omega) = \left\{ v \in (H^1(\Omega))^2: v_i = 0 \text{ on } \Gamma_u \right\}$$

supposing that the continuous medium is fixed in Γ_u. $L^2(\Omega)$ is the space of functions defined on Ω, which are square-integrable over Ω, in the Lebesgue sense. $H^1(\Omega)$ is the Sobolev space of functions with one square-integrable partial derivative, defined by

$$H^1(\Omega) = \left\{ v = (v_i): v_i, \frac{\partial v_i}{\partial x_j} \in L^2(\Omega) \right\}$$

Assuming that the geometric boundary conditions are satisfied, the Hellinger-Reissner functional (1) for isotropic and linear elastic materials can be written in matrix form:

$$V_R(u,\sigma) = \int_\Omega \left(\sigma^T \Delta u - \frac{1}{2} \sigma^T C \sigma - u^T b \right) d\Omega - \int_{\Gamma_t} u^T t \, d\Gamma \qquad (2)$$

where

$$u = \begin{bmatrix} u_1 & u_2 \end{bmatrix}^T \qquad \text{is the vector of displacements}$$

$$b = \begin{bmatrix} b_1 & b_2 \end{bmatrix}^T \qquad \text{is the vector of body forces}$$

$$t = \begin{bmatrix} t_1 & t_2 \end{bmatrix}^T \qquad \text{is the vector of tractions on the boundary}$$

$$\sigma = \begin{bmatrix} \sigma_{11} & \sigma_{22} & \sigma_{12} \end{bmatrix}^T \qquad \text{is the vector of stresses}$$

$$\Delta = \begin{bmatrix} \partial/\partial x_1 & 0 \\ 0 & \partial/\partial x_2 \\ \partial/\partial x_2 & \partial/\partial x_1 \end{bmatrix} \qquad \text{is a differential operator}$$

$$C = \frac{1}{E^{\bullet}} \begin{vmatrix} 1 & -v^{\bullet} & 0 \\ -v^{\bullet} & 1 & 0 \\ 0 & 0 & 2(1+v^{\bullet}) \end{vmatrix}$$ is the matrix of elastic properties of material

with

$$E^{\bullet} = E \qquad\qquad v^{\bullet} = v \qquad\qquad (3)$$

in plane stress and

$$E^{\bullet} = \frac{E}{1-v^2} \qquad\qquad v^{\bullet} = \frac{v}{1-v} \qquad\qquad (4)$$

in plane strain; E is the modulus of elasticity and v is Poisson's ratio of the material. We represent displacement and stress fields by

$$u = N\ q_e \qquad\qquad \sigma = L\ m_e \qquad\qquad (5)$$

where q_e and m_e are the displacement and stress degrees of freedom of the element and N and L are the shape function matrices, given by

$$N = \begin{bmatrix} N_1 & N_2 & \cdots & N_8 \end{bmatrix}$$
$$L = \begin{bmatrix} L_1 & L_2 & \cdots & L_8 \end{bmatrix} \qquad\qquad (6)$$

with

$$N_i = \begin{bmatrix} N_i & 0 \\ 0 & N_i \end{bmatrix}$$

$$L_i = \begin{bmatrix} N_i & 0 & 0 \\ 0 & N_i & 0 \\ 0 & 0 & N_i \end{bmatrix}$$

where N_i are the corresponding element interpolation functions.
Introducing (5) into the functional (2) gives

$$V_R = \sum_e \left(m_e^T \int_{\Omega_e} L^T \Delta\ N\ d\Omega\ q_e - \frac{1}{2}\ m_e^T \int_{\Omega_e} L^T C\ L\ d\Omega\ m_e \right.$$

$$\left. - q_e^T \int_{\Omega_e} N^T b\ d\Omega\ - q_e^T \int_{\Gamma_e} N^T t\ d\Gamma \right) \qquad\qquad (7)$$

or

$$V_R = \sum_e \left(m_e^T H_e q_e - \frac{1}{2} m_e^T G_e m_e - q_e^T P_{v_e} - q_e^T P_{s_e} \right) \qquad (8)$$

where

$$G_e = \int_{\Omega_e} L^T C L \, d\Omega \qquad \text{is the element flexibility matrix}$$

$$H_e = \int_{\Omega_e} L^T \Delta N \, d\Omega \qquad \text{is the element flexibility / stiffness matrix}$$

$$(9)$$

$$P_{v_e} = \int_{\Omega_e} N^T b \, d\Omega \qquad \text{is the element vector of body forces}$$

$$P_{s_e} = \int_{\Gamma_e} N^T t \, d\Gamma \qquad \text{is the element vector of boundary forces}$$

Representing by f_e the element force vector

$$f_e = P_{v_e} + P_{s_e} \qquad (10)$$

equation (8) can be written as

$$V_R = \sum_e \left(m_e^T H_e q_e - \frac{1}{2} m_e^T G_e m_e - q_e^T f_e \right) \qquad (11)$$

The stationarity condition of Reissner's functional (11) leads to an equation at the element level:

$$\begin{bmatrix} -G_e & H_e \\ H_e^T & 0 \end{bmatrix} \begin{bmatrix} m_e \\ q_e \end{bmatrix} = \begin{bmatrix} 0 \\ f_e \end{bmatrix} \qquad (12)$$

Assembling equations (12) for all elements, the system equations are

$$\begin{bmatrix} -G & H \\ H^T & 0 \end{bmatrix} \begin{bmatrix} m \\ q \end{bmatrix} = \begin{bmatrix} 0 \\ f \end{bmatrix} \qquad (13)$$

where m is the global vector of stress degrees of freedom and q is the global vector of displacement degrees of freedom.

The mixed element used in this paper has 8 nodes with 2 displacements and 3 stress degrees of freedom per node. This mixed isoparametric quadratic element is represented in Fig. 1.

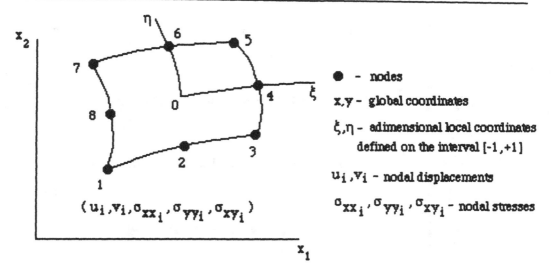

Fig. 1 - Isoparametric quadratic mixed element.

The corresponding element interpolation functions are:

$$N_i = \frac{1}{4} (1 + \xi_0) (1 + \eta_0) (\xi_0 + \eta_0 - 1) \qquad (14)$$

for the nodes 1, 3, 5 and 7,

$$N_i = \frac{1}{2} (1 - \xi^2) (1 + \eta_0) \qquad (15)$$

for the nodes 4 and 8, and

$$N_i = \frac{1}{2} (1 + \xi_0) (1 - \eta^2) \qquad (16)$$

for the nodes 2 and 6 .

In these expressions we have

$$\xi_0 = \xi_i \, \xi \qquad\qquad\qquad \eta_0 = \eta_i \, \eta \qquad (17)$$

where ξ_i and η_i are the local coordinates of node i.

The element matrices are calculated by substituting equations (6) and (14)-(17) into equations (9). The domain integrals are integrated numerically using a 3*3 Gaussian mesh and the boundary integrals are integrated numerically using 3 Gaussian points.

3 SHAPE OPTIMAL DESIGN

For the two dimensional linear elastic structure described in Fig. 2, the objective is to determine the domain Ω such that the objective function $J(\Omega, u, \sigma)$ is minimized under the conditions that the total volume of the body does not exceed a prescribed value A, $\Omega \in \Omega_{adm}$, where Ω_{adm} is the set of geometrical or technological constraints and the variables u and σ are the solution of the state equation.

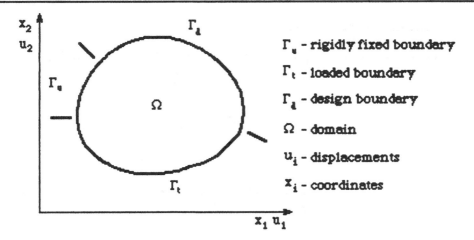

Fig.2 - Two dimensional elastic structure

A general objective function $J(\Omega,u,\sigma)$ can be defined as

$$J(\Omega,u,\sigma) = J_\Omega(\Omega,u,\sigma) + J_\Gamma(\Gamma,u,\sigma) \qquad (18)$$

where

$$J_\Omega(\Omega,u,\sigma) = \int_\Omega f(\Omega,u,\sigma)\ d\Omega \qquad (19)$$

$$J_\Gamma(\Gamma,u,\sigma) = \int_\Gamma g(\Gamma,u,\sigma)\ d\Gamma \qquad (20)$$

Supposing that the objective function and constraints are differentiable with respect to the design, the mathematical formulation of the problem is

$$\min_{\Omega \in \Omega_{adm}} \quad J(\Omega,u,\sigma) \qquad (21)$$

subject to the area constraint

$$\int_\Omega d\Omega\ -\ A\ \leq\ 0 \qquad (22)$$

where A is the pre-defined area and subject to the condition for equilibrium which is defined by the stationarity conditions of Reissner's functional (1)

$$\int_\Omega \tau_{ij} u_{ij}\ d\Omega\ -\ \int_\Omega b_{ijkl}\ \sigma_{ij}\tau_{kl}\ d\Omega\ =\ 0 \qquad \forall\ \tau \in T \qquad (23)$$

$$\int_\Omega \sigma_{ij}\ v_{i,j}\ d\Omega\ -\ \int_\Omega \bar{b}_i\ v_i\ d\Omega\ -\ \int_{\Gamma_t} \bar{t}_i\ v_i\ d\Gamma\ =\ 0 \qquad \forall\ v \in V \qquad (24)$$

The Lagrangian associated with this minimization problem is

$$L(\Omega,u,v,\sigma,\tau,\Lambda) = \int_\Omega f(\Omega,u,v)\,d\Omega + \int_\Gamma g(\Gamma,u,v) + \Lambda\left(\int_\Omega d\Omega - A\right)$$

$$+ \int_\Omega \tau_{ij}\,u_{i,j}\,d\Omega - \int_\Omega b_{ijkl}\,\sigma_{ij}\,\tau_{kl}\,d\Omega + \int_\Omega \sigma_{ij}\,v_{i,j}\,d\Omega$$

$$- \int_\Omega \bar{b}_i\,v_i\,d\Omega - \int_\Gamma \bar{t}_i\,v_i\,d\Gamma \qquad\qquad (25)$$

where

$$u, v \in V(\Omega) \qquad\qquad \sigma, \tau \in T(\Omega)$$

and the Lagrangian multiplier $\Lambda \geq 0$.

The perturbed domain Ω^* is defined, relative to the reference domain Ω, by the mapping $T: R^2 \rightarrow R^2$, in such a manner that:

$$\Omega^* = T(\Omega,t) \qquad\qquad (26)$$

with

$$(\Omega^*)_{t-0} = \Omega = T(\Omega,0) \qquad\qquad (27)$$

Assuming sufficient regularity on T and neglecting second and higher order terms, this mapping can be expressed, for small enough $t > 0$ by

$$T(\Omega,t) = T(\Omega,0) + t\,V(\Omega) \qquad\qquad (28)$$

where $V(\Omega)$ is the domain perturbation or velocity field. Denoting by

$$J = \begin{bmatrix} 1 + t\,V_{1,1} & t\,V_{1,2} \\ t\,V_{2,1} & 1 + t\,V_{2,2} \end{bmatrix} \qquad\qquad (29)$$

the Jacobian of T, we have

$$|J| = \det(I + tV) = 1 + t\,\mathrm{div}\,V \qquad\qquad (30)$$

Considering the augmented functional L^* defined on the perturbed domain Ω^*

$$L^*(\Omega^*,u^*,v^*,\sigma^*,\tau^*,\Lambda^*) = \int_{\Omega^*} f^*(\Omega^*,u^*,\sigma^*)\,d\Omega^* + \int_{\Gamma^*} g^*(\Gamma^*,u^*,\sigma^*)\,d\Gamma^* +$$

$$+ \Lambda^*\left(\int_{\Omega^*} d\Omega^* - A^*\right) + \int_{\Omega^*} \tau_{ij}^*\,u_{i,j^*}^*\,d\Omega^* - \int_{\Omega^*} b_{ijkl}\,\sigma_{ij}^*\,\tau_{kl}^*\,d\Omega^* +$$

$$+ \int_{\Omega^*} \sigma_{ij}^*\,v_{i,j^*}^*\,d\Omega^* - \int_{\Omega^*} \bar{b}_i^*\,v_i^*\,d\Omega^* - \int_{\Gamma_t^*} \bar{t}_i^*\,v_i^*\,d\Gamma^* \qquad\qquad (31)$$

in which the prescribed value A^* and Λ^* are independent of domain variation. The following first-order expansions for the functions, where the superscript point denotes the material derivative (e.g. Haug $et\ al.$, 1986), are used

$$u_i^* = u_i + t\,\dot{u}_i \qquad\qquad \sigma_{ij}^* = \sigma_{ij} + t\,\dot{\sigma}_{ij}$$

$$v_i^* = v_i + t\,\dot{v}_i \qquad\qquad \tau_{ij}^* = \tau_{ij} + t\,\dot{\tau}_{ij}$$

$$f^* = f + t\left.\frac{df}{dt}\right|_{t-0} \qquad\qquad g^* = g + t\left.\frac{dg}{dt}\right|_{t-0}$$

$$\bar{b}_i^* = \bar{b}_i + t\,\dot{\bar{b}}_i \qquad\qquad \bar{t}_i^* = \bar{t}_i + t\,\dot{\bar{t}}_i$$

$$u_{i,j^*}^* = u_{i,j} + t\left(\dot{u}_{i,j} - u_{i,m}\,V_{m,j}\right)$$

$$v_{i,j^*}^* = v_{i,j} + t\left(\dot{v}_{i,j} - v_{i,m}\,V_{m,j}\right) \tag{32}$$

and considering that

$$d\Omega^* = |J|\,d\Omega = (1 + t\operatorname{div}V)\,d\Omega$$

$$d\Gamma^* = |J|\,\|J^{-T}n\|\,d\Gamma = (1 + t\operatorname{div}V)\,\alpha(t)\,d\Gamma \tag{33}$$

where $\|A\| = (A^T . A)^{1/2}$ is the Euclidean norm and

$$\alpha(t) = (1 - 2\,t\,n_i\,n_j\,V_{i,j})^{1/2} \tag{34}$$

The Lagrangian defined on Ω^*, and referred to the initial domain Ω, can be expressed as

$$L^*(\Omega^*, u^*, v^*, \sigma^*, \tau^*, \Lambda^*) = L(\Omega, u, v, \sigma, \tau, \Lambda) + \left(\int_\Gamma g(\Gamma, u, \sigma)\,\alpha(t)\,d\Gamma \right. -$$

$$\left. - \int_\Gamma g(\Gamma, u, \sigma)\,d\Gamma\right) - \left(\int_{\Gamma_t}\bar{t}_i\,v_i\,\alpha(t)\,d\Gamma - \int_{\Gamma_t}\bar{t}_i\,v_i\,d\Gamma\right) +$$

$$+ t\left\{\int_\Omega \dot{f}(\Omega, u, \sigma)\,d\Omega + \int_\Omega \frac{\partial f}{\partial u_i}\,\dot{u}_i\,d\Omega + \int_\Omega \frac{\partial f}{\partial \sigma_{ij}}\,\dot{\sigma}_{ij}\,d\Omega + \right.$$

$$+ \int_\Omega f(\Omega, u, \sigma)\operatorname{div}V\,d\Omega + \int_\Gamma \dot{g}(\Gamma, u, \sigma)\,\alpha(t)\,d\Gamma + \int_\Gamma \frac{\partial g}{\partial u_i}\,\dot{u}_i\,\alpha(t)\,d\Gamma +$$

$$+ \int_\Gamma \frac{\partial g}{\partial \sigma_{ij}} \dot{\sigma}_{ij} \, \alpha(t) \, d\Gamma + \int_\Gamma g\,(\Gamma,u,\sigma)\, \alpha(t) \, \text{div}\, \mathbf{V} \, d\Gamma + \int_\Omega \text{div}\, \mathbf{V} \, d\Omega +$$

$$+ \int_\Omega \tau_{ij} \dot{u}_{i,j} \, d\Omega - \int_\Omega \tau_{ij} \, u_{i,m} \, V_{m,j} \, d\Omega + \int_\Omega \dot{\tau}_{ij} \, u_{i,j} \, d\Omega + \int_\Omega \tau_{ij} \, u_{i,j} \, \text{div}\, \mathbf{V} \, d\Omega -$$

$$- \int_\Omega b_{ijkl} \, \dot{\tau}_{ij} \, \sigma_{kl} \, d\Omega - \int_\Omega b_{ijkl} \, \tau_{ij} \, \dot{\sigma}_{kl} \, d\Omega - \int_\Omega b_{ijkl} \, \tau_{ij} \, \sigma_{kl} \, \text{div}\, \mathbf{V} \, d\Omega +$$

$$+ \int_\Omega \sigma_{ij} \, \dot{v}_{ij} \, d\Omega - \int_\Omega \sigma_{ij} \, v_{i,m} \, V_{m,j} \, d\Omega + \int_\Omega \dot{\sigma}_{ij} \, v_{i,j} \, d\Omega + \int_\Omega \sigma_{ij} \, v_{i,j} \, \text{div}\, \mathbf{V} \, d\Omega -$$

$$- \int_\Omega \bar{b}_i \, \dot{v}_i \, d\Omega - \int_\Omega \dot{\bar{b}}_i \, v_i \, d\Omega - \int_\Omega \bar{b}_i \, v_i \, \text{div}\, \mathbf{V} \, d\Omega - \int_{\Gamma_t} \dot{\bar{t}}_i \, v_i \, \alpha(t) \, d\Gamma -$$

$$- \int_{\Gamma_t} \bar{t}_i \, \dot{v}_i \, \alpha(t) \, d\Gamma - \int_{\Gamma_t} \bar{t}_i \, v_i \, \alpha(t) \, \text{div}\, \mathbf{V} \, d\Gamma \Big\} \qquad\qquad (35)$$

The total material derivative of Lagrangian L is defined by

$$\frac{dL}{d\Omega} (\Omega,u,v,\sigma,\tau,\Lambda) = \frac{dL^*}{dt}\Big|_{t-0} = \lim_{t \to 0^+} \frac{L^* - L}{t} \qquad\qquad (36)$$

Considering that

$$\frac{d\alpha}{dt}\Big|_{t-0} = \frac{1}{2} \frac{-2\,n_i\,n_j\,V_{i,j}}{(1-2\,t n_i\,n_j\,V_{i,j})^{1/2}}\Big|_{t-0} = -n_i\,n_j\,V_{i,j}$$

$$\frac{d}{dt}(t\alpha)\Big|_{t-0} = t\frac{d\alpha}{dt}\Big|_{t-0} + \alpha\Big|_{t-0} = 1 \qquad\qquad (37)$$

the following is obtained

$$\frac{dL}{d\Omega}(\Omega,u,v,\sigma,\tau,\Lambda) = \int_\Omega \dot{f}\,(\Omega,u,\sigma)\, d\Omega + \int_\Omega f\,(\Omega,u,\sigma)\, \text{div}\, \mathbf{V}\, d\Omega +$$

$$+ \int_\Gamma \dot{g}\,(\Gamma,u,\sigma)\, d\Gamma + \int_\Gamma g\,(\Gamma,u,\sigma)\, \text{div}_\Gamma\, \mathbf{V}\, d\Gamma + \Lambda \int_\Omega \text{div}\, \mathbf{V}\, d\Omega -$$

$$- \int_\Omega \tau_{ij} \, u_{i,m} \, V_{m,j} \, d\Omega + \int_\Omega \tau_{ij} \, u_{i,j} \, \text{div} \, \mathbf{V} \, d\Omega - \int_\Omega b_{ijkl} \, \tau_{ij} \, \sigma_{kl} \, \text{div} \, \mathbf{V} \, d\Omega -$$

$$- \int_\Omega \sigma_{ij} \, v_{i,m} \, V_{m,j} \, d\Omega + \int_\Omega \sigma_{ij} \, v_{i,j} \, \text{div} \, \mathbf{V} \, d\Omega - \int_\Omega \bar{b}_i \, v_i \, d\Omega - \int_\Omega \bar{b}_i v_i \, \text{div} \, \mathbf{V} \, d\Omega -$$

$$- \int_{\Gamma_t} \bar{t}_i \, v_i \, d\Gamma - \int_{\Gamma_t} \bar{t}_i \, v_i \, \text{div}_\Gamma \, \mathbf{V} \, d\Gamma + \int_\Omega \frac{\partial f}{\partial u_i} \, \dot{u}_i \, d\Omega + \int_\Gamma \frac{\partial g}{\partial u_i} \, \dot{u}_i \, d\Gamma +$$

$$+ \int_\Omega \tau_{ij} \, \dot{u}_{i,j} \, d\Omega + \int_\Omega \frac{\partial f}{\partial \sigma_{ij}} \, \dot{\sigma}_{ij} \, d\Omega + \int_\Gamma \frac{\partial g}{\partial \sigma_{ij}} \, \dot{\sigma}_{ij} \, d\Gamma - \int_\Omega b_{ijkl} \, \tau_{ij} \, \dot{\sigma}_{kl} \, d\Omega +$$

$$+ \int_\Omega \dot{\sigma}_{ij} \, v_{i,j} \, d\Omega + \int_\Omega \dot{\tau}_{ij} \, u_{i,j} \, d\Omega - \int_\Omega b_{ijkl} \, \dot{\tau}_{ij} \, \sigma_{kl} \, d\Omega + \int_\Omega \sigma_{ij} \, \dot{v}_{i,j} \, d\Omega -$$

$$- \int_\Omega \bar{b}_i \, \dot{v}_i \, d\Omega - \int_{\Gamma_t} \bar{t}_i \, \dot{v}_i \, d\Gamma \qquad\qquad (38)$$

where

$$\text{div}_\Gamma \, \mathbf{V} = \text{div} \, \mathbf{V} - n_i \, n_j \, V_{k,l} = H \, (\mathbf{V.n}) \qquad\qquad (39)$$

Equating to zero terms of the equation containing material derivatives of τ and $\mathbf{v}$, the state equations are obtained:

$$\int_\Omega \sigma_{ij} \, \dot{v}_{i,j} \, d\Omega - \int_\Omega \bar{b}_i \, \dot{v}_i \, d\Omega - \int_{\Gamma_t} \bar{t}_i \, \dot{v}_i \, d\Gamma = 0 \qquad\qquad (40)$$

$$\int_\Omega \dot{\tau}_{ij} \, u_{i,j} \, d\Omega - \int_\Omega b_{ijkl} \, \dot{\tau}_{ij} \, \sigma_{kl} \, d\Omega = 0 \qquad\qquad (41)$$

Equating to zero terms of the equation containing material derivatives of σ and u, the adjoint equations are obtained:

$$\int_\Omega \tau_{ij} \, \dot{u}_{i,j} \, d\Omega + \int_\Omega \frac{\partial f}{\partial u_i} \, \dot{u}_i \, d\Omega + \int_\Gamma \frac{\partial g}{\partial u_i} \, \dot{u}_i \, d\Gamma = 0 \qquad\qquad (42)$$

$$\int_\Omega \dot{\sigma}_{ij} \, v_{i,j} \, d\Omega - \int_\Omega b_{ijkl} \, \dot{\sigma}_{ij} \, \tau_{kl} \, d\Omega + \int_\Omega \frac{\partial f}{\partial \sigma_{ij}} \, \dot{\sigma}_{ij} \, d\Omega + \int_\Gamma \frac{\partial g}{\partial \sigma_{ij}} \, \dot{\sigma}_{ij} \, d\Gamma = 0$$
$$(43)$$

Finally, the sensitivity of the Lagrangian to the domain perturbation is obtained, substituting (u, σ), the solution to the state equations, and (v, τ), the solution to the adjoint equations, on the total material derivative (38):

$$\frac{dL}{d\Omega} (\Omega, u, v, \sigma, \tau, \Lambda) = \int_\Omega \dot{f} (\Omega, u, \sigma) \, d\Omega + \int_\Omega f (\Omega, u, \sigma) \, div \, V \, d\Omega +$$

$$+ \int_\Gamma \dot{g} (\Gamma, u, \sigma) \, d\Gamma + \int_\Gamma g (\Gamma, u, \sigma) \, div_\Gamma V \, d\Gamma + \Lambda \int_\Omega div \, V \, d\Omega -$$

$$- \int_\Omega \tau_{ij} \, u_{i,m} \, V_{m,j} \, d\Omega + \int_\Omega \tau_{ij} \, u_{i,j} \, div \, V \, d\Omega - \int_\Omega b_{ijkl} \, \tau_{ij} \, \sigma_{kl} \, div \, V \, d\Omega -$$

$$- \int_\Omega \sigma_{ij} \, v_{i,m} \, V_{m,j} \, d\Omega + \int_\Omega \sigma_{ij} \, v_{i,j} \, div \, V \, d\Omega - \int_\Omega \dot{b}_i \, v_i \, d\Omega - \int_\Omega \bar{b}_i v_i \, div \, V \, d\Omega -$$

$$- \int_{\Gamma_t} \dot{\bar{t}}_i \, v_i \, d\Gamma - \int_{\Gamma_t} \bar{t}_i \, v_i \, div_\Gamma V \, d\Gamma \qquad (44)$$

If the objective is to find the domain $\Omega \in \Omega_{adm}$ such that the compliance

$$\Psi_0 = J (\Omega, u, \sigma) = \frac{1}{2} \left(\int_\Omega \bar{b}_i \, u_i \, d\Omega + \int_{\Gamma_t} \bar{t}_i \, u_i \, d\Gamma \right) \qquad (45)$$

is minimized with a constraint on volume equal to A:

$$\Psi_1 = \int_\Omega d\Omega - A \leq 0 \qquad (46)$$

and observing that

$$J_\Omega(\Omega, u, \sigma) = \int_\Omega f (\Omega, u, \sigma) \, d\Omega = \frac{1}{2} \int_\Omega \bar{b}_i \, u_i \, d\Omega \qquad (47)$$

$$J_\Gamma (\Gamma, u, \sigma) = \int_\Gamma g (\Gamma, u, \sigma) \, d\Gamma = \frac{1}{2} \int_{\Gamma_t} \bar{t}_i \, u_i \, d\Gamma \qquad (48)$$

and using (42)-(43), we obtain the adjoint equations

$$\int_\Omega \tau_{ij} \, \dot{u}_{i,j} \, d\Omega + \frac{1}{2} \int_\Omega \bar{b}_i \, \dot{u}_i \, d\Omega + \frac{1}{2} \int_{\Gamma_t} \bar{t}_i \, \dot{u}_i \, d\Gamma = 0 \qquad (49)$$

$$\int_\Omega \dot{\sigma}_{ij} \, v_{i,j} \, d\Omega - \int_\Omega b_{ijkl} \, \dot{\sigma}_{ij} \, \tau_{kl} \, d\Omega = 0 \qquad (50)$$

These equations are similar to the state equations with body forces and tractions related to the real problem by

$$(\bar{b}_i)_a = -\frac{1}{2}\bar{b}_i \qquad\qquad (\bar{t}_i)_a = -\frac{1}{2}\bar{t}_i \qquad\qquad (51)$$

where the subscript a denotes adjoint state. So the adjoint variables are related to the state variables by

$$v_i = -\frac{1}{2}u_i \qquad\qquad \tau_{ij} = -\frac{1}{2}\sigma_{ij} \qquad\qquad (52)$$

Considering that $f(\Omega, u, \sigma) = f(u)$, $g(\Gamma, u, \sigma) = g(u)$, and that body forces and tractions are not variables with the domain variation, the Lagrangian sensitivity is

$$\frac{dL}{d\Omega}(\Omega, u, v, \sigma, \tau, \Lambda) = \int_\Omega \bar{b}_i\, u_i\, \mathrm{div}\, V\, d\Omega + \int_{\Gamma_t} \bar{t}_i\, u_i\, \mathrm{div}_\Gamma V\, d\Gamma +$$

$$+ \Lambda \int_\Omega \mathrm{div}\, V\, d\Omega + \int_\Omega \sigma_{ij}\, u_{i,m}\, V_{m,j}\, d\Omega - \frac{1}{2}\int_\Omega \sigma_{ij}\, u_{i,j}\, \mathrm{div}\, V\, d\Omega \quad (53)$$

Using Green's Theorem the correspondent boundary expression is:

$$\frac{dL}{d\Omega}(\Omega, u, v, \sigma, \tau, \Lambda) = \int_\Gamma \bar{b}_i\, u_i\, V.n\, d\Gamma + \int_{\Gamma_t}\left[\nabla(\bar{t}_i\, u_i)^T.n + H\bar{t}_i u_i\right] d\Gamma +$$

$$+ \Lambda \int_\Gamma V.n\, d\Gamma - \int_\Omega \sigma_{ij,j}\, u_{i,m}\, V_m\, d\Omega - \frac{1}{2}\int_\Gamma \sigma_{ij}\, u_{i,j}\, V.n\, d\Gamma +$$

$$+ \int_\Gamma \sigma_{ij}\, n_j\, u_{i,m}\, V_m\, d\Gamma \qquad\qquad (54)$$

If it is assumed that body forces and boundary tractions on Γ_d are zero, the domain equation is:

$$\frac{dL}{d\Omega}(\Omega, u, v, \sigma, \tau, \Lambda) = \Lambda \int_\Omega \mathrm{div}\, V\, d\Omega + \int_\Omega \sigma_{ij}\, u_{i,m}\, V_{m,j}\, d\Omega -$$

$$- \frac{1}{2}\int_\Omega \sigma_{ij}\, u_{i,j}\, \mathrm{div}\, V\, d\Omega \qquad\qquad (55)$$

or the corresponding boundary expression is

$$\frac{dL}{d\Omega}(\Omega, u, v, \sigma, \tau, \Lambda) = \Lambda \int_\Gamma V.n\, d\Gamma - \frac{1}{2}\int_\Gamma \sigma_{ij}\, u_{i,j}\, V.n\, d\Gamma \qquad (56)$$

Using this last equation, the first variation of the compliance functional Ψ_0 (45) can be efficiently obtained with mixed finite elements. Since mixed elements can provide better boundary results than displacement finite elements, we may expect more accurate sensitivity values. It is important to note that

$$\frac{dL}{d\Omega} (\Omega, u, v, \sigma, \tau, \Lambda) = \Lambda \, \delta \Psi_1 + \delta \Psi_0 \qquad (57)$$

where $\delta \Psi_0$ and $\delta \Psi_1$ are the first variations of the compliance (45) and of the area constraint (46).

4 NUMERICAL MODEL

The first variation of the compliance is

$$\delta \Psi_0 = - \frac{1}{2} \int_{\Gamma_d} U \, V_n \, d\Gamma \qquad (58)$$

where

$$U = \sigma_{ij} \, \varepsilon_{ij} = \frac{1}{2} \sigma_{ij} \, (u_{i,j} + u_{j,i}) = \sigma_{ij} u_{i,j} \qquad (59)$$

is the strain energy density and $V_n = (\mathbf{V} \cdot \mathbf{n})$ is the normal perturbation field of the domain, defined on Γ_d.

The specific strain energy at one boundary point of an unloaded boundary is given by

$$U = \sigma_{ss} \, \varepsilon_{ss} = \frac{1}{E^*} \, \sigma_{ss}^2 \qquad (60)$$

where σ_{ss} is the tangential stress, which is the first invariant of the stress tensor

$$\sigma_{ss} = \sigma_{11} + \sigma_{22} \qquad (61)$$

The mixed element used is an isoparametric eight node mixed element with two displacement and three stress degrees of freedom per node. The boundary geometry is described by linear design elements. The nodes of each design element coincide with the extreme nodes of one side of the mixed element, as shown in Fig. 3. The design variables are defined as the norm of the position vector of the interpolation nodes with respect to a pre-defined origin O. The tangential stresses on the boundary are quadratic.

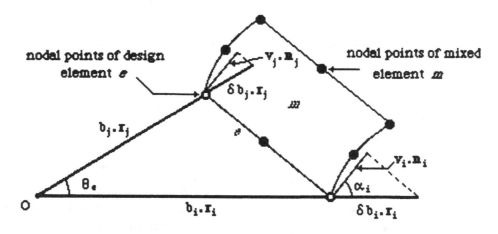

Fig.3 - Design variables and their correlation with the mixed finite element.

The tangential stress σ_{ss} on the side of the mixed element m, can be described by the quadratic shape functions L_i as

$$\sigma_{ss} = L_i \; \sigma_{ss_i} \tag{62}$$

where σ_{ss_i} is the nodal tangential stress and the shape functions L_i are:

$$L_1(\xi) = \frac{1}{2} \; (\xi^2 - \xi \,)$$

$$L_2(\xi) = (\, 1 - \xi^2 \,) \tag{63}$$

$$L_3(\xi) = \frac{1}{2} \; (\, \xi^2 + \xi \,)$$

where ξ is the tangential adimensional coordinate of the side of the element.

The normal boundary perturbation V_n on a linear geometric design element e, can be described by the linear shape functions N_i^L as

$$V_n = N_i^L \; v_i \tag{64}$$

where v_i is the normal perturbation of each nodal point and the shape functions N_i^L are:

$$N_1^L (\xi) = \frac{1}{2} \; (\, 1 - \xi \,) \qquad\qquad N_2^L (\xi) = \frac{1}{2} \; (\, 1 + \xi \,) \tag{65}$$

As shown in Fig. 3

$$v_i = \delta \, b_i \; (\, r_i \cdot n_i \,) \tag{66}$$

where r_i and b_i are, respectively, the unit vector and the norm of the position vector of the corresponding interpolation node i, δb_i is the variation of design variable b_i and n_i is the unit normal vector to the boundary in node i. The normal perturbation field is

$$V_n = N_i^L \; \delta \, b_i \; \cos \alpha_i \tag{67}$$

Introducing (60) and (67) into expression (58) we obtain, for the geometric element e, in matrix representation

$$\delta \, \Psi_{0_e} = - \; \frac{\ell_e}{2 \, E^{\bullet}} \; (\, s^T \; M \; v \,) \tag{68}$$

where ℓ_e is the element lenght and

$$s = \left[\; \sigma_{ss_1}^2 \quad \sigma_{ss_2}^2 \quad \sigma_{ss_3}^2 \quad 2 \, \sigma_{ss_1} \sigma_{ss_2} \quad 2 \, \sigma_{ss_1} \sigma_{ss_2} \quad 2 \, \sigma_{ss_2} \sigma_{ss_3} \; \right]^T$$

$$M = \int_{-1}^{+1} L^{\bullet} \; N^{L^T} \; d\xi \tag{69}$$

with

$$\mathbf{L}^{\bullet} = \left[\begin{array}{cccccc} L_1^2 & L_2^2 & L_3^2 & L_1 L_2 & L_1 L_3 & L_2 L_3 \end{array} \right]^T$$

$$\mathbf{N}^L = \left[\begin{array}{cc} N_1^L & N_2^L \end{array} \right]^T$$

and, finally,

$$\mathbf{v} = \left[\begin{array}{cc} \delta b_i \ \cos \alpha_i & \delta b_j \ \cos \alpha_j \end{array} \right]^T$$

where i and j represent nodal points of element e. Equation (69) is numerically integrated by 3 Gaussian points.

The first variation of the area constraint (46),

$$\delta \Psi_1 = \int_{\Gamma_d} V_n \ d\Gamma \tag{70}$$

can be obtained for element e, as shown in Fig. 4, by the expression:

$$\delta \Psi_{1_e} = \int_{\Gamma_e} V_n \ d\Gamma = \frac{1}{2} \ \sin \theta_e \ \left(b_i \ \delta b_j + b_j \ \delta b_i \right) \tag{71}$$

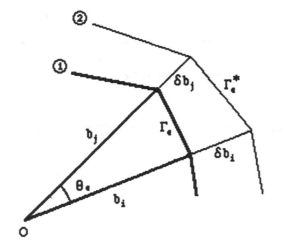

① - initial boundary

② - perturbed boundary

Γ_e - initial boundary of element

Γ_e^* - perturbed boundary of element

b_i - design variable

δb_i - perturbation of design variable

Fig. 4 - Influence of the perturbation of design variables on the variation of the area

The nonlinear programming problem (45)-(46) is solved using the method of sequential convex programming (Fleury and Braibant, 1986) and the modified method of feasible directions, available in the commercial programme ADS (Automated Design Synthesis) (Vanderplaats, 1987).

4 APPLICATIONS

Consider the problem represented in Fig. 5.

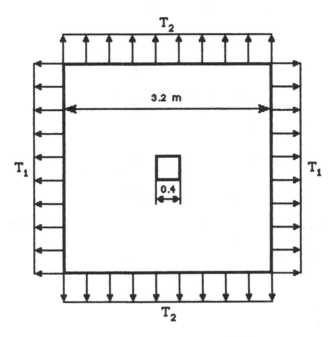

Fig. 5 - Square plate with hole.

For the infinite plate the analytical solution is a circular hole (Banichuk, 1983), when the load $T_1 - T_2$ and an elliptical hole with a semi-axis ratio equal to the ratio of the external applied forces when $T_1 \neq T_2$. Since the analytical solution of the problem is known, this example is a good test to check the numerical procedure developed.

The problem data is the following:
- plane stress
- material elastic properties:

  $E - 200$ GPa

  $\upsilon - 0.3$

- applied forces:

  first load case: $T_1 = T_2 = 100$ MPa

  second load case $T_1 - 75$ MPa $T_2 - 100$ MPa

- the objective function is the compliance
- the maximum admissible area is equal to the initial area of the hole.

The problem is solved by modelling 1/4 of the plate with 16 mixed quadratic elements. Five design variables, as represented in Fig. 6, are used.

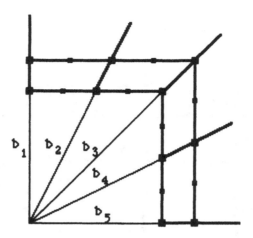

Fig. 6 - Design variables.

For the first load case, $T_1 = T_2$, results are presented in Figs. 7 to 9.

The evolution of the compliance and the area constraint are shown in Fig. 7. The constraint value presented here is obtained by the expression:

$$\Psi_{1_{rep}} = \frac{\Psi_1}{\int_{\Omega} d\Omega} * 10^6 \qquad (72)$$

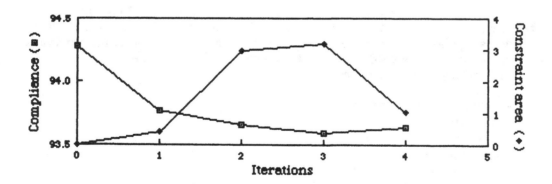

Fig. 7 - Evolution of objective and constraint functions. First load case.

In Fig. 8, initial and final meshes are shown. The mesh regenerator updates only 8 elements of the sub-domain near the design boundary.

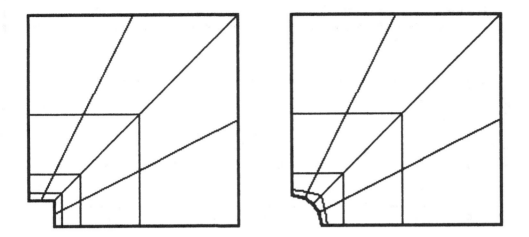

Fig. 8 - Initial and final mesh. First load case.

The evolution of the hole design is represented in Fig. 9.

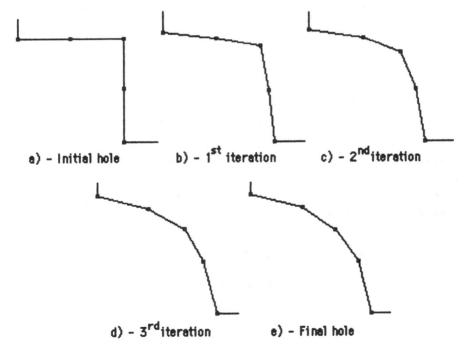

a) – Initial hole   b) – 1st iteration   c) – 2nd iteration

d) – 3rd iteration   e) – Final hole

Fig. 9 - Evolution of hole design. First load case.

For the second load case, $T_1 = 0.75 \, T_2$, results are presented in Figs. 10 to 12. In Fig. 10 the evolution of the compliance and area constraint are shown. The constraint value presented is obtained by the expression (72).

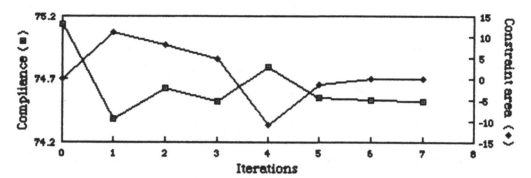

Fig. 10 - Evolution of objective and constraint functions. Second load case.

In Fig.11 initial and final meshes are shown. The mesh regenerator updates only 8 elements of the sub-domain near the design boundary.

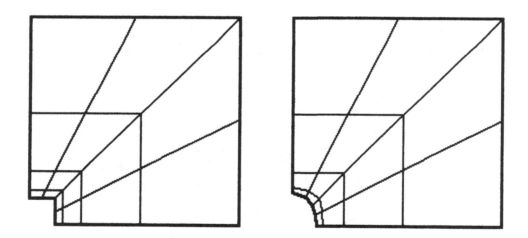

Fig. 11 - Initial and final mesh. Second load case.

The evolution of hole design is represented in Fig. 12.

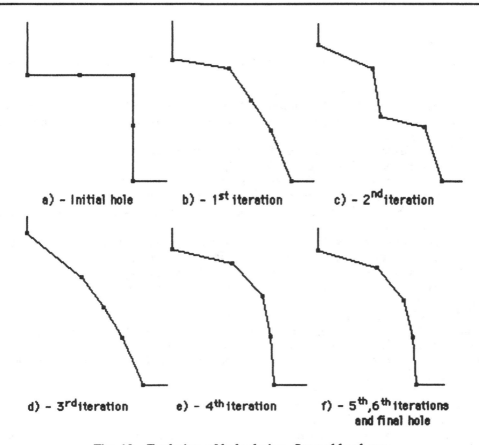

a) – Initial hole   b) – 1st iteration   c) – 2nd iteration

d) – 3rd iteration   e) – 4th iteration   f) – 5th,6th iterations
                                                                                 and final hole

Fig. 12 - Evolution of hole design. Second load case.

5 CONCLUDING REMARKS

In the applications, the final designs are almost identical with the analytical solutions for infinite plates (Banichuk, 1983). For bi-axial equal loads, the final design is an excellent approximation of a circle, with a radius difference of 0.16 %. The mean stress intensity factor on the boundary of the hole is 2.025 which has 1% discrepancy compared to the analytical solution for the infinite plate.

For bi-axial unequal loads, the final design is a polygon having a strong resemblance to an ellipse. The ratio of the semi-axis is 0.73, which compares favourably with 0.75 given by the analytical solution.

In the applications, optimal design is achieved in a few iterations. Considering that the model used is very simple, only 5 design variables and 4 linear elements describing the design boundary, the results are excellent. It can be concluded that mixed elements can be efficient in the shape optimal design of structures.

Mixed elements may offer some advantages over displacement finite elements since the stresses and sensitivities are more accurate.

Further applications are required to demonstrate the efficiency and accuracy of the domain formulation for sensitivities developed in this paper.

REFERENCES

Banichuk, N.V., *Problems and Methods of Optimal Structural Design*, Plenum Press, New York, 1983

Braibant, V., Shape sensitivity by finite elements, *J. Stuct. Mech.* **14**, 209-228, 1986

Dems, K. and Mroz, Z., Variational approach by means of adjoint systems to structural optimization and sensitivity analysis - II Stucture shape variation, *Int. J. Num. Meth. Engrg.* **20**, 527-552, 1984

Fleury, C. and Braibant, V., Structural optimization: a new dual method using mixed variables, *Int. J. Num. Meth. Engrg.* **23**, 409-428, 1986

Francavilla, A., Ramakrishnan, C.V. and Zienkiewicz, O.C., Optimization of shape to minimize stress concentration, *J. Strain Analysis* **20**, 63-70, 1975

Haug, E.J., Choi, K.K. and Komkov, V., *Design Sensitivity Analysis of Stuctural Systems*, Academic Press, New York, 1986

Leal, R.P. and Mota Soares,C.A., Mixed elements in design sensitivity analysis of plates with dynamic and stability constraints, In: *Proc.Third International Conference on Recent Advances in Structural Dynamics* 1,123-132, Southampton, 1988

Leal, R.P. and Mota Soares,C.A., Mixed elements in the optimal design of plates, *Structural Optimization* **1**, 127-136, 1989

Leal, R.P. and Mota Soares,C.A., Shape optimal stuctural design using mixed elements and minimum compliance techniques, In: S. Saigal and S. Mukherjee, eds., *Sensitivity Analysis and Optimization with Numerical Methods*, 79-93, ASME, 1990

Mota Soares,C.A. and Leal, R.P., Mixed elements in the sensitivity analysis of structures, *Eng. Opt.* **11**, 227-237, 1987

Mota Soares,C.A., Leal, R.P. and Choi, K.K., Boundary elements in shape optimal design of stuctural components, In: C.A. Mota Soares, ed., *Computer Aided Optimal Design: Structural and Mechanical Systems*, 605-631, Springer, Berlin, 1987

Rodrigues, H.C., *Shape optimal design of elastic bodies using a mixed variational formulation*, Ph. D. Thesis, The University of Michigan, Ann Arbor, MI, 1988a

Rodrigues, H.C., Shape optimal design of elastic bodies using a mixed variational formulation, *Comput. Meth. Appl. Mech. Eng.* **69** , 29-44, 1988b

Vanderplaats, G.N., *ADS - A Fortran program for Automated Design Synthesis - Version 2.01* , Engineering Design Optimization,Inc., 1987

Washizu, K., *Variational Methods in Elasticity and Plasticity*, Pergamon, 1974

Zienkiewicz, O.C. and Campbell, J.S., Shape optimization and sequential linear programming, In: R.H. Gallagher and O.C. Zienkiwicz, eds., *Optimum Structural Design*, Wiley, London, 1973

Chapter 15

SHAPE OPTIMAL DESIGN OF
AXISYMMETRIC SHELL STRUCTURES

C.A. Mota Soares
CEMUL, Lisbon, Portugal

J.I. Barbosa
ENIDH, Oeiras, Portugal

C.M. Mota Soares
CEMUL, Lisbon, Portugal

1 - INTRODUCTION

Structural optimization   using finite element techniques requires the sequential use of structural and sensitivity analyses combined with a numerical optimizer. The success of the structural optimization process depends on the proper choices with respect to the finite element model, sensitivity analysis, objective function, constraints, design variables and method of solution of the nonlinear mathematical problem.

In this section the structural and sensitivity analyses of thin axisymmetric shell structures are presented using a frustum-cone finite element with 8 degrees of freedom, based on Love-Kirchhof assumptions. In the structural analysis program, the calculation of the sensitivities of displacements, stresses and natural frequency, with respect to perturbations in the design variables, are included. These sensitivities are evaluated using the semi-analytical method as presented by Gendong and Yingwei (1987) and Barthelemy et al. (1988). In this method the sensitivities of the stiffness or mass matrices or load vector, are obtained by a finite difference approach at element level considering a small perturbation of the design variables. Alternatively the sensitivities are evaluated analytically, using a symbolic manipulator to overcome the difficulty of obtaining the sensitivities when the design variables are nodal coordinates. It is also possible to calculate the sensitivities using a global finite difference technique. In this case two complete structural analyses for each design variable are necessary and they are shown here only for comparison purposes.

The formulation presented is applied to the minimum weight design of thin axisymmetric shell structures subjected to constraints on displacements, stresses, natural frequency, volume of the shell material and enclosed volume of the structure. The maximization of a natural frequency of a specified mode shape can be also carried out. For static constraints the adjoint structure technique is used as presented in Haftka and Kamat (1987). The design variables are thicknesses and/or radial nodal coordinates. The ADS (Automated Design Synthesis) program of Vanderplaats (1984) is used to solve the nonlinear mathematical programming problem.

2 - THIN AXISYMMETRIC SHELLS

2.1 - KINEMATICS

Structural analysis of axisymmetric shells using finite element methods requires a discretized model where the complete shell can be idealized as a series of shell ring elements joined at their nodal point circles. Its behaviour will be characterized by the displacements of these nodal circles which are described in terms of a finite number of displacement variables or generalized displacements.

For an arbitrary shell the strain-displacement relations for small displacements in an orthogonal curvilinear system are given by Kraus (1967) :

$$\epsilon_i = \frac{\partial}{\partial \alpha_i}\left(\frac{u_i}{\sqrt{g_i}}\right) + \frac{1}{2\,g_i} \sum_{k=1}^{3} \frac{\partial g_i}{\partial \alpha_k}\frac{u_k}{\sqrt{g_k}} \qquad\qquad (i=1,2,3) \qquad\qquad (1)$$

$$\gamma_{ij} = \frac{1}{\sqrt{g_i\,g_j}}\left[g_i\frac{\partial}{\partial \alpha_j}\left(\frac{u_i}{\sqrt{g_i}}\right) + g_j\frac{\partial}{\partial \alpha_i}\left(\frac{u_j}{\sqrt{g_j}}\right)\right] \qquad (i, j = 1,2,3 \; ; \; i \neq j) \qquad (2)$$

where ϵ_i are the normal strains, α_i the curvilinear coordinates, u_i the displacement components, g_i the first fundamental magnitudes and γ_{ij} the shear strains.

Considering the Love-Kirchhoff approximation of the theory of thin elastic shells which is based on the postulation that the shell is thin, the deflections of the shell are small, the transverse normal stress is negligible and normals to the reference surface of the shell remain normal to it and undergo no change in length during deformation. For a conical shell (Fig. 1) represented by its reference surface of revolution, the displacement components are represented as :

$$u_1 = U(S, \theta, \xi) \qquad ; \qquad u_2 = V(S, \theta, \xi) \qquad ; \qquad u_3 = W(S, \theta, \xi)$$

For this particular situation and assuming for thin shells $\xi/R_i \simeq 0$, where $1/R_i$ are the principal curvatures, equations (1) and (2) can be represented as :

$$\epsilon_{SS} = \frac{\partial U}{\partial S}$$

$$\epsilon_{\theta\theta} = \frac{1}{r}\left(\frac{\partial V}{\partial \theta} + U \cos \phi + W \sin \phi\right)$$

$$\epsilon_{\xi\xi} = \frac{\partial W}{\partial \xi} \qquad\qquad (3)$$

$$\gamma_{S\theta} = \frac{\partial V}{\partial S} + \frac{1}{r}\left(\frac{\partial U}{\partial \theta} - V \cos \phi\right)$$

$$\gamma_{S\xi} = \frac{\partial W}{\partial S} + \frac{\partial U}{\partial \xi}$$

$$\gamma_{\theta\xi} = \frac{1}{r}\frac{\partial W}{\partial \theta} + r\frac{\partial}{\partial \xi}\left(\frac{V}{r}\right)$$

$$U = U(S, \theta, \xi)$$
$$V = V(S, \theta, \xi)$$
$$W = W(S, \theta, \xi)$$

where U, V, W are the components of the displacement vector of a spatial point and S, θ, ξ are, respectively, the coordinates along the meridian, parallel circle and normal to the reference surface of the shell (Fig. 1).

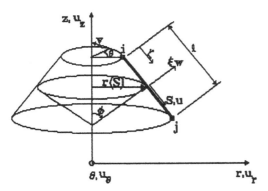

Fig. 1 – Frustum–cone finite element
Geometry and displacements

Assuming the following displacement distribution :

$$U(S,\theta,\xi) = u(S,\theta) + \xi \frac{\partial U}{\partial \xi}\Big|_{\xi=0}$$

$$V(S,\theta,\xi) = v(S,\theta) + \xi \frac{\partial V}{\partial \xi}\Big|_{\xi=0} \qquad\qquad (4)$$

$$W(S,\theta,\xi) = w(S,\theta)$$

where $u(S,\theta)$, $v(S,\theta)$ and $w(S,\theta)$ represent the components of the displacement vector of a point on the reference middle surface of the shell and $\frac{\partial U}{\partial \xi}|_{\xi=0}$ and $\frac{\partial V}{\partial \xi}|_{\xi=0}$ represent the rotations of tangents to the reference surface oriented along the lines S and θ, respectively. Let $\frac{\partial U}{\partial \xi}|_{\xi=0} = \beta_S(S,\theta)$ and $\frac{\partial V}{\partial \xi}|_{\xi=0} = \beta_\theta(S,\theta)$ and using the Love-Kirchhoff assumptions ($\gamma_{S\xi} = \gamma_{\theta\xi} = 0$), one obtains for the frustum cone :

$$\beta_S = -\frac{\partial w}{\partial S}$$

$$\beta_\theta = \frac{v}{r}\sin\phi - \frac{1}{r}\frac{\partial w}{\partial \theta}$$

(5)

The displacement vector $\underline{U} = [\ U\ \ V\ \ W\ ]^T$ of a given point (S, θ, ξ) can be expressed in terms of the displacement vector $U = [u, v, w]^T$ of the reference surface and the rotations of tangents (β_S, β_θ) of the same reference surface oriented along the parametric lines as :

$$\underline{U} = \begin{bmatrix} u \\ v \\ w \end{bmatrix} + \begin{bmatrix} \xi & 0 \\ 0 & \xi \\ 0 & 0 \end{bmatrix} \begin{bmatrix} \beta_S \\ \beta_\theta \end{bmatrix}$$

(6)

Substituting relations (6) into the remainder of equations (3), the non-vanishing strains in the thin elastic shell are given by :

$$\epsilon = \epsilon^\circ + \xi\,\chi$$

(7)

with:

$$\epsilon = [\ \epsilon_{SS}\ \ \epsilon_{\theta\theta}\ \ \gamma_{S\theta}\ ]^T\quad ;\quad \epsilon^\circ = [\ \epsilon_{SS}^\circ\ \ \epsilon_{\theta\theta}^\circ\ \ \gamma_{S\theta}^\circ\ ]^T\quad ;\quad \chi = [\ \chi_{SS}\ \ \chi_{\theta\theta}\ \ \chi_{S\theta}\ ]^T$$

and:

$$\epsilon^\circ = \Delta_m\,U$$

(8)

$$\chi = \Delta_f\,U$$

(9)

where the operators Δ_m and Δ_f are, respectively :

$$\Delta_m = \begin{bmatrix} \dfrac{\partial}{\partial S} & 0 & 0 \\[2mm] \dfrac{\cos\phi}{r} & \dfrac{1}{r}\dfrac{\partial}{\partial\theta} & \dfrac{\sin\phi}{r} \\[2mm] \dfrac{1}{r}\dfrac{\partial}{\partial\theta} & \left(\dfrac{\partial}{\partial S} - \dfrac{\cos\phi}{r}\right) & 0 \end{bmatrix}$$

$$\Delta_f = \begin{bmatrix} 0 & 0 & -\dfrac{\partial^2}{\partial S^2} \\[3mm] 0 & \dfrac{\sin\phi}{r^2}\dfrac{\partial}{\partial\theta} & -\dfrac{1}{r^2}\left(\dfrac{\partial^2}{\partial\theta^2} + r\cos\phi\,\dfrac{\partial}{\partial S}\right) \\[3mm] 0 & \dfrac{2\sin\phi}{r^2}\left(r\dfrac{\partial}{\partial S} - \cos\phi\right) & \dfrac{2}{r^2}\left(\cos\phi\,\dfrac{\partial}{\partial\theta} - r\dfrac{\partial^2}{\partial S\,\partial\theta}\right) \end{bmatrix}$$

The quantities ϵ_{SS}^o, $\epsilon_{\theta\theta}^o$, $\gamma_{S\theta}^o$ represent, respectively, the meridional, circumferential and shearing strains of the reference surface. The quantities χ_{SS} and $\chi_{\theta\theta}$ represent the changes in the curvature of the reference surface and $\chi_{S\theta}$ represents the torsion of the same surface during deformation.

Assuming that the displacements u, v, w can be expanded in Fourier series of the type :

$$U = \sum_{n=0}^{N} \left(C_n U_n + \hat{C}_n \hat{U}_n \right) \tag{10}$$

where:

$$C_n = \begin{bmatrix} \cos n\theta & 0 & 0 \\ 0 & \sin n\theta & 0 \\ 0 & 0 & \cos n\theta \end{bmatrix} \quad ; \quad \hat{C}_n = \begin{bmatrix} \sin n\theta & 0 & 0 \\ 0 & \cos n\theta & 0 \\ 0 & 0 & \sin n\theta \end{bmatrix}$$

$$U_n = [\, u_n \ v_n \ w_n \,]^T \quad ; \quad \hat{U}_n = [\, \hat{u}_n \ \hat{v}_n \ \hat{w}_n \,]^T$$

The first and second terms of equation (10) represent these components of displacements which are, respectively, symmetric and antisymmetric with respect to the plane passing through $\theta=0$, u_n, v_n and w_n being the amplitudes of symmetric part and $\hat{u}_n$, $\hat{v}_n$ and $\hat{w}_n$ the amplitudes of antisymmetric part for the nth harmonic and N is the number of terms in the truncated Fourier series.

Since the angular dependence of displacement components is expressed in terms of trigonometric functions, the orthogonality properties of such functions yields a formulation of the problem as a series of uncoupled quasi two-dimensional problems, in which the displacement amplitudes u_n, v_n, w_n, $\hat{u}_n$, $\hat{v}_n$ and $\hat{w}_n$ are the unknowns. The meridional dependence of these displacements amplitudes along the frustum cone represented here only by the symmetric part, for sake of simplicity, can be assumed as :

$$U_n = N\, q_{e_n}^l \tag{11}$$

where :

$$N = \begin{bmatrix} N_1 & 0 & 0 & 0 & N_2 & 0 & 0 & 0 \\ 0 & N_1 & 0 & 0 & 0 & N_2 & 0 & 0 \\ 0 & 0 & N_3 & N_4 & 0 & 0 & N_5 & N_6 \end{bmatrix} \quad ; \quad \begin{aligned} N_1 &= (1-\zeta) \\ N_2 &= \zeta \\ N_3 &= (1-3\,\zeta^2+2\,\zeta^3) \\ N_4 &= (\zeta-2\,\zeta^2+\zeta^3)\,\ell \\ N_5 &= (3\,\zeta^2-2\,\zeta^3) \\ N_6 &= (-\zeta^2+\zeta^3)\,\ell \end{aligned} \tag{12}$$

$\zeta = \frac{S}{\ell}$ - Coordinate over the length of the frustum-cone element.

$\ell = \sqrt{(r^j-r^i)^2+(z^j-z^i)^2}$ - Length of the frustum-cone element

$r = (1-\zeta)\, r^i + \zeta\, r^j \quad ; \quad r^i, r^j$ - Radial coordinates of nodes i and j

$$q^l_{e_n} = [\; u^i_n \;\; v^i_n \;\; w^i_n \;\; \frac{dw^i_n}{dS} \;\; u^j_n \;\; v^j_n \;\; w^j_n \;\; \frac{dw^j_n}{dS} \;]^T$$ - Vector of amplitudes displacement components in local coordinates

Substitution of equations (10) and (11) into equation (8) yields for the nth harmonic :

$$\epsilon_{m_n} = B^*_{m_n} \; q^l_{e_n} \tag{13}$$

being :

$$\epsilon_{m_n} = \begin{bmatrix} \overset{\circ}{\epsilon}_{SS_n} \\ \overset{\circ}{\epsilon}_{\theta\theta_n} \\ \overset{\circ}{\gamma}_{S\theta_n} \end{bmatrix} \quad ; \quad B^*_{m_n} = \begin{bmatrix} B_{11} & 0 & 0 & 0 & -B_{11} & 0 & 0 & 0 \\ B_{21} & B_{22} & B_{23} & B_{24} & B_{25} & B_{26} & B_{27} & B_{28} \\ B_{31} & B_{32} & 0 & 0 & B_{35} & B_{36} & 0 & 0 \end{bmatrix} \tag{14}$$

where $\overset{\circ}{\epsilon}_{SS_n}$, $\overset{\circ}{\epsilon}_{\theta\theta_n}$ and $\overset{\circ}{\gamma}_{S\theta_n}$ represent, respectively, the amplitudes for the nth harmonic of the meridional, circumferential and shearing strains of the reference surface. The $B^*_{m_n}$ are derived by applying the strain operator Δ_m to the displacement shape functions yielding :

$$B_{11} = -\frac{1}{\ell} \cos n\theta \qquad ; \qquad B_{21} = \frac{N_1}{r} \cos \phi \cos n\theta$$

$$B_{22} = \frac{N_1}{r} n \cos n\theta \qquad ; \qquad B_{23} = \frac{N_3}{r} \sin \phi \cos n\theta$$

$$B_{24} = \frac{N_4}{r} \sin \phi \cos n\theta \qquad ; \qquad B_{25} = \frac{N_2}{r} \cos \phi \cos n\theta$$

$$B_{26} = \frac{N_2}{r} n \cos n\theta \qquad ; \qquad B_{27} = \frac{N_5}{r} \sin \phi \cos n\theta \tag{15}$$

$$B_{28} = \frac{N_6}{r} \sin \phi \cos n\theta \qquad ; \qquad B_{31} = -\frac{N_1}{r} n \sin n\theta$$

$$B_{35} = -\frac{N_2}{r} n \sin n\theta \qquad ; \qquad B_{32} = -(\frac{1}{\ell} + \frac{N_1}{r} \cos \phi) \sin n\theta$$

$$B_{36} = (\frac{1}{\ell} - \frac{N_2}{r} \cos \phi) \sin n\theta$$

Considering the transformation matrix L relating nodal local coordinates (S, θ, ξ) to nodal global coordinates (r, θ, z) (Fig. 1), one obtains the relationship between the displacements amplitudes in the local referential $(q^l_{e_n})$ and the displacements amplitudes in the global referential (q_{e_n}) :

$$q^l_{e_n} = L \; q_{e_n} \tag{16}$$

where :

$$q_{e_n} = [\; u^i_{r_n} \;\; u^i_{z_n} \;\; u^i_{\theta_n} \;\; \frac{dw^i_n}{dS} \;\; u^j_{r_n} \;\; u^j_{z_n} \;\; u^j_{\theta_n} \;\; \frac{dw^j_n}{dS} \;]^T$$

$$L = \begin{bmatrix} [\; A \;] & [\; 0 \;] \\ [\; 0 \;] & [\; A \;] \end{bmatrix}$$

$$
[A] = \begin{bmatrix} \cos\phi & -\sin\phi & 0 & 0 \\ 0 & 0 & 1 & 0 \\ \sin\phi & \cos\phi & 0 & 0 \\ 0 & 0 & 0 & 1 \end{bmatrix} \quad ; \quad [0] = \begin{bmatrix} 0 & 0 & 0 & 0 \\ 0 & 0 & 0 & 0 \\ 0 & 0 & 0 & 0 \\ 0 & 0 & 0 & 0 \end{bmatrix}
$$

Substituting equation (16) into the strain-displacement relation (eq. 13), one obtains the membrane terms of the strain-displacement in terms of the element degrees of freedom of the nth harmonic as:

$$
\epsilon_{m_n} = B_{m_n} q_{e_n} \tag{17}
$$

where :

$$
B_{m_n} = B^*_{m_n} L \tag{18}
$$

For the bending terms the procedure is identical, yielding :

$$
\chi_{f_n} = B^*_{f_n} q^l_{e_n} \tag{19}
$$

where :

$$
\chi_{f_n} = \begin{bmatrix} \chi_{SS_n} \\ \chi_{\theta\theta_n} \\ \chi_{S\theta_n} \end{bmatrix} \quad ; \quad B^*_{f_n} = \begin{bmatrix} 0 & 0 & B_{13} & B_{14} & 0 & 0 & -B_{13} & B_{18} \\ 0 & B_{22} & B_{23} & B_{24} & 0 & B_{26} & B_{27} & B_{28} \\ 0 & B_{32} & B_{33} & B_{34} & 0 & B_{36} & B_{37} & B_{38} \end{bmatrix}
$$

and :

$$
B_{13} = \frac{(6-12\zeta)}{\ell^2}\cos n\theta \qquad ; \quad B_{14} = \frac{(4-6\zeta)}{\ell}\cos n\theta
$$

$$
B_{18} = \frac{(2-6\zeta)}{\ell}\cos n\theta \qquad ; \quad B_{22} = \frac{N_1\, n\sin\phi}{r^2}\cos n\theta
$$

$$
B_{23} = \left(\frac{n^2}{r^2}N_3 + \frac{\cos\phi}{\ell\, r}(6\zeta-6\zeta^2)\right)\cos n\theta \qquad ; \quad B_{24} = \left(\frac{n^2\,N_4}{r^2} + \frac{(-1+4\zeta-3\zeta^2)\cos\phi}{r}\right)\cos n\theta
$$

$$
B_{26} = \frac{N_2\, n\sin\phi}{r^2}\cos n\theta \qquad ; \quad B_{27} = \left(\frac{n^2}{r^2}N_5 - \frac{\cos\phi}{\ell\, r}(6\zeta-6\zeta^2)\right)\cos n\theta
$$

$$
B_{28} = \left(\frac{n^2\,N_6}{r^2} + \frac{(2\zeta-3\zeta^2)\cos\phi}{r}\right)\cos n\theta \qquad ; \quad B_{32} = \left(-\frac{2\sin\phi}{r\,\ell} - \frac{2\,N_1\cos\phi\sin\phi}{r^2}\right)\sin n\theta
$$

$$
B_{33} = \left(-\frac{2\,n}{r^2}N_3\cos\phi - \frac{2\,n}{\ell\, r}(6\zeta-6\zeta^2)\right)\sin n\theta \quad ; \quad B_{36} = \left(\frac{2\sin\phi}{r\,\ell} - \frac{2\,N_2\sin\phi\cos\phi}{r^2}\right)\sin n\theta
$$

$$
B_{34} = \left(-\frac{2\,n\,N_4}{r^2}\cos\phi + \frac{2\,n\,(1-4\zeta+3\zeta^2)}{r}\right)\sin n\theta \quad ; \quad B_{37} = \left(\frac{2\,n}{\ell\, r}(6\zeta-6\zeta^2) - \frac{2\,n}{r^2}N_5\cos\phi\right)\sin n\theta
$$

$$
B_{38} = \left(\frac{2\,n\,(-2\zeta+3\zeta^2)}{r} - \frac{2\,n\,N_6}{r^2}\cos\phi\right)\sin n\theta
$$

Similarly if one considers the transformation of coordinates (L) of the local referential (S, θ, ξ) to the global referential (r, θ, z), the changes of curvatures for the nth harmonic can be represented as :

$$\chi_{f_n} = B_{f_n} \, q_{e_n} \tag{20}$$

where :

$$B_{f_n} = B_{f_n}^* \, L \tag{21}$$

2.2 - CONSTITUTIVE EQUATIONS

For a linear elastic solid, considering infinitesimal deformation and orthotropic materials the constitutive equation can be written as :

$$\underset{\sim}{\tau} = \underset{\sim}{C} \, \underset{\sim}{\epsilon} \tag{22}$$

where :

$$\underset{\sim}{\tau} = [\, \tau_{SS} \; \tau_{\theta\theta} \; \tau_{\xi\xi} \; \tau_{S\theta} \; \tau_{S\xi} \; \tau_{\theta\xi} \,]^T$$

$$\underset{\sim}{\epsilon} = [\, \epsilon_{SS} \; \epsilon_{\theta\theta} \; \epsilon_{\xi\xi} \; \epsilon_{S\theta} \; \epsilon_{S\xi} \; \epsilon_{\theta\xi} \,]^T$$

and $\underset{\sim}{C}$ is the constitutive matrix for the three-dimensional linear elastic solid.

As a consequence of the Love theory expressed by $\gamma_{S\xi}=\gamma_{\theta\xi}=\epsilon_{\xi\xi}=\tau_{\xi\xi}=0$, the system of stress-strain relations for thin orthotropic axisymmetric shells can be reduced to :

$$\tau = C \, \epsilon \tag{23}$$

where:

$$\tau = [\tau_{SS} \; \tau_{\theta\theta} \; \tau_{S\theta} \,]^T$$

$$C = \begin{bmatrix} E_S^* & \nu_{\theta S} \, E_\theta^* & 0 \\ \nu_{S\theta} \, E_\theta^* & E_\theta^* & 0 \\ 0 & 0 & G_{S\theta} \end{bmatrix}$$

$$E_\theta^* = \frac{E_\theta}{1 - \nu_{S\theta} \, \nu_{\theta S}} \quad ; \quad E_S^* = \frac{E_S}{1 - \nu_{S\theta} \, \nu_{\theta S}}$$

where E_S, E_θ , $G_{S\theta}$, $\nu_{S\theta}$, $\nu_{\theta S}$, are Young's moduli, shear modulus and Poisson's ratio for the material referred to the S and θ directions. Substitution of equations (7) in equations (23) yields :

$$\tau = C \, \epsilon^\circ + \xi \, C \, \chi \tag{24}$$

Integrating the stress distribution across the thickness of the shell by neglecting $\xi/R_i \simeq 0$ one obtains :

$$\tau = \frac{1}{h} \mathcal{N} + \xi \frac{12}{h^3} \mathcal{M} \tag{25}$$

where:

$$\mathcal{N} = [\ \mathcal{N}_{SS} \ \mathcal{N}_{\theta\theta} \ \mathcal{N}_{S\theta} \ ]^T \qquad \text{(Membrane resultants)}$$

$$\mathcal{M} = [\ \mathcal{M}_{SS} \ \mathcal{M}_{\theta\theta} \ \mathcal{M}_{S\theta} \ ]^T \qquad \text{(Bending moments)}$$

and :

$$\mathcal{N} = \int_{-h/2}^{h/2} \tau \ d\xi \ = D_m \sum_{n=0}^{N} \epsilon_{m_n} \tag{26}$$

$$\mathcal{M} = \int_{-h/2}^{h/2} \tau \ \xi \ d\xi \ = D_f \sum_{n=0}^{N} \chi_{f_n} \tag{27}$$

where D_m and D_f are the membrane and bending constitutive matrices given by :

$$D_m = h \ C \qquad ; \qquad D_f = \frac{h^3}{12} C$$

2.3 - KINETIC ENERGY

The kinetic energy T^e of the eth element is given by the expression :

$$T^e = \frac{1}{2} \int \rho \ \underline{\dot{U}}^T \ \underline{\dot{U}} \ d\Omega = \frac{1}{2} \int_0^{2\pi} \int_0^1 \int_{-h/2}^{h/2} \rho \ \underline{\dot{U}}^T \ \underline{\dot{U}} \ \ell \ r \ d\xi \ d\zeta \ d\theta \tag{28}$$

where $\underline{\dot{U}} = \frac{d\underline{U}}{dt}$, $d\Omega$ is the elementary volume and ρ is the mass per unit of volume. Substituting equation (5) into equation (6) and integrating over the thickness of the element one obtains :

$$T^e = \frac{1}{2} \rho \int_0^{2\pi} \int_0^1 h \ \underline{\dot{U}}^T \ \underline{\dot{U}} \ \ell \ r \ d\zeta \ d\theta + \frac{1}{2} \frac{\rho}{12} \int_0^{2\pi} \int_0^1 h^3 \ \dot{\kappa}^T \ \dot{\kappa} \ \ell \ r \ d\zeta \ d\theta \tag{29}$$

where :

$$\underline{\dot{U}} = \frac{d}{dt}[\ u \quad v \quad w\ ]^T$$

$$\dot{\kappa} = \frac{d}{dt} [\ -\frac{\partial w}{\partial S} \quad \frac{1}{r}(v \sin \phi - \frac{\partial w}{\partial \theta}) \quad 0\ ]^T$$

In equation (29) the first term represents the translational kinetic inertia and the second term represents the rotational kinetic inertia. Its important to note that the coupled terms of

Fourier series of the type ($\sin m\theta \; \sin n\theta$) and ($\cos m\theta \; \cos n\theta$) for $m \neq n$ are not considered by making use of the orthogonality of the harmonic functions in the interval $0 \leq \theta \leq 2\pi$ and because the thickness is uniform along θ, ($h(\theta)$=constant), those terms are equal to zero resulting for the integration in θ :

$$\int_0^{2\pi} \cos n\theta \; \cos m\theta \; d\theta = \mathcal{A}_1$$

$$\int_0^{2\pi} \cos n\theta \; \sin m\theta \; d\theta = 0 \qquad \begin{array}{ll} \mathcal{A}_1 = \mathcal{A}_2 = 0 & \text{for } m \neq n \\ \mathcal{A}_1 = 2\pi \; ; \; \mathcal{A}_2 = 0 & \text{for } m=n=0 \\ \mathcal{A}_1 = \mathcal{A}_2 = \pi & \text{for } m=n>0 \end{array} \qquad (30)$$

$$\int_0^{2\pi} \sin n\theta \; \sin m\theta \; d\theta = \mathcal{A}_2$$

Substituting equations (6), (10) and (11) in the equation (29) for the frustum-cone finite element one obtains :

$$T^e = \tfrac{1}{2}\rho \sum_{n=0}^{N} \int_0^{2\pi} \int_0^1 h \, (N_n^* \, L \, \dot{q}_{e_n})^T \, (N_n^* \, L \, \dot{q}_{e_n}) \, \ell \, r \, d\zeta \, d\theta +$$

$$+ \tfrac{1}{2} \tfrac{\rho}{12} \sum_{n=0}^{N} \int_0^{2\pi} \int_0^1 h^3 \, (R_n^* \, L \, \dot{q}_{e_n})^T \, (R_n^* \, L \, \dot{q}_{e_n}) \, \ell \, r \, d\zeta \, d\theta \qquad (31)$$

where :

$$\dot{q}_{e_n} = \tfrac{d}{dt} \, q_{e_n} \; ; \quad N_n^* = C_n \, N \quad \text{and} \quad R_n^* = C_n \, R_n$$

in which :

$$R_n = \begin{bmatrix} 0 & 0 & \dfrac{-\partial N_3}{\partial S} & \dfrac{-\partial N_4}{\partial S} & 0 & 0 & \dfrac{-\partial N_5}{\partial S} & \dfrac{-\partial N_6}{\partial S} \\[2mm] 0 & \dfrac{N_1}{r}\sin\phi & \dfrac{n\,N_3}{r} & \dfrac{n\,\ell\,N_4}{r} & 0 & \dfrac{N_2}{r}\sin\phi & \dfrac{n\,N_5}{r} & \dfrac{n\,\ell\,N_6}{r} \\[2mm] 0 & 0 & 0 & 0 & 0 & 0 & 0 & 0 \end{bmatrix}$$

The kinetic energy can be represented in a simplified form as :

$$T^e = \tfrac{1}{2} \sum_{n=0}^{N} \left(\dot{q}_{e_n}^T \left(M_{n_T}^e + M_{n_R}^e \right) \dot{q}_{e_n} \right) = \tfrac{1}{2} \sum_{n=0}^{N} \left(\dot{q}_{e_n}^T \, M_n^e \, \dot{q}_{e_n} \right) \qquad (32)$$

being M_n^e the element mass matrix for the nth harmonic represented by :

$$M_n^e = \rho \int_0^{2\pi} \int_0^1 h \, (N_n^* \, L)^T \, (N_n^* \, L) \, \ell \, r \, d\zeta \, d\theta + \tfrac{\rho}{12} \int_0^{2\pi} \int_0^1 h^3 \, (R_n^* \, L)^T \, (R_n^* \, L) \, \ell \, r \, d\zeta \, d\theta \qquad (33)$$

The element mass matrices M_n^e are full (8x8) matrices which are evaluated taking into consideration the orthogonality properties of the trigonometric functions (eq. 30). The simplest mathematical model for inertia properties of structural elements is the lumped-mass representation. In this model the mass properties of the system are separated from the elastic properties and equivalent concentrated masses are placed at the nodal points to represent the inertia forces in the direction of the assumed element degree of freedom. These masses refer to both translation and rotational inertia of the element. This assumption excludes dynamic coupling between the element displacement and the resulting mass matrix is purely diagonal, thus implying a computer time reduction and a decrease in computer storage. The diagonal mass matrix is obtained from the full consistent mass matrix by adding the off diagonal terms to the appropriate diagonal elements (e.g. Desai and Abel, 1972).

2.4 - STRAIN ENERGY

The strain energy of the eth element is represented by the expression :

$$E^e = \frac{1}{2} \int_\Omega \tau^T \, \epsilon \, d\Omega = \frac{1}{2} \int_0^{2\pi} \int_0^1 \int_{-h/2}^{h/2} \epsilon^T \, C \, \epsilon \; r \, \ell \, d\xi \, d\zeta \, d\theta \tag{34}$$

Introducing equation (7) in the above equation and taking advantage of the orthogonality properties of the trigonometric functions one obtains :

$$E^e = \frac{1}{2} \sum_{n=0}^{N} q_{e_n}^T \, K_n^e \, q_{e_n} \tag{35}$$

with :

$$K_n^e = K_{m_n}^e + K_{f_n}^e \tag{36}$$

$$K_{m_n}^e = \int_0^{2\pi} \int_0^1 B_{m_n}^T \, D_m \, B_{m_n} \; r \, \ell \, d\zeta \, d\theta \tag{37}$$

$$K_{f_n}^e = \int_0^{2\pi} \int_0^1 B_{f_n}^T \, D_f \, B_{f_n} \; r \, \ell \, d\zeta \, d\theta \tag{38}$$

where $K_{m_n}^e$ and $K_{f_n}^e$ are the membrane and bending terms of the element stiffness matrix for the nth harmonic. These matrices are evaluated analytically in the θ direction taking into consideration the orthogonality properties of the trigonometric functions. In the ζ direction the integration is carried out by Gaussian formulae.

2.5 - EXTERNAL WORK

Expanding the surface loads $\underline{p} = [p_u \ p_\theta \ p_w]^T$ in Fourier series and using the assumption of orthogonality of the trigonometric functions on the interval $0 \le \theta \le 2\pi$, the external work becomes :

$$W^e = \int_0^{2\pi} \int_0^1 \mathbf{U}^T \underline{p} \ r \ \ell \ d\zeta \ d\theta \tag{39}$$

where:

$$\underline{p} = \sum_{n=0}^N C_n \underline{p}^n \quad ; \quad \underline{p}^n = [\ p_u^n \ p_\theta^n \ p_w^n\ ]^T$$

Assuming a linear dependence between the above magnitudes of the forces vector and the components of these vector on the nodes of the frustum-cone finite element, one obtains :

$$\underline{p}^n = \mathfrak{N} \ p_{e_n}^l \tag{40}$$

where :

$$p_{e_n}^l = [\ p_{u_i}^n \ p_{\theta_i}^n \ p_{w_i}^n \ M_i^n \ p_{u_j}^n \ p_{\theta_j}^n \ p_{w_j}^n \ M_j^n\ ]^T$$

$$\mathfrak{N} = \begin{bmatrix} N_1 & 0 & 0 & 0 & N_2 & 0 & 0 & 0 \\ 0 & N_1 & 0 & 0 & 0 & N_2 & 0 & 0 \\ 0 & 0 & N_1 & 0 & 0 & 0 & N_2 & 0 \end{bmatrix} \tag{41}$$

Considering the equations (6), (10), (11) and (16) and substituting in equation (39) on obtains :

$$W^e = \sum_{n=0}^N \int_0^{2\pi} \int_0^1 \left((C_n \ N \ L \ q_{e_n})^T \ C_n \ \mathfrak{N} \ p_{e_n}^l \right) r \ \ell \ d\zeta \ d\theta \tag{42}$$

Using $N_n^* = C_n \ N$ and making $\mathfrak{N}_n^* = C_n \ \mathfrak{N}$, yields :

$$W^e = \sum_{n=0}^N q_{e_n}^T \ Q_n^e \tag{43}$$

where :

$$Q_n^e = \int_0^{2\pi} \int_0^1 L^T \ N_n^{*T} \ \mathfrak{N}_n^* \ p_{e_n}^l \ r \ \ell \ d\zeta \ d\theta \tag{44}$$

is the load vector of the element for the nth harmonic which is consistent with the assumed displacement field used in deriving mass and stiffness matrices. Vectors Q_n^e are evaluated analytically in the θ direction taking into consideration the orthogonality properties of the trigonometric functions. In the ζ direction the integration may be carried out by Gaussian formulae or alternatively by using a symbolic manipulator.

2.6 - EQUATIONS OF MOTION

Problems of static equilibrium are governed by the variational principles such a minimum total potential energy, while the dynamic equilibrium are most simply formulated in terms of Hamilton's variational principle, defining a Lagrangian function through the expression :

$$\mathcal{L} = T^e - V^e \tag{45}$$

with :

$$V^e = E^e - W^e \tag{46}$$

Substituting the values of T^e, E^e and W^e given by the equations (32), (35) and (43) in the Lagrangian function $\mathcal{L}$, one obtains :

$$\mathcal{L} = \sum_{n=0}^{N} \left(\tfrac{1}{2}\left(\dot{q}_{e_n}^T\, M_n^e\, \dot{q}_{e_n} - q_{e_n}^T\, K_n^e\, q_{e_n} \right) + q_{e_n}^T\, Q_n^e \right) \tag{47}$$

Applying the appropriate Lagrange equations of the motion for equilibrium yields for the element and for the nth harmonic :

$$M_n^e\, \ddot{q}_{e_n} + K_n^e q_{e_n} = Q_n^e \tag{48}$$

where:

$$\ddot{q}_{e_{n'}} = \frac{d^2(q_{e_n})}{dt^2}$$

For the system these equations become :

$$M\,\ddot{q} + K\,q = Q \tag{49}$$

where M, K, q and Q are, respectively, the mass and stiffness matrices, the displacement vector and the load vector of the nth harmonic.

Assuming free undamped vibration the static and dynamic equilibrium equations for the nth harmonic can be represented symbolically, respectively, as :

$$K\,q = Q \tag{50}$$

$$K\,q = \omega^2\, M\, q \tag{51}$$

where ω is the natural frequency.

3 - SENSITIVITY ANALYSIS

3.1 - INTRODUCTION

The evaluation of sensitivities of structural response to changes in design variables is a crucial stage in the optimal design of complex structures, representing a major factor with regard to the computer time required for the optimization process. Hence it is important to have efficient techniques to calculate these derivatives. The simplest technique of evaluating sensitivities of response with respect to changes in design variables is through the finite difference approximation, here called global finite difference, which is computationaly expensive or through the use of semi-analytical or analytical methods as described in the next section. These latter methods can both be applied through the direct or adjoint structure technique (Haftka and Kamat, 1987). In this section the formulation of the evaluation of sensitivities of axisymmetric shells is presented for the particular case of axisymmetric loading and boundary conditions. For this particular situation the element degrees of freedom of the frustum element are reduced from 8 to 6 degrees of freedom, since $n=0$ and $v_n = 0$.

More numerically based solutions are reported by Marcelin and Trompette (1988) using a finite element with a two node straight element and/or a three node parabolic element also based on Love-Kirchhoff shell theory and a semi-analytical method to evaluate the sensitivities. Other authors, such a Plaut et al (1984) and Chenais (1987), present alternative theories and models for the optimization of shell structures. More recently, Mehrez and Rousselet (1989) evaluate the analysis and optimization of shells of revolution using Koiter's model with the implementation of B-Splines for the middle surface and finite element for displacements.

3.2 - ADJOINT STRUCTURE METHOD

3.2.1. - STATICS

A typical optimization constraint such as a limit on a displacement, stress component or effective stress can be represented by :

$$g_j = g_j (q ; b) \leq 0 \tag{52}$$

where b is the vector of design variables and $j \in (1, ... ,m)$, m being the number of constraints. The sensitivity of the constraint g_j is :

where :

$$\frac{dg_j}{db_i} = \frac{\partial g_j}{\partial b_i} + Z_j^T \frac{dq}{db_i} \tag{53}$$

$$Z_j = \frac{\partial g_j}{\partial q} \tag{54}$$

is the vector of adjoint forces.

Thus in the method of the adjoint structure a virtual structure is defined that satisfies the equilibrium equation :

$$K \lambda_j = Z_j \tag{55}$$

where λ_j is the system adjoint degrees of freedom for the constraint g_j . The solution of the system equation (55) gives λ_j. It should be noted that the adjoint structure is identical to the real structure, but subjected to a different load. To increase computational efficiency, the already factorized form of the stiffness matrix should be used.

Considering the static equilibrium equations (50) and differentiating these with respect to a design variable b_i :

$$K \frac{\partial q}{\partial b_i} + \frac{\partial K}{\partial b_i} q = \frac{\partial Q}{\partial b_i} \tag{56}$$

Premultiplying by Z_j^T one obtains :

$$Z_j^T \frac{\partial q}{\partial b_i} = Z_j^T K^{-1} \left(\frac{\partial Q}{\partial b_i} - \frac{\partial K}{\partial b_i} q \right) \tag{57}$$

The inversion of the stiffness matrix K is easily avoided using the adjoint structure method through the solution of equation (55). Thus, the sensitivities given by equation (53) can be evaluated for the harmonic n=0, as :

$$\frac{dg_j}{db_i} = \frac{\partial g_j}{\partial b_i} + \lambda_j^T \left(\frac{\partial Q}{\partial b_i} - \frac{\partial K}{\partial b_i} q \right) \tag{58}$$

where $\frac{\partial K}{\partial b_i}$ is the sensitivity of the system stiffness matrix and $\frac{\partial Q}{\partial b_i}$ is the sensitivity of the system load vector. When the forces are independent of the design variables the sensitivity of the system load vector is zero and then equation (58) simplifies to :

$$\frac{dg_j}{db_i} = \frac{\partial g_j}{\partial b_i} - \lambda_j^T \frac{\partial K}{\partial b_i} q \tag{59}$$

The term $\frac{\partial g_j}{\partial b_i}$ is usually zero or can easily be obtained.

3.2.2. - DYNAMICS

Considering the mode of vibration q_k which corresponds to the natural frequency ω_k, the eigenvalue problem, equation (51), is represented for the system as :

$$K\, q_k = \omega_k^2\, M\, q_k \tag{60}$$

Differentiating the above equation with respect to a design variable b_i and premultiplying by q_k^T one obtains :

$$2\,\omega_k\, \frac{\partial \omega_k}{\partial b_i}\, q_k^T\, M\, q_k = q_k^T \left(\left(\frac{\partial K}{\partial b_i} - \omega_k^2 \frac{\partial M}{\partial b_i} \right) q_k + (K - \omega_k^2\, M)\, \frac{\partial q_k}{\partial b_i} \right) \tag{61}$$

Considering the modal normalization $q_k^T\, M\, q_k = 1$, the sensitivity of the natural frequency corresponding to mode k with respect to changes in design variables is given by :

$$\frac{\partial \omega_k}{\partial b_i} = \frac{1}{2\omega_k}\, q_k^T \left(\frac{\partial K}{\partial b_i} - \omega_k^2 \frac{\partial M}{\partial b_i} \right) q_k \tag{62}$$

where $\frac{\partial M}{\partial b_i}$ is the system mass sensitivity matrix. Thus to evaluate the sensitivity of natural frequencies with respect to changes in the design variables there is no need to define an adjoint structure.

3.3 - SENSITIVITY ANALYSIS OF AXISYMMETRIC SHELLS

3.3.1 - Analytical Method

The analytical derivative of the element stiffness matrix (eqs. 37 and 38) with respect to a variable b_i^* (not necessary a design variable) can be represented in a symbolic form as :

$$\frac{\partial K^e}{\partial b_i^*} = 2\pi \int_0^1 \left\{ \left(\left(B^T D \frac{\partial B}{\partial b_i^*} \right)^T + \left(B^T D \frac{\partial B}{\partial b_i^*} \right) + \left(B^T \frac{\partial D}{\partial b_i^*} B \right) \right) r\,\ell + \left(B^T D B \right) \left(\ell \frac{\partial r}{\partial b_i^*} + r \frac{\partial \ell}{\partial b_i^*} \right) \right\} d\zeta \tag{63}$$

where:
$$D = \begin{bmatrix} D_m & 0 \\ 0 & D_f \end{bmatrix} \quad ; \quad B = \begin{bmatrix} B_{m_n} \\ B_{f_n} \end{bmatrix}$$

The derivative of the element force vector is :

$$\frac{\partial Q^e}{\partial b_i^*} = 2\pi \int_0^1 \left\{ \left(\left(\frac{\partial L^T}{\partial b_i^*} N^{*T} \underline{p} \right) + \left(L^T \frac{\partial N^{*T}}{\partial b_i^*} \underline{p} \right) \left(L^T N^{*T} \frac{\partial \underline{p}}{\partial b_i^*} \right) \right) r\,\ell + \left(L^T N^{*T} \underline{p} \right) \left(\frac{\partial r}{\partial b_i^*} \ell + r \frac{\partial \ell}{\partial b_i^*} \right) \right\} d\zeta \tag{64}$$

The derivatives of the stiffness matrix are evaluated at each Gaussian point, separately for membrane and bending. When the design variables are radial coordinates, the derivatives of equation (63) and the integration of equations (64) and their derivatives $\partial Q^e / \partial b_i^*$ are carried out using the symbolic manipulator MATHEMATICA (e.g. Wolfram, 1988). Full details are presented by Barbosa (1990).

For the particular case of a frustum-cone finite element the sensitivities for the harmonic $n=0$ of stiffness matrix K^e or mass matrix M^e, are obtained easily when the design variables are thicknesses, using equations (33), (37) and (38). In fact the mass matrix M^e depends explicitly on the thickness while for the stiffness matrix K^e the dependence is only on constitutive matrices D_m and D_f When the thickness is assumed constant within the element one obtains :

$$\frac{\partial K^e}{\partial h} = 2\pi \int_0^1 \left(B_m^T \frac{\partial D_m}{\partial h} B_m + B_f^T \frac{\partial D_f}{\partial h} B_f \right) r \; \ell \; d\zeta \tag{65}$$

$$\frac{\partial M^e}{\partial h} = 2\pi \int_0^1 \left(\left(\tfrac{1}{h}\right) \rho \, h \, (N^* L)^T \, (N^* L) + \left(\tfrac{3}{h}\right) \frac{\rho \, h^3}{12} (R^* L)^T \, (R^* L) \right) r \; \ell \; d\zeta \tag{66}$$

or :

$$\frac{\partial K^e}{\partial h} = \tfrac{1}{h} K_m^e + \tfrac{3}{h} K_f^e \tag{67}$$

$$\frac{\partial M^e}{\partial h} = \tfrac{1}{h} M_T^e + \tfrac{3}{h} M_R^e \tag{68}$$

Considering that the load vector Q^e is independent of the thickness, it becomes :

$$\frac{\partial Q^e}{\partial h} = 0 \tag{69}$$

For nodal coordinates or when the thickness distribution varies within the element, the shape of the model is related through the linking relation (Vanderplaats, 1984) :

$$F = F^c + T b \tag{70}$$

where F is the vector of dependent variables (thicknesses and/or radial nodal coordinates of the finite element model), T the linking matrix which relates the vector of shape design variables b with the dependent variables and F^c a vector of constant terms.

The derivatives of the stiffness matrix with respect to a shape design variable required by equation (58) can be evaluated using relations (63) and the chain rule of differentiation as :

$$\frac{\partial K^e}{\partial b_i} = \frac{\partial K^e}{\partial b_j^*} \frac{\partial b_j^*}{\partial b_i} \qquad i = 1, n \quad ; \quad j = 1, 2 \tag{71}$$

where b_j^* is the value of the element nodal variable concerned, yielding :

$$\frac{\partial K^e}{\partial b_i} = \frac{\partial K^e}{\partial b_j^*} T_{ij}^e \qquad i = 1, n \quad ; \quad j = 1, 2 \tag{72}$$

where T_{ij}^e is related to the linking matrix T (eq. 70) through the topological finite element code procedure. The sensitivities of the element mass matrix or load vector are obtained in a similar way.

The sensitivities of a constraint function with respect to a design variable are evaluated at element level using equations (58) and (62), as :

$$\frac{dg_j}{db_i} = \frac{\partial g_j}{\partial b_i} + \sum_{\ell \in E} \lambda_j^{(\ell)T} \left(\frac{\partial Q^{e(\ell)}}{\partial b_i} - \frac{\partial K^{e(\ell)}}{\partial b_i} q^{e(\ell)} \right) \tag{73}$$

$$\frac{\partial g_j}{\partial b_i} = -\frac{1}{2\omega_o \omega_k} \sum_{\ell \in E} q_k^{e(\ell)T} \left(\frac{\partial K^{e(\ell)}}{\partial b_i} - \omega_k^2 \frac{\partial M^{e(\ell)}}{\partial b_i} \right) q_k^{e(\ell)} \tag{74}$$

being E the set of elements which are affected by the design variable b_i, $\lambda_j^{(\ell)}$ the vector of the adjoint displacement of the element (ℓ), ω_k the natural frequency of the vibrating mode q_k and ω_o the limiting natural frequency.

3.3.2 - Semi-Analytical Method

In this technique the vector of adjoint forces Z_j is obtained analytically and the gradients of equations (58) and (62), with terms of the type $\frac{\partial F}{\partial b_i}$, are evaluated by finite difference approximation, F being a function dependent on the design variables b.

The gradients $\frac{\partial F}{\partial b_i}$ can be evaluated through the approximations :

$$\frac{\partial F}{\partial b_i} \approx \frac{F(b + \Delta b) - F(b)}{\delta b_i} \qquad \text{(Forward difference - FFD)} \tag{75}$$

$$\frac{\partial F}{\partial b_i} \approx \frac{F(b + \Delta b) - F(b - \Delta b)}{2\delta b_i} \qquad \text{(Central difference - CFD)} \tag{76}$$

where $\Delta b = [0, ..., \delta b_i, ..., 0]$ and δb_i is a small perturbation. It should be noted that to evaluate $F(b + \Delta b)$, the coordinates perturbations δR due to a design perturbation δb_i must be calculated, which is done through the linking relation.

With regard to shape design variables and considering the linking relation (70), the sensitivities of the element stiffness, mass or load vector can also be obtained analytically through:

$$\frac{\partial F^e}{\partial b_i} = \frac{\partial F^e}{\partial r_1^e} \frac{\partial r_1^e}{\partial b_i} + \frac{\partial F^e}{\partial r_2^e} \frac{\partial r_2^e}{\partial b_i} = \frac{\partial F^e}{\partial r_1^e} T_{i1}^e + \frac{\partial F^e}{\partial r_2^e} T_{i2}^e \tag{77}$$

where F^e can be the stiffness and mass matrices or load vector of the eth element and r_j^e, z_j^e (j=1,2) are the coordinates of the two ring nodes of the frustum conical element.

For the semi-analytical method the gradients with respect to changes in nodal radius are evaluated at element level through forward finite difference :

$$\frac{\partial K^e}{\partial b_i} \approx \frac{K^e(r_1^e + \delta r_1^e , z_1^e ; r_2^e + \delta r_2^e , z_2^e) - K^e(r_1^e , z_1^e ; r_2^e , z_2^e)}{\delta b_i} \qquad \text{a)}$$

$$\frac{\partial M^e}{\partial b_i} \approx \frac{M^e(r_1^e + \delta r_1^e , z_1^e ; r_2^e + \delta r_2^e , z_2^e) - M^e(r_1^e , z_1^e ; r_2^e , z_2^e)}{\delta b_i} \qquad \text{b)} \qquad (78)$$

$$\frac{\partial Q^e}{\partial b_i} \approx \frac{Q^e(r_1^e + \delta r_1^e , z_1^e ; r_2^e + \delta r_2^e , z_2^e) - Q^e(r_1^e , z_1^e ; r_2^e , z_2^e)}{\delta b_i} \qquad \text{c)}$$

3.3.3 - Finite Difference Techniques

Alternatively a global finite difference approach can be used. In this case the sensitivity of a constraint with respect to a change δb_i in a design variable is given by :

$$\frac{dg_j}{db_i} \approx \frac{g_j(b_1, \dots , b_i + \delta b_i, \dots , b_m) - g_j(b)}{\delta b_i} \qquad (79)$$

which needs one extra structural analysis for each design variable, increasing the computational effort.

3.3.4 - Limit on Displacements

A constraint in a displacement is represented in normalized form by :

$$g_j = \frac{q_f}{q_o} - 1 \leq 0 \qquad (80)$$

where q_f is the real generalized displacement corresponding to the system degree of freedom f and q_o is the maximum admissible generalized displacement.

The vector of adjoint forces is :

$$Z_j = \left[\frac{\partial g_j}{\partial q_1} \dots \frac{\partial g_j}{\partial q_f} \dots \frac{\partial g_j}{\partial q_\eta} \right]^T = \left[0 \dots \frac{1}{q_o} \dots 0 \right]^T \qquad (81)$$

where η is the total number of degrees of freedom. It should be noted that the adjoint structure is identical with the real structure and it is subjected to a force or moment of intensity $1/q_o$ on the corresponding degree of freedom where the displacement or rotation is limited.

3.3.5 - Limit on Stresses

Limit on a stress component of the stress tensor or an effective stress is represented by

$$g_j = \bar{\sigma}/\sigma_o - 1 \leq 0 \tag{82}$$

where σ_o is the maximum allowable stress, which may be different for tension and compression, and $\bar{\sigma}$ is the stress component or the effective stress. For instance for a Von Mises effective stress, $\bar{\sigma}$ is defined by the following relation :

$$\bar{\sigma} = \left[(\tau_{SS}^m)^2 + (\tau_{\theta\theta}^m)^2 - \tau_{SS}^m \tau_{\theta\theta}^m + 3\tau_{S\theta}^m + (\tau_{SS}^f)^2 + (\tau_{\theta\theta}^f)^2 - \tau_{SS}^f \tau_{\theta\theta}^f + 3\tau_{S\theta}^f \right]^{\frac{1}{2}} \tag{83}$$

where :

$$\sigma^m = \frac{1}{h} D_m B_m q_e = A q_e \qquad \text{(Membrane and shear stresses)}$$

$$\sigma^f = \frac{6}{h^2} D_f B_f q_e = C q_e \qquad \text{(Bending stresses)}$$

$$\sigma^m = \left[\tau_{SS}^m \quad \tau_{\theta\theta}^m \quad \tau_{S\theta}^m \right]^T \quad ; \qquad \sigma^f = \left[\tau_{SS}^f \quad \tau_{\theta\theta}^f \quad \tau_{S\theta}^f \right]^T$$

$$A = \left[A_1 \quad A_2 \quad A_3 \right]^T \quad ; \qquad C = \left[C_1 \quad C_2 \quad C_3 \right]^T$$

For a stress constraint at a Gaussian point of the element (ℓ), which corresponds to adjoint load $z_j^{(\ell)}$, the system adjoint force vector is then :

$$z_j^T = \left[0 \cdots z_j^{(\ell)} \cdots 0 \right]^T \tag{84}$$

where :

$$z_j^{(\ell)} = \frac{1}{\sigma_o \bar{\sigma}} \left[A_1^T A_1 + A_2^T A_2 - \frac{1}{2} A_1^T A_2 - \frac{1}{2} A_2^T A_1 + 3 A_3^T A_3 + \right.$$

$$\left. + C_1^T C_1 + C_2^T C_2 - \frac{1}{2} C_1^T C_2 - \frac{1}{2} C_2^T C_1 + 3 C_3^T C_3 \right] q_e^{(\ell)} \tag{85}$$

Thus, for a pointwise limit on a stress, such as defined by equation (83), the design sensitivity of stress to thickness variation, evaluated by equation (59), yields :

$$\frac{dg_j}{dh_i} = \frac{1}{\sigma_o \bar{\sigma} h^{(\ell)}} \left[(\tau_{SS}^f)^2 + (\tau_{\theta\theta}^f)^2 - \tau_{SS}^f \tau_{\theta\theta}^f + 3\tau_{S\theta}^f \right] - \lambda_j^T \frac{\partial K}{\partial h_i} q \tag{86}$$

where $h^{(\ell)}$ is the thickness of the element (ℓ), where the constraint g_j has been imposed.

3.3.6 - Limit on Natural Frequencies

A constraint on the natural frequency of mode k can be easily evaluated once the eigenvalue problem, equation (62), is solved. There is no need to assemble the sensitivities of the system stiffness or mass matrices. Following the procedure described for the static case, one obtains for a normalized constraint $g_j = 1 - \omega_k/\omega_o \leq 0$ and for each element:

$$\frac{dg_j}{db_i} = - \frac{1}{2\,\omega_o\,\omega_k} \sum_{\ell \in E} q_k^{T(\ell)} \left(\frac{\partial K^{(\ell)}}{\partial b_i^{(\ell)}} - \omega_k^2 \frac{\partial M^{(\ell)}}{\partial b_i^{(\ell)}} \right) q_k^{(\ell)} \tag{87}$$

where ω_o is the limiting natural frequency for mode k.

4 - OPTIMAL DESIGN

The objective is to find b to minimize the volume V of the material of the shell structure. Alternatively, the objective is to maximize the natural frequency of the mode k of the structure assuming constant volume. The problem is stated as:

$$\min V\,(b) \qquad \text{or} \qquad \max \omega_k = \omega_k(q_k, b) \tag{88}$$

subjected to constraints :

$$g_j\,(b) \leq 0 \quad ; \quad j=1,2, \dots ,m \quad \text{or} \quad V-V_o = 0 \qquad \text{or} \qquad C-C_o = 0 \tag{89}$$

$$b_i^\ell \leq b_i \leq b_i^u \quad ; \quad i=1,2, \dots ,n \tag{90}$$

where b_i^ℓ and b_i^u are, respectively, the lower and upper limiting values of the design variables, V_o and V the initial and final volume of the shell material, C_o and C the initial and final enclosed volume of the structure and ω_k is the natural frequency of mode k. The number of design variables is designated by n. The gradients of the constraints are evaluated using the analytical sensitivity analysis formulation for limits on displacements, stresses or natural frequencies when the design variables are thicknesses and the analytical, or alternatively the semi-analytical, formulation when design variables are the radial coordinates. The optimal design problem is solved by the techniques of nonlinear mathematical programming described by Vanderplaats (1984) and the algorithms of the ADS program.

5 - APPLICATIONS

A computer program for personal computers has been developed based on the formulation presented and using the modified feasible direction method of the ADS program for the optimization. As examples circular plates and shells structures are optimized.

5.1 - Clamped Circular Plate - Weight minimization

The load, geometric and material properties of the initial design (Fig. 2) are :

p_z = 0.0689 MPa (normal pressure) ; a = 1 m (radius) ; h = 25 mm (thickness);
E = 200 GPa (Elasticity modulus) ; ν = 0.3 (Poisson's ratio) ;
ρ=7800 Kg.m^{-3} (mass density).

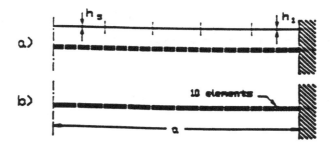

Fig. 2 - Initial design

Five thicknesses are considered as design variables . The plate has been modelled with 20 finite elements. Predictions of design sensitivities due to thickness variation and constraints of displacement at r=0, stress τ^f_{SS} at r=0.995m and fundamental frequency are shown in Table I, for design variables 1 and 5. They are compared with global finite differences solutions obtained by the same finite element model considering a slightly perturbed design. The difference in the value of the constraint function divided by Δh is the approximate sensitivity. The agreement between the proposed solutions and the alternative results is quite favourable.

TABLE I - Sensitivities due to Thickness variation

	Deflection at r= 0		Stress τ_{SS}^{f} at r=0.995m		Fundamental natural frequency	
	dg_j/dh_1	dg_j/dh_5	dg_j/dh_1	dg_j/dh_5	$\partial\omega/\partial h_1$	$\partial\omega/\partial h_5$
Analytical	- 0.2065	- 0.0742	- 4.1484 - 4.7859*	- 0.2585 - 0.2578*	11171	820
Finite differences Δh=0.25 mm	- 0.2038	-0.0735	- 4.9600 - 4.9648*	- 0.2488 - 0.2540*	11134	829

* - Model (b)

The objective of the design is to minimize the volume of the material of the circular plate. The plate has been firstly optimized considering deflection and stress limits, respectively, of w_o =5.5 mm and σ_o =120 MPa. The upper limiting bound of the design variables has been set at h $\leq$ 25 mm. The optimal design shape using model (a) is shown in Fig. 3. The number of iterations is 13. The number of function evaluations is 145 and the gradient evaluations are 13. With regard to the initial design, a reduction in volume of 20% was found. Only the displacement constraint is activated. The maximum τ_{SS}^{f} reached 112.3 MPa at r=0.425 m.

The plate has been also optimized with regard to a limit (ω_o = 40 Hz) in the natural frequency of the fundamental mode. The optimal design is shown in Fig. 4. The numerical solution has been obtained in 4 iterations, 16 function evaluations and 2 gradient evaluations. The volume reduction is 31% with reference to the initial design.

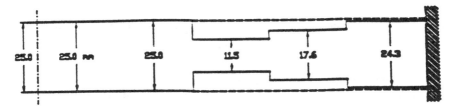

Fig. 3 - Optimal design - Static constraints

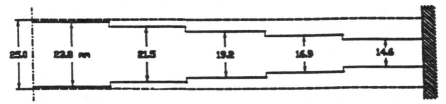

Fig. 4 - Optimal design - Dynamic constraint

5.2 - Clamped Circular Plate - Natural Frequency Maximization

The geometric and material properties of the initial design (Fig. 5) are :

a = 1 m (radius) ; h = 25 mm (thickness); E = 200 GPa (Elasticity modulus) ;
ν = 0.3 (Poisson's ratio) ; ρ=7800 Kg.m^{-3} (mass density).

The objective is to maximize the fundamental natural frequency without increasing the mass
of the plate. Five design variables (thicknesses) are considered. The plate has been modelled with
20 finite elements. The optimal thicknesses of the plate are shown in Fig. 5.

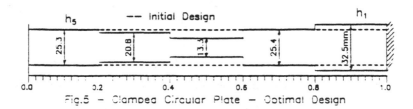

Fig.5 - Clamped Circular Plate - Optimal Design

The number of iterations, function evaluations and gradient evaluations are 5, 16, 5,
respectively. The natural frequency of the fundamental mode is ω_0 = 62.2 Hz in the initial
design. In the optimal design it is found to be ω_0= 72.5 Hz and the volume of the material is the
same.

5.3 - Cylindrical Tank

A cylindrical tank clamped at the bottom is submitted to the action of a liquid. The
geometric and material properties of the initial design are :

a = 20 m ; L = 10 m (Depth of tank) ; h = 0.813 m (initial constant thickness);
E = 28 GPa ; ν = 1/6 ; γ = 9.81 KN.m^{-3} (liquid weight per unit volume).

A finite element model (Fig. 6) with 46 elements and 14 design variables has been
considered. The meridional stress τ_{ss} and the membrane circumferential stress $\tau_{\theta\theta}^m$ has been
constrained to σ_0=3.6 MPa and σ_0 = 1.5 MPa.

The objective of the design is to minimize the volume of the material of the cylindrical tank.
The number of iterations, function evaluations and gradient evaluations were 10, 79 and 8,
respectively. The results of the optimum design are shown in Fig. 6. For the optimal design the
maximum stresses are in the bottom of the tank for τ_{ss} and at the height of 0.525m for $\tau_{\theta\theta}^m$. A

reduction in volume of 27% has been reached between the initial and optimal designs.

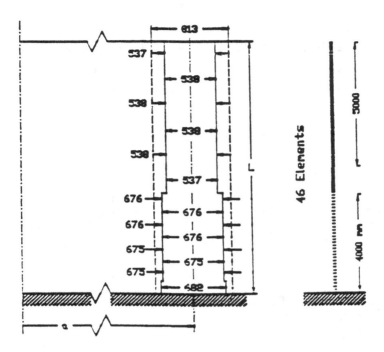

Fig. 6 - Cylindrical Tank - Model and Optimal Design

5.4 - Conical Structure Clamped at Lower End

The geometric and material properties of the initial design (fig. 7) are :

$R_1 = 1.25$ m ; $R_2 = 2.768$ m ; $L = 4.17$ m ;
$h = 62.5$ mm (Thickness) ; $E = 30$ GPa (Young's modulus) ;
$\nu = 0.15$ (Poisson's ratio) ; $\rho = 2410$ Kg.m^{-3} (mass per unit volume).

The objective of the design is to maximize the fundamental natural frequency of the structure without increasing its mass. As design variables 6 thicknesses (h_1-h_6) are considered. The structure is discretized using three models with 30, 60 and 120 frustum-cone finite elements, respectively. In Table II the analytical sensitivities of the design variables h_1 to h_6 are presented, in the initial design with constant thickness for the three models used.

In Fig. 8 the optimal design for model with 60 elements is represented and in Table III the thicknesses found for the three models are shown. The number of iterations, function evaluations and gradient evaluations are, respectively, 5, 22 and 3 for the models with 30 and 60 elements and 4, 19 and 2 for the model with 120 elements. A rapid convergence as shown in Fig.9 is verified, where for the model with 60 elements the final value of the objective function is found almost in the first iteration. In the initial design and for the three models the fundamental natural frequency is $\omega_o = 926$ rad/s . For the models of 30, 60 and 120 elements in the optimal design the frequency is, respectively, 1135 rad/s, 1184 rad/s and 1197 rad/s. The volume of the material of the structure is constant. There is a good agreement between the models of 60 and 120 elements and an alternative model using the 3D axisymmetric shell finite element presented by Thambiratnam et al (1989). The model with 30 elements can not represent the structure adequately.

Table II - Sensitivities due to thickness variation

Model	Design Variables					
	h_1	h_2	h_3	h_4	h_5	h_6
30 Elements	-1928.5	-1277.5	-398.3	546.0	1428.7	2124.1
60 Elements	-1928.2	-1276.0	-397.7	547.7	1431.7	2128.6
120 Elements	-1926.9	-1274.8	-397.0	548.4	1432.6	2129.9

Table III - Thicknesses (mm) in optimal design

Model	Design Variables					
	h_1	h_2	h_3	h_4	h_5	h_6
30 Elements	34.8	28.6	54.7	79.8	101.4	117.0
60 Elements	15.7	26.7	54.7	85.1	113.3	135.5
120 Elements	8.6	29.1	56.9	86.9	115.0	136.9

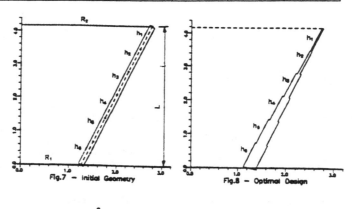

Fig.7 — Initial Geometry   Fig.8 — Optimal Design

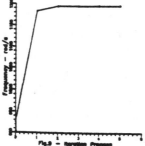

Fig.9 — Iteration Process

5.5 - Simply Supported Cone-Cylinder Connection with Internal Pressure

The geometric and material properties are:

R=1.0m (Cylinder radius),   H=0.6m (Height of cylinder),
h=0.010m (thickness),   E=200GPa, ν=0.3.

A finite element model with 50 elements is considered (Fig. 10). The design variables are 6 radial coordinates (b_1, ... , b_6). Tables IV and V show the sensitivities for the initial design with respect to changes in the radial coordinates for a displacement radial constraint in the junction (r=1.0m , z=0.6m) and a meridional stress constraint τ_{SS} in the Gaussian point of the cylindrical element adjacent to the junction. The cone-cylinder connection is submitted to an internal pressure of p=0.2MPa. The global sensitivities are calculated using central finite difference (CFD) with a perturbation of Δb_i=0.001b_i. From Tables IV and V it is observed that the analytical sensitivities for shape design sensitivities compare very favourably with the global finite element sensitivities obtained with the same model. It is also seen that the semi-analytical sensitivities only compare favourably for very small perturbations in the design variables, being highly influenced by the perturbation used, due to the truncation on the finite difference method.

Table IV : Sensitivities due to a radial displacement constraint

Source	Perturbation	Sensitivities					
		b_1	b_2	b_3	b_4	b_5	b_6
CFD	0.001b_i	-0.0430	0.6042	4.6708	-12.7946	0.0926	5.7383
Analytical		-0.0430	0.6041	4.6703	-12.7940	0.0922	5.7387
Semi-Analytical	0.000000001b_i	-0.0430	0.6042	4.6708	-12.7946	0.0922	5.7387
	0.0000001b_i	-0.0430	0.6042	4.6709	-12.7942	0.0928	5.7390
	0.00001b_i	-0.0430	0.6044	4.6811	-12.7507	0.1541	5.7679
	0.001b_i	-0.0428	0.6289	5.7062	-8.4132	6.2830	8.6519
	0.01b_i	-0.0410	0.8691	15.9369	29.7409	61.1342	34.3096

Table V : Sensitivities due to a meridional stress constraint

Source	Perturbation	Sensitivities					
		b_1	b_2	b_3	b_4	b_5	b_6
CFD	0.001b_i	-0.0077	0.8343	-6.5946	16.7119	-7.8595	-1.8374
Analytical		-0.0077	0.8342	-6.5940	16.7115	-7.8591	-1.8375
Semi-Analytical	0.000000001b_i	-0.0077	0.8343	-6.5946	16.7121	-7.8591	-1.8375
	0.0000001b_i	-0.0077	0.8343	-6.5947	16.7114	-7.8598	-1.8376
	0.00001b_i	-0.0077	0.8344	-6.6034	16.6458	-7.9269	-1.8488
	0.001b_i	-0.0077	0.8480	-7.4871	10.0850	-14.6369	-2.9660
	0.01b_i	-0.0077	0.9802	-16.3546	-48.1310	-74.5055	-12.9122

For the initial design the maximum meridional stress is τ_{SS}=105.4 MPa at the cylindrical element adjacent to the junction. The model is optimized considering a radial deflection of

u_{r_o}=0.3mm, a meridional stress limit of σ_o=120 MPa and the enclosed volume of the initial design at 2.932 m^3 (equality constraint). The optimal design (Fig. 10) is obtained in 5 iterations, 12 function evaluations and 5 gradient evaluations, with a reduction in the material of the shell of 1.4%, however, for the optimal design the maximum meridional stress in the model decreases to σ_ϕ=47.5 MPa. During the iteration process only the enclosed volume of the initial design is an active constraint. Increasing the internal pressure to p=0.28MPa the meridional stress is 147.6 MPa in the initial design, which gives a violated constraint, but the optimal design is the same and is obtained in 6 iterations, 20 function evaluations and 6 gradient evaluations. The meridional stress decreases to 70.3 MPa.

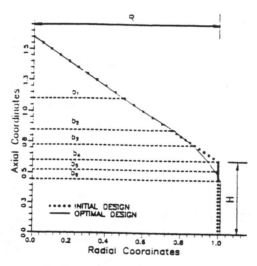

Fig.10 — Cone—Cylinder Connection

5.6 - Conclusions

The analytical sensitivities for shape design are more accurate than the semi-analytical sensitivities. Hence analytical sensitivities should be recommended for the optimization of axisymmetric type structures although they are more difficult to obtain. This problem can be overcome using a symbolic manipulator. However the analytical sensitivities take more computer time than the semi-analytical ones. In these applications a CPU ratio of 2.5 is achieved between the analytical/semi-analytical evaluation of sensitivities.

Shell structures are very sensitive to small imperfections, consequently the semi-analytical techniques for sensitivities cannot be very accurate and it will require a very small perturbation to obtain acceptable results. However this perturbation can create problems of numerical stability. In these applications a perturbation of 0.00001% gives acceptable results.

REFERENCES

Barbosa, J. I., 1990, "Analytical Sensitivities for Axisymmetric Shells Using Symbolic Manipulator MATHEMATICA", CEMUL Report, July 1990.

Barthelemy, B., Chon, C. T. and Haftka, R. T., 1988, "Accuracy Problems Associated with Semi-Analytical Derivatives of Static Response", Journal of Finite Elements in Analysis and Design, Vol.4, pp. 249-265.

Chenais, D., 1987, "Shape Optimization in Shell Theory : Design Sensitivity of the Continuous Problem", Eng. Opt, 11, pp. 289-303.

Desai, C. S. and Abel, J. F., 1972, Introduction To The Finite Element Method, A Numerical Method For Engineering Analysis. Van Nostrand Reinhold Company, New York.

Gendong, C. and Yingwei, L., 1987, "A New Computation Scheme for Sensitivity Analysis", Eng. Opt., Vol. 12, pp. 219-234.

Haftka, R. T. and Kamat, M. P., 1987, "Finite Elements in Structural Design", Computer Aided Optimal Design : Structural and Mechanical Systems (Ed. Mota Soares, C. A.), Springer-Verlag, pp. 241-270, Berlin.

Kraus, H., 1967, Thin Elastic Shells, John Wiley & Sons, Inc., New York

Marcelin, J. L. and Trompette Ph., 1988, "Optimal Shape Design of Thin Axisymmetric Shells", Eng. Opt., Vol. 13, pp. 108-117.

Mehrez, S. and Rousselet, B., 1989, "Analysis and Optimization of a Shell of Revolution", Computer Aided Optimum Design of Structures : Applications. Ed. C. A. Brebbia and S. Hernandez, Computational Mechanics Publications, Springer-Verlag, pp. 123-133.

Plaut, R. H., Johnson, L. W. and Parbery, R., 1984, "Optimal Form of Shallow Shells with Circular Boundary", Transactions of the ASME, Vol. 51, pp. 526-538.

Thambiratnam, D. P., Thevendran, V. and Lee, S. L., 1989, "Computer Aided Optimum Design of Structures for Vibration Isolation", Computer Aided Optimum Design of Structures: Recent Advances, (Ed. C. A. Brebbia and S. Hernandez), Computational Mechanics Publications, Springer-Verlag, pp. 49-59. U. K.

Vanderplaats, G. N., 1984, Numerical Optimization Techniques for Engineering Design, McGraw-Hill, New York.

Wolfram, S., 1988, MATHEMATICA - A System for Doing Mathematics by Computer, Addison - Wesley Publishing Company, Inc.

Chapter 16

APPLICATIONS OF ARTIFICIAL NEURAL NETS IN STRUCTURAL MECHANICS

L. Berke

NASA Lewis Research Center, Cleveland, Ohio, USA

P. Hajela

Rensselaer Polytechnic, Troy, New York, USA

ABSTRACT

A brief introduction to the fundamentals of Neural Nets is given first, followed by two applications in structural optimization. In the first case the feasibility of simulating with neural nets the many structural analyses performed during optimization iterations was studied. In the second case the concept of using neural nets to capture design expertese was investigated.

1. INTRODUCTION

Considerable activity can be observed in the development and application of a certain class of trainable network paradigms, namely the biologically motivated Artificial Neural Nets (ANN). This upsurge of developmental activities is expected to contribute to the availability of powerful new capabilities in the near future. It is this expectation that motivated the examination of the usefulness of ANN in structural optimization as one of the many potential applications in structural analysis and design.

It appears that only a few structural problems were investigated as demonstrations of neural net capabilities. References 1 and 2 are examples dealing with simple oscillators and a beam design problem, respectively. As will be discussed in subsequent sections, there are many potentially productive applications.

Developments in Computational Structures Technology (CST) are closely linked to developments in computational capabilities. ANN technology, and its exploitation for CST falls in the same category and an initial investigation, as reported here, appeared warranted. CST is very demanding on computer resources and new approaches for their utilization are continuously being explored. For example it is not widely known that the first engineering application of Artificial Intelligence (AI) in the form of a rule based expert systems was in structural analysis. This first expert system capability, Structural Analysis Consultant (SACON) became somewhat of a classic model for similar applications (refs. 3,4). AI is now a widely accepted technology for engineering applications as illustrated by the large number of conferences and publications.

Artificial neural nets are also a class of AI paradigms, and provide new opportunities for applications of computer science developments in CST. This brief note proposes a few potentially profitable applications and presents results of feasibility studies associated with automated structural design.

Only the basic ideas of artificial neural nets will be provided in this brief study. References 5 and 6 are recommended for those interested in a detailed introduction to neural nets. Reference 6 also describes an extension of neural nets called "functional links." This represents an extension of the biologically motivated approaches towards more specific mathematical functionalities present in engineering or other applications.

The feasibility studies presented here were conducted using well known small "toy" problems. Research is underway to explore the limits of applicability. This includes increasing problem complexity and dimensionality as indicated by the number of input-output variables, and the nature of nonlinearities in their functional dependence. As will be discussed next the "training" of neural nets involve the minimization of some error measure. Consequently, limitations on dimensionality of problems associated with optimization techniques also apply, at least in the case of ANN utilized in this study. ANN hardware is the subject of vigorous developments and eventually may provide considerable increase in processing power.

2. BASIC CONCEPTS OF ARTIFICIAL NEURAL NETS

As stated earlier, only a brief and incomplete introduction is given here, and References 5 and 6 are suggested as introductory reading. Many other texts are available, and the body of publications is increasing very rapidly.

Neural nets can be viewed in different ways. The original motivation came from creating computer models that can mimic certain brain functions, and the word neural was attached to the designation of this class of models. For computer specialists ANN are a class of parallel distributed processors with some particular processing capability in the artificial neurons and modification of data during communication among them. The particular class of ANN utilized in this study can be viewed either as brain function models if one is romantically inclined, or simply as a flexible technique for creating nonlinear regression models for data with complex interdependencies.

Figure 1a shows a simplistic representation of a neuron with the following components of interest: a cell body with the mechanism which controls cell activity, the "axon" that transmits stimulus from one neuron to others, the "dendrites" which also receive electrical signals from connected neurons or from an external source, and the "synapses" which define interconnections and their respective strengths. In a human brain the number of neurons approaches a trillion, each connected perhaps to tens of thousands of other neurons forming an immense network. Figure 1b shows a small segment of this network in the cerebral cortex.

Artificial neural nets were conceived as very simple models of certain brain activities. Of interest for us here are those aspects of biological neural net activities that are associated with learning, memory, and generalization from accumulated experience. Learned information is thought to be represented by a pattern of synaptic connection strengths that modify the incoming stimuli, strengthening or inhibiting them. When the accumulation of the received stimuli in the neuron reaches a certain threshold, it "fires," sending out an electrical stimulus to all connected neurons. Learning in turn is thought to be associated with the development and retention of a pattern of the connection strengths in various regions of this immense network. It has been suggested that such retained patterns are somewhat similar in nature to holograms that also contain complex information in a vast arrangement of simple patterns.

Artificial neural nets simulate the above activities in brain tissue through very simple concepts. An artificial neuron receives information labeled x_i from the incoming n connections from other neurons as indicated in Figure 2a. Such neurons and their connections can be assembled in principle into any architecture of connectivities as indicated in Figure 2b. The information x_i sent out by the connecting neurons and received by the jth neuron of a net are modified by connection strengths w_{ij}. The jth neuron performs a summation of the modified information as also indicated in Figure 2a., and processes the value r of the sum through an activation function producing an output z_j. This output is then sent as a stimulus to all connecting neurons, and determines in turn, the activity of those neurons. Figure 3 shows a few activation functions, with the sigmoid function being the most popular. More complex neuron activation functions can be devised for various special purposes.

Training of neural nets of interest here involves the evolution of the connection strengths w_{ij} everywhere in the net through "training". Sets of known input and associated output values are presented to the net and the w_{ij} are adjusted during an iterative procedure to minimize a selected error measure between the desired output and the one produced by the net. Once trained, the network responds to a new input within the domain of its training by "propagating" it through the net and producing an output. This output is an estimate within certain error, of the output that the actual computational or physical process would have produced.

Several neural net paradigms have emerged as a result of over four decades of research, each with its own purpose and capabilities. The particular class of neural nets that are of interest to us here fall in the category of "feed forward" nets because the input data given to the network is propagated forward towards the output nodes. The "delta-error back propagation" algorithm (see Refs. 5 or 6) is used usually for their "supervised" learning. It is essentially a special purpose steepest descent algorithm to adjust the w_{ij} connection strengths, and other additional internal parameters that are sometimes added to increase flexibility. In principle other optimization methods can also be used, and the development of efficient learning algorithms is an active area of research.

Most currently available neural net capabilities are simulations of the distributed parallel processing concept on serial machines, and such simulations were also used in this study. Neural nets present premier applications for parallel machines or for the developments of special purpose hardware. These approaches are all being investigated, and neural nets enjoy vigorous funding and developments worldwide. As mentioned before, it is this fact that served as motivation for the present study. Other CST applications are also being investigated in view of expected increases in capabilities.

To start out with an application one requires a set of known input and output pairs that must be generated by the "real" process one is planning to simulate. The number of training pairs, and how they span the intended domain of training, is part of what is. still an art in ANN requiring experimentation and experience. The same statement is also valid for the architecture of the neural net one intends to use. The examples given later will provide some idea of what is required for a successful application. For the engineering applications presented here, it is perhaps worthwhile to think of neural nets as a peculiar automated multidimensional surface fitting or nonlinear regression capability. What one would accept as a representative input-output set to produce a useful surface fit, is most likely a good start to determine the training pairs for the neural nets.

For the present application it is sufficient to discuss the simplest forms of net architectures. A single layer net is called a "flat" net and is of little interest here in its basic form. It has limited capabilities to represent nonlinearities unless these are specifically captured in the input. An example of this is the use of reciprocal variables in problems involving structural stiffness, a case to be discussed later. Reference 6 provides a powerful generalization of this concept referred to as "functional links". In general, nets have an input and output layer with the number of neurons in each of these matching the number of input and output variables, respectively, and one or more "hidden layers." As an example, Figure 4 has two nodes in its input layer, three nodes in its single hidden layer, and one node in its output layer. In later discussions such a net will be designated a (2,3,1) net signifying the number of nodes in its three layers. A net can provide an n-to-m mapping, which, for the case of Figure 4, is a 2-to-1 mapping.

The number n and m of nodes in the input and output layer is determined by the number of input and output variables in the training set. It is however, important to determine the necessity of one or more "hidden" layers in the network. A single hidden layer with nodes numbering somewhere between the average and the sum of the input and output nodes is suggested in the literature as a good first start. To add more layers for added flexibility is a temptation which must be resisted in the simple cases addressed in this note. A general suggestion is to try to use as few nodes as possible. As in any optimization problem one should avoid needless increase in the number of optimization variables.

The n-to-m mappings discussed here can also be separated into m n-to-1 mappings or a number of mappings involving groups of output variables. Very large problems may have to be separated that way to keep the number of learning parameters within a practical range for any of the single mappings. For the small problems discussed here, and based on limited experimentation, the training times for one n-to-m mapping appeared to be slightly more favorable than for the equivalent m n-to-1 mappings to equal accuracy.

Once an architecture has been selected, the training starts out with a random set of connection weights w_{ij} usually generated automatically by the particular capability used. These connection weights are then adjusted by a learning algorithm to minimize the difference between the training output values and the values produced by "propagating" the associated input through the net. The training is sensitive to the choices of the various net learning parameters. The principal parameters are the "learning rate" which essentially governs the "step size," a concept familiar to the optimization community, and the "momentum coefficient" which forces the search to continue in the same direction to aid numerical stability, and to go over local minima encountered in the search.

During supervised learning these parameters are adjusted periodically based on the changing convergence trend during iterations. In the "Ten Bar Truss Optimum Design Expert" example discussed later, a publicly available NASA developed capability NETS 2.0 was used. Its user manual, Reference 7, provides a good introduction for someone who would like to experiment with neural nets. NETS 2.0 has a number of other learning parameters and provides good default values for them, including some adaptive features during training iterations. A few possible applications within CST are suggested next, followed by a representative set of the results obtained in preliminary feasibility studies.

3. NEURAL NETS IN COMPUTATIONAL STRUCTURES TECHNOLOGY

The history of the exploitation of computer technology by CST can be viewed, even if somewhat romantically, as attempted simulations of the brain processes of an expert designer at higher and higher levels of abstraction. Procedural codes, expert systems, and neural nets represent this higher and higher levels of abstraction from "number crunching," "expert judgments" and finally a "feel" for a problem area, respectively. These three levels represent increasing intellectual levels and ability to provide quick expert estimates for solutions with less and less participation required of the human user. The final aspiration of researchers in CST is the development of automated expert design capability; neural nets perhaps provide an approach towards that goal.

As described earlier, artificial neural nets perform their functions by developing specific "patterns" of their connection weights. These patterns, and not any individual value serves as the storage of the knowledge. It would be naive to make much of this supposed similarity to brain functions. Much has been learned about the electrochemical activities of brain cells and of the vast neural nets they form. What all that means is poorly understood if at all, and the functioning of the brain remains largely unknown. A more realistic view of the class of neural nets employed in the present study would be that they are essentially glorified surface fitting capabilities. The important consideration is that neural net research activities are expected to result in major novel hardware and software capabilities when compared to other mathematical procedures for nonlinear regression.

The major advantage of a trained neural net over the original (computational) process is that results can be produced in a few clock cycles representing orders of magnitude less computational effort than the original process. This processing time, once the net is trained, is also insensitive to the effort it takes to generate an output by the original process. Consequently benefits can be higher for those problem areas that are computationally very intensive, such as optimization, especially in multidisciplinary settings. There is of course a catch, namely that in those cases the generation of sufficient training data is also costlier. Practical applications can be envisioned where a problem is frequently solved within limited variations of the input parameters. Organizations with specific products for slightly changing applications could develop or evolve trained neural nets based on sets of past solutions. New solutions could then be obtained with negligible efforts. Machine components that are of a certain basic configuration slightly changing from application to application could be good practical examples.

The w_{ij} weights, as they develop, may contain information concerning hidden functional relationships between the variables for some of the applications providing "feature extraction" capabilities. A version of neural nets designated as "unsupervised learning" has such feature extraction capabilities. Training pairs can be preprocessed to be grouped into clusters based on similarity of features within a certain selected radius. Training effort is then reduced by using the "centers" of such clusters for generalization.

Multidisciplinary design optimization provides particularly intriguing possibilities. For example, nets could be trained for each of the participating disciplines, and integrated to represent appropriate coupling or to use an additional net that develops the important coupling functionalities through feature extraction.

The feasibility of two particular applications at two distinct levels of abstraction were studied in some detail and the results are presented next. The first one involved training a neural net to replace analyses of given structural configurations during optimization iterations. The second exercise was to train a neural net to provide estimates of the actual optimum structures directly totally avoiding the conventional analysis and optimization iterations.

4. NEURAL NET ASSISTED OPTIMIZATION.

This first feasibility study to simulate analysis with the quick response of neural nets was motivated by the approximation concepts in structural optimization. The idea here was to train a neural net to provide computationally inexpensive estimates of analysis output needed for sensitivity evaluations, which in turn is needed by most optimization codes. The numerical experimentation also served to gain initial experience with neural nets.

The familiar five bar and ten bar truss "toy" problems, shown in Figures 5 and 6 respectively, were used for this initial feasibility study. First, various sets of input-output training pairs and network configurations were examined to find the combination that reduced the training effort and produced trained nets which yielded good results as measured by their ability to generalize.

Once an acceptable trained neural net was obtained, it was attached to an optimizer, and all analysis information was obtained from it instead of invoking a conventional analysis capability. Mixing neural net predictions with occasional conventional analyses was not explored, but it is an approach that could possibly exploit the advantages of both.

To create the training sets conventional optimum designs had to be created for two reasons. First, optimum designs were required for comparisons with designs obtained using neural net simulation of the analyses. The second was to perform analyses with random sets of the values of the design variables, in this case the bar areas, within certain preset variations of their optimum values. The optimization involved constraints on the nodal displacements. Consequently, the input-output training pairs for analysis simulation consisted of the bar areas as inputs and nodal displacements as analysis output variables respectively. How many pairs to use, and within what range of variations, is itself a research question. Because of the nature of the sigmoid function at least the output variables are to be scaled by the user or automatically by the neural net code, to within the most active range of the sigmoid function. Scaling minimum and maximum values to 0.1 and 0.9 is usually suggested.

At this point one has to prescribe the number of iterations for which the network must be trained to obtain desired levels of accuracy. A number from a few hundred to tens of thousands is routinely accepted in neural net applications, even for small nets as in this study. For this level of experimentation one often initiates a run on a PC or a work station and lets it run to a large number overnight in somewhat of an overkill.

Some of the net configurations examined for the five bar truss exercise are shown in Figure 7. As the first attempt a 5-to-4 mapping with a (5,4) net was tried with no hidden layer as shown in Figure 7a. The four output variables were the four nodal displacements indicated in Figure 5. Since the active constraints were essentially related to displacements d_2 and d_4, the rest of the nets considered only these two displacements as output. Figure 7b consequently is a (5,2) net with reduced training effort. Figure 7c is a (10,2) net with the reciprocals of the bar areas also included to help the net capture without a hidden layer the inverse relation between bar areas and nodal displacements. Table 1 contains data on the results of these initial training efforts with other functional relationships also included to try to capture nonlinearities. These attempts without a hidden layer were not totally satisfactory in terms of obtained accuracy or number of required training cycles.

Including a hidden layer, as shown in Figure 7d, produced acceptable results. Table 2 presents the results of optimization using various net and training set combinations. Using the (5,7,2) net and scaled variables, an optimum design was obtained within 2.4% of the exact optimum design proving the feasibility of the basic concept of neural net assisted optimization. Table 3 presents the results of similar experimentation for the ten bar truss supporting the same conclusion. Similar results were obtained for a higher dimensionality wing box problem.

5. NEURAL NETS AS EXPERTS FOR DIRECT OPTIMUM DESIGN ESTIMATES

The next set of experiments were conducted to explore the idea of training a neural net to estimate optimum designs directly for given design conditions and bypass all the analyses and optimization iterations of the conventional approach. It is conceivable in practice that successful similar designs could be collected within some domain of design conditions, input-output pairs defined, and then a neural net trained to serve as "intelligent corporate memory" that can provide a new design for different design requirements instantaneously.

Now let us suppose that we work in a company that markets equipment that is mounted in all cases on ten bar trusses as shown in Figure 6. These trusses have to carry the equipment weight (2 X 100 K) at the two lower free nodes while these support points cannot deflect more than 2 in. The dimensions L_1 and L_2 and H of the trusses can vary between 300 and 400 inches, depending on the particular installation. The engineer who was designing the trusses for the past 30 years and could simply tell the optimum bar areas for any combination of those dimensions has just retired. Can we create an accurate simulation of this departed expert? Yes we can, and rather simply!

To experiment with various training sets, optimum designs were generated by conventional methods for varying first only H in 5 inch increments. The results are given in Table 4. There are three kinds of output numbers in the set. These are the bar areas that change, areas that are at the preselected minimum value of 0.1 for all designs, and the weight, which is of a different order of magnitude. A representative A_1 and A_2, and the optimum weight Wt were considered in the first numerical experiments. The neural net code NETS 2.0 (Ref. 7) was used for all of the direct optimum ten bar truss design exercises.

A number of small net configurations were tried for these 1-to-3 mappings more or less as a learning exercise . Table 5 shows the results of training with a (1,6,3) net, a probable overkill with too many nodes in the hidden layer. Table 6 gives the results of design estimates of the trained net for the remaining check cases of Table 4 that were not included in the training set. The training was performed to 1% RMS accuracy within 200 iterations. As can be seen, both the training accuracy and the estimates for the new cases is around a third of one percent for the individual values and can be considered quite satisfactory. It is also of some interest to note that the net had to evolve its w_{ij} connection weights and other internal parameters provided by NETS 2.0 in such a manner that it could also reproduce a constant .1 value for any input while also producing accurate values of variables of different orders of magnitude for the same inputs.

After the above limited exercises the 3-to-11 mappings were performed between L_1, L_2, H, and the ten bar areas $A_1,...,A_{10}$ and the optimum weight Wt. A rather limited training set of only ten input-output patterns was created using ten random sets of L_1, L_2 and H and the corresponding ten optimum designs. It is interesting to note that ten training pairs did quite well in this case versus the hundreds of training pairs used for the neural net assisted optimization study. Of course, in this problem only three variables are varied to cover a domain. The net used for this exercise is shown in Fig. 8. The ten optimum designs used for training the net are given in Table 7. Optimum designs were then obtained for another seven random sets of L_1, L_2, and H, as checks on the estimates obtained from the trained network. Table 8 shows the seven design conditions and the optimum designs.

The (3,14,11) net given in Figure 8 was used with the nodes in the hidden layer taken as a sum of the input and output nodes. The training was repeated with a (3,6,11) net which also was successful but the learning parameters had to be adjusted after about 100 iterations. During experimentation with various options during training, it was found that it is beneficial to code the 0.1 minimum sizes as 0.5. The active midpoint of the sigmoid activation function is the explanation. This value also represented net accuracy in better detail. NETS 2.0 worked very well, and 1% RMS accuracy was obtainable with 200 iterations in around 30 seconds on a SUN 386i, and using only the default values for the learning parameters for the (3,14,11) net. Because of this good performance, exercises were conducted to overtrain the net. Letting it run for 5000 iterations an RMS accuracy of 0.006% was obtained. Over-training is to be avoided because the neural net at that point becomes a memory with lessened ability to generalize. The overtrained net actually reproduced the training results exactly, but it did a little worse if anything against the seven check conditions than the net trained only for 1% RMS accuracy. The results are shown for one of the check cases in Fig. 9 where the first bar is the desired result, the second is obtained with 1% RMS accuracy and the third is obtained with 0.006% RMS accuracy of training. As can be seen nothing has been gained in accuracy of the estimated results.

Table 9 shows comparisons and the percent error of net estimates for the seven check cases of Table 8 for the net trained for 1% accuracy. As can be seen, the results are quite satisfactory and certainly would be good enough information to produce the ten bar trusses to support the equipment at minimum weight and 2.0 inches maximum deflections. The net produced its estimates by computing a few sums of products in practically no computer time. The mental activities our retired expert designer employed to come up with his optimum designs have been replaced by a trained (3,14,11) neural net of similar capability for this limited task.

6. CLOSING REMARKS

It has been shown that artificial neural nets have intriguing applications in Computational Structural Technology. What has been presented here are two of the possibilities. There are now efforts underway to explore multidisciplinary design applications, to "package" composite material property generation codes as quick response neural net simulations, to develop structural component life prediction capabilities and to capture constitutive material relationships both from theoretical codes and directly from test data. Integrating neural net capabilities with expert systems and optimization algorithms into an automated capability to generate trained neural nets once the domain of interest is defined for often occurring structural components in a design office is also being explored. There are many other applications possible. Controls is one of the most successful applications of neural nets. Investigations dealing with control of large space structures and with smart structures could also prove profitable.

REFERENCES

1. Rehak, D. R.; Thewalt, C. R.; and Doo, L. L.:   Neural Network Approaches in Structural Mechanics Computations. Computer Utilization in Structural Engineering, ed. J. J. Nelson, Jr., ASCE PRoceedings from Structures Congress 1989.
2. McAulay, A. D.:   Optical Neural Network for Engineering Design. NAECON, 1988.
3. Bennet, J.; Creary, L.; Engelmore, R.; and Melosh, R.:   SACON: A Knowledge-Based Consultant for Structural Analysis. Stanford Heuristic Programming Project, Computer Science Department Report No. STAN-CS-78-699, September 1978.
4. Melosh, R. J.; Berke, L.; and Marcal, P. V.:   Knowledge-Based Consultation for Selecting a Structural Analysis Strategy, NASA Conference Publication 2059, November 1978.
5. Rummelhart, D. E., and McCleland, J. L.:   Parallel Distributed Processing, Volume 1: Foundations. The MIT Press, Cambridge, MA, 1988.
6. Yoh-Han Pao: Adaptive Pattern Recognition and Neural Networks. Addison-Wesley Publishing Co., 1989.
7. Baffes, P. T.:   NETS 2.0 User's Guide. LSC-23366, NASA Lyndon B. Johnson Space Center, September 1989.

Table 1. Summary of Training Results with no Hidden Layer

Number of Training Sets	Network Description	Cycles of Training	Error Description
50	(5,2) - five areas as inputs, two vertical displacements as outputs.	1500	ϵ = 0.087 Error not decreasing
50	(10,2) - five areas and five reciprocal areas as inputs - two vertical displacements as outputs.	50000	ϵ = 0.03927 Error decreasing slowly
50	(15,2) - five areas and ten area products of type $A_i A_j$ ($i \neq j$) as inputs - two vertical displacements as outputs.	50000	ϵ = 0.05235 Error decreasing slowly
50	(20,2) - five areas, five reciprocal areas and ten values of type $sin(A_i/A_{max})$, $cos(A_i/A_{max})$ used as input - two vertical displacements as output.	50000	ϵ = 0.0398 Error decreasing slowly

(xx,yy) denotes xx input nodes and yy output nodes.

Table 2. Optimal Design for Five Bar Truss Using Trained Neural Net for Analysis

Network Description		Design Variables					Objective Functions
		x_1	x_2	x_3	x_4	x_5	
(5-7-4) 200 training sets, all four output displacements mapped.	Initial	1.0	1.0	1.0	1.0	1.0	58.28
	Final	2.131	2.032	2.679	2.766	1.0	128.626
(5-7-4) 500 training sets, all four output displacements mapped.	Initial	1.0	1.0	1.0	1.0	1.0	58.28
	Final	1.952	2.013	2.763	2.760	1.0	127.759
(5-7-2) 100 training sets, two vertical displacements mapped.	Initial	1.0	1.0	1.0	1.0	1.0	58.28
	Final	1.535	1.778	2.29	2.265	1.0	107.56
(5-7-2) 100 training sets, two vertical displacements scaled as constraints and mapped.	Initial	1.0	1.0	1.0	1.0	1.0	58.28
	Final	1.505	1.584	2.138	2.211	1.0	102.399
	Exact Solution	1.5	1.5	2.121	2.121	1.0	100.0

(xx-yy-zz) denotes a three layer architecture with xx input layer nodes, yy hidden layer nodes, and zz output layer nodes.

Table 3. Optimal Design for Ten Bar Truss Using Trained Neural Net for Analysis

Design Variables	Network Description			
	(10-6-6-2)* 100 Training Sets Used in a Range of ±25% About Optimum	(10-6-6-2)* 400 Training Sets Used in a Range of ±25% About Optimum	(10-6-6-2) 100 Training Sets Used in a Range of 0.01-55.0 in² Output Scaled to Reduce Range of Variation	Solution From Exact Analysis
X_1	30.774	30.967	30.508	30.688
X_2	0.112	0.100	0.100	0.100
X_3	17.40	19.136	26.277	23.952
X_4	11.425	14.279	11.415	15.461
X_5	0.108	0.100	0.100	0.100
X_6	0.407	0.434	0.413	0.552
X_7	5.593	5.593	5.593	8.421
X_8	22.953	20.031	21.434	20.606
X_9	20.886	19.966	22.623	20.554
X_{10}	0.100	0.100	0.100	0.100
Objective Function	4692.49	4666.71	5010.22	5063.81

* Lower bound of design variables used as initial design - was infeasible.

Table 4. Ten Bar Truss Optimum Designs with H as Design Condition

H	A1	A2	A3	A4	A5	A6	A7	A8	A9	A10	WT
300	9.53	.1	9.67	4.73	.1	.1	6.34	6.15	6.15	.1	1749.6
305	9.37	.1	9.51	4.65	.1	.1	6.28	6.09	6.09	.1	1733.2
310	9.22	.1	9.36	4.57	.1	.1	6.22	6.03	6.03	.1	1717.4
315	9.07	.1	9.21	4.50	.1	.1	6.16	5.98	5.98	.1	1702.3
320	8.93	.1	9.07	4.43	.1	.1	6.11	5.92	5.92	.1	1687.9
325	8.79	.1	8.92	4.36	.1	.1	6.05	5.87	5.87	.1	1673.5
330	8.66	.1	8.79	4.29	.1	.1	6.00	5.83	5.83	.1	1660.9
335	8.53	.1	8.66	4.23	.1	.1	5.96	5.78	5.78	.1	1648.3
340	8.40	.1	8.54	4.16	.1	.1	5.92	5.73	5.73	.1	1635.9
345	8.28	.1	8.41	4.10	.1	.1	5.87	5.68	5.68	.1	1624.5
350	8.16	.1	8.29	4.05	.1	.1	5.82	5.65	5.65	.1	1613.5
355	8.05	.1	8.17	4.00	.1	.1	5.78	5.60	5.60	.1	1603.2
360	7.93	.1	8.06	3.93	.1	.1	5.74	5.68	5.68	.1	1593.1
365	7.83	.1	7.95	3.88	.1	.1	5.70	5.53	5.53	.1	1583.5
370	7.72	.1	7.84	3.83	.1	.1	5.66	5.50	5.50	.1	1574.3
375	7.61	.1	7.73	3.77	.1	.1	5.63	5.45	5.45	.1	1565.2
380	7.51	.1	7.63	3.73	.1	.1	5.60	5.42	5.42	.1	1557.1
385	7.42	.1	7.53	3.68	.1	.1	5.56	5.39	5.39	.1	1549.1
390	7.32	.1	7.44	3.63	.1	.1	5.53	5.36	5.36	.1	1541.4
395	7.23	.1	7.34	3.58	.1	.1	5.49	5.32	5.32	.1	1534.1
400	7.14	.1	7.25	3.54	.1	.1	5.46	5.30	5.30	.1	1527.1

Table 5. Training Accuracy with (1,6,3) Net

Training Pairs				Training Accuracy (RMS = 0.9%)				
Input	Output							
H	A1	A2	WT	A1	%	A2	WT	%
300	9.53	.1	1749.6	9.543	.14	.101	1744.4	.30
310	9.22	.1	1717.4	9.240	.23	.101	1721.0	.21
320	8.93	.1	1687.9	8.955	.28	.100	1693.4	.33
330	8.66	.1	1660.9	8.668	.11	.100	1665.6	.28
340	8.40	.1	1635.9	8.401	.01	.100	1640.3	.27
350	8.16	.1	1613.5	8.159	.012	.100	1618.1	.29
360	7.93	.1	1593.1	7.930	.00	.100	1597.2	.26
370	7.72	.1	1574.3	7.711	.12	.100	1577.2	.19
380	7.51	.1	1557.1	7.507	.07	.100	1558.2	.07
390	7.32	.1	1541.4	7.326	.08	.100	1542.1	.05
400	7.16	.1	1527.1	7.187	.38	.100	1529.3	.14

Table 6. Test Set of Neural Net Estimates

Input	Optimum			N-N Estimates				
H	A1	A2	WT	A1	%	A2	WT	%
305	9.37	.1	1733.2	9.364	.11	.101	1733.5	.00
315	9.07	.1	1702.3	9.097	.30	.100	1707.1	.28
325	8.79	.1	1673.5	8.801	.12	.100	1679.0	.33
335	8.53	.1	1648.3	8.529	.01	.100	1652.8	.27
345	8.28	.1	1624.5	8.283	.04	.100	1629.5	.31
355	8.05	.1	1603.2	8.043	.09	.100	1607.5	.27
365	7.83	.1	1583.5	7.818	.15	.100	1587.0	.22
375	7.61	.1	1565.2	7.604	.08	.100	1567.4	.20
385	7.42	.1	1549.1	7.361	.79	.100	1550.2	.07
395	7.23	.1	1534.1	7.252	.30	.100	1535.4	.08

Table 7. Ten Bar Truss Optimum Designs for Training with L_1, L_2, and H as Design Conditions

Input			Output										
L1	L2	H	A1	A2	A3	A4	A5	A6	A7	A8	A9	A10	WT
310	350	380	6.89	.1	7.00	3.62	.1	.1	5.24	5.08	5.35	.1	1356.7
345	326	360	7.40	.1	7.50	3.56	.1	.1	5.62	5.45	5.31	.1	1456.4
371	329	310	8.95	.1	9.10	4.17	.1	.1	6.33	6.14	5.74	.1	1684.5
360	300	340	7.69	.1	7.83	3.47	.1	.1	5.73	5.25	5.56	.1	1492.6
315	340	340	7.64	.1	7.76	3.93	.1	.1	5.53	5.36	5.56	.1	1407.6
380	355	390	7.47	.1	7.60	3.58	.1	.1	5.67	5.50	5.32	.1	1605.7
322	319	400	6.35	.1	6.46	3.13	.1	.1	5.21	5.05	5.03	.1	1314.2
400	300	400	9.25	.1	9.41	3.93	.1	.1	6.77	6.56	5.56	.1	1780.9
300	400	300	9.27	.1	9.39	5.25	.1	.1	5.73	5.57	6.57	.1	1593.8
311	350	315	8.33	.1	8.45	4.36	.1	.1	5.70	5.53	5.88	.1	1464.6

Table 8. Ten Bar Truss Optimum Designs for Checking Estimates of the Trained Neural Net

Input			Optimum Solutions										
L1	L2	H	A1	A2	A3	A4	A5	A6	A7	A8	A9	A10	WT
342	351	383	7.18	.1	7.29	3.60	.1	.1	5.44	5.27	5.34	.1	1466
360	360	360	7.93	.1	8.06	3.93	.1	.1	5.74	5.56	5.56	.1	1593
320	350	360	7.38	.1	7.50	3.82	.1	.1	5.43	5.26	5.49	.1	1417
340	370	340	8.29	.1	8.41	4.28	.1	.1	5.74	5.56	5.82	.1	1578
310	350	380	6.89	.1	6.99	3.62	.1	.1	5.24	5.08	5.35	.1	1356
345	326	360	7.39	.1	7.51	3.56	.1	.1	5.62	5.45	5.30	.1	1456
371	329	310	8.95	.1	9.16	4.17	.1	.1	6.33	6.14	5.74	.1	1684

Table 9. Trained Neural Net Estimates of Ten Bar Truss Optimum Designs

A1	7.138	8.000	7.365	8.398	6.908	7.310	8.916
%	0.580	0.780	0.200	1.300	0.260	1.080	0.380
A2	0.503	0.505	0.502	0.507	0.502	0.501	0.504
A3	7.245	8.160	7.485	8.558	7.013	7.448	9.089
%	0.620	1.240	0.200	1.760	0.330	0.820	0.770
A4	3.554	3.957	3.777	4.354	3.551	3.533	4.177
%	1.230	0.690	1.130	1.730	1.900	0.760	0.170
A5	0.499	0.499	0.500	0.499	0.500	0.500	0.500
A6	0.501	0.501	0.500	0.501	0.500	0.500	0.500
A7	5.456	5.783	5.461	5.781	5.309	5.586	6.359
%	0.290	0.750	0.570	0.710	1.320	0.610	0.460
A8	5.293	5.594	5.301	5.598	5.164	5.416	6.157
%	0.470	0.610	0.780	0.680	1.650	0.620	0.280
A9	5.326	5.574	5.477	5.863	5.336	5.311	5.740
%	0.260	0.250	0.240	0.740	0.260	0.210	0.000
A10	0.499	0.500	0.499	0.499	0.497	0.503	0.500
WT	1466.000	1598.000	1417.000	1585.000	1362.000	1451.000	1693.000
%	0.000	0.310	0.000	0.440	0.440	0.340	0.530

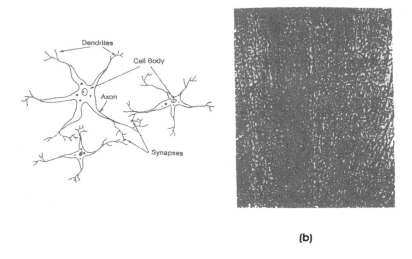

(b)

Figure 1. Biological Neuron and Neural Net

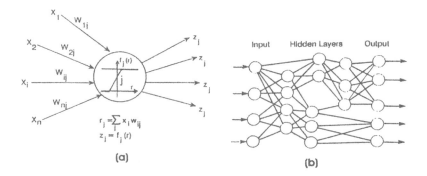

(a)

(b)

Figure 2. Artificial Neuron and Neural Net

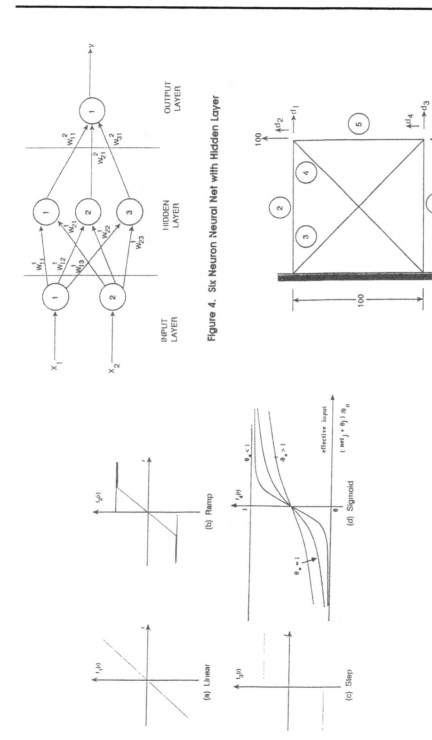

Figure 4. Six Neuron Neural Net with Hidden Layer

Figure 5. Five Bar Truss

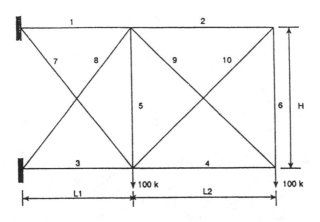

Figure 6. Ten Bar Truss

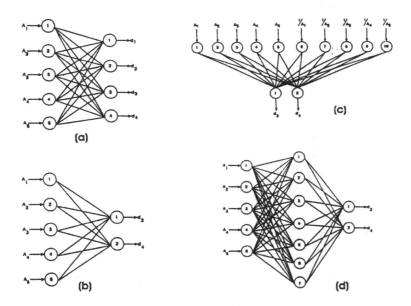

Figure 7. Neural Nets for Five Bar Truss Problem

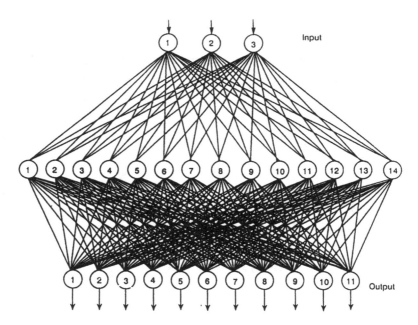

Figure 8. Neural Net for Ten Bar Truss Optimization Expert

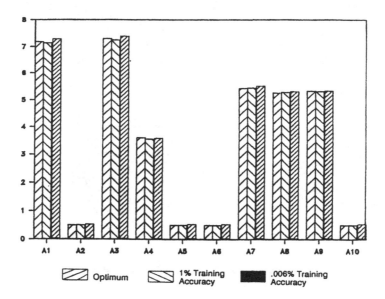

Figure 9. Optimum Design versus Neural Net Estimates.

Chapter 17

MATHEMATICAL PROGRAMMING TECHNIQUES FOR
SHAPE OPTIMIZATION OF SKELETAL STRUCTURES

B.H.V. Topping
Heriot-Watt University, Riccarton, Edinburgh, UK

1   Introduction

This chapter presents a review of mathematical programming methods used in the design of skeletal elastic structures in which the possibility of altering the shape, position or layout of the members is considered. Virtually every type of optimization procedure including linear, non-linear, and dynamic programming has been applied to this design problem. These methods have been implemented using three main approaches. The first, referred to as the 'ground structure' approach, is one in which members are removed from a highly connected structure to derive an optimum subset of bars. In the second approach the co-ordinates of the joints of the structure are treated as design variables and moved during the optimization procedure to enable an optimum layout to be designed. The third type of method includes those which allow for topological considerations at certain points during the design process and generally keeps the design variables in two separate groups. The paper discusses the way in which each of the mathematical programming methods has been applied to these approaches.

Since the advent of computer methods of analysis and design, considerable academic interest has been shown in the application of mathematical programming techniques to the optimization of engineering structures. These methods appear to have been little understood and rarely used by practising civil engineers who often consider them to be of no relevance to practical design situations. This is probably due to the apparent mathematical complexity and, in general, the limited scope of the methods.

Most of these methods seek to minimize the weight or volume of the structure considering design limits or constraints on the member stresses and the deflection of the structure. The design variables for skeletal structures are usually the cross sectional properties of the members, i.e., the area and second moments of area. These methods, however, represent a limited and rather narrow approach to the design of most structures. The possibility of altering the shape, position or layout of the structural members to improve the design during the optimization process represents a significant advance on techniques which only consider the effects of varying the cross sectional properties of the members. Solutions derived using techniques which allow for topological changes provide valuable yardsticks for designers and in recent years practical design using these methods has become more of a reality.

This paper presents a review of methods of topological design applied to skeletal elastic structures and seeks to highlight the main areas of development. The paper is based on two earlier of reviews of this topic [95, 96]. Two further reviews have been published that are of interest in this context. The first by Spillers [92] discusses shape optimization of structures in general including finite element structures. The second by Kirsch [26] relates analytical and mathematical programming methods applied to the shape optimization of skeletal structures.

One of the earliest approaches to the optimization of structural layout was developed by Michell [48] in 1904, using a theorem previously developed by Clerk Maxwell [46]. Michell's theorem states that *"a frame attains the limit of economy of material possible in any frame-structure under the same applied forces, if the space occupied by it can be subjected to an appropriate small deformation, such that the strain in all the bars of the frame are increased by equal fractions of their lengths, not less than the fractional change of length of any element of the space."* Michell structures are designed for only one loading case and depend on an appropriate specification of the strain field. Unfortunately, these structures are statically determinate and impractical, consisting of non-standard lengths and an infinite number of bars.

This early work was discussed by many other [8, 50, 51] and Chan [3] developed techniques for graphical construction of suitable strain fields. Nonetheless, the work still suffered from the same impracticalities. Prager [62] presented techniques for assessing the efficiency of near optimal trusses and the modified design criteria for members in a discretized Michell structure. Parkes [52] has published a theoretical investigation in which he considered the effect of the cost of joints when deriving the optimum form of networks under a simple single load case. Prager [63] also gave the problem of joints his consideration and Prager and Rozvany [64, 65, 71] have developed an optimal theory for the layout of grillages using an analogy of the Michell Theory.

These papers represent the earliest research into the optimization of structural layout and the subsequent theoretical developments. Although these techniques are completely different in philosophy to mathematical programming techniques they provided yardsticks which the first mathematical programmers found invaluable.

2 Introduction of Mathematical Programming Techniques

In 1968 Hemp and Chan [3, 18, 19, 20], and in parallel work Dorn, Gomory and Greenberg [10], overcame the impracticalities of Michell structures by considering a 'ground structure' of a grid of points which include the structural joints, supports and load positions. The grid was connected by many potential members. They showed that a pin-jointed structure subjected to a virtual displacement field which maximizes the external virtual work and complies with strain constraints is an optimum. This form of solution is referred to as the dual method. The virtual displacements of the points were varied using linear programming methods to make the virtual work a maximum and so a strain field was derived in which all permitted members achieved the maximum virtual strain. Other members were then removed. Some of the remaining members and joints can usually be removed by considering equilibrium, but the reduced structure may still, in some cases, be indeterminate. However, for most problems with a single loading case the structure is or can be made determinate. Consequently, the statically determinate form is an optimal structure with respect to elastic and plastic safety conditions despite the fact that compatibility conditions are ignored in the primal formulation and equilibrium conditions in the dual formation.

The original form of the linear programming problem called the primal is:
Minimize Objective Function:

$$Volume = \sum_{j=1}^{m} A_j L_j \tag{1}$$

Subject to:
   Stress Constraints:

$$-\sigma_p A_j \leq P_j \leq \sigma_p A_j \qquad (j = 1\ to\ m)$$

Equilibrium Conditions:

$$\sum P_j K_{ij} = F_i \qquad (i = 1\ to\ n)$$

in which A, L, P are the member cross-sectional areas, lengths and forces respectively; $\sigma = $ the permissible stress; $K = $ a matrix of member direction cosines; $F = $ the applied load vector; $m = $ the number of structural members; and $n = $ number of joints in the structure. Using duality principles [66] the problem may be restated as:
Maximize Virtual Work:

$$Virtual\ Work = \sum_{i=1}^{n} F_i v_i \tag{2}$$

Subject to:
  Strain Constraints:

$$-\epsilon L_j \prec h L_j \prec \epsilon L_j \qquad (j = 1\ to\ m)$$

  Compatibility Conditions:

$$\sum K_{ij} v_i = \epsilon L_j \qquad (i = 1\ to\ n)$$

in which v = a virtual nodal displacement vector; ϵ = a vector of virtual member strains; and h = a virtual strain parameter.

The fact that the optimum layout for a structure subject to a single load case is statically determinate was generally appreciated long before 1968. In 1956 Barta [2] published an investigation of determinate subsets of a redundant truss to prove a theorem discussed by Sved [93] in 1954 in which it was stated that *"by removing properly chosen redundant bars from the given network it is possible to obtain a statically determinate structure which yields the least weight of the structure"* for a single loading case. In 1958 Pearson [54] derived solutions for ground structures using the equilibrium conditions alone. A random number generator was used to vary the redundant member forces until the optimum solution has been derived. He noted that for a single loading case *"normally only a statically determine set of links have non-zero areas."* However, he was only considering a much reduced 'ground structure' with little choice of alternative members.

Fleron [13] developed an iterative procedure for minimizing the weight of a truss 'ground structure' in which each step corresponded to a different statically determinate sub-structure. He arrived at two conclusions for a 'ground structure' which is subject to one loading case and where the permissible stresses for compression and tension members are fixed but not necessarily equal:

1. If one structure has a weight less than all others, this structure will be statically determinate.

2. If a 'p' times statically indeterminate structure is an optimum one there exists at least 'p + 1' statically determinate optimum trusses.

Further discussion of the linear programming approach to optimization of topology may be found in papers by Pope [59] and Richards and Chan [68].

The 'ground structure' approach, in which a mathematical programming technique is used to remove members during an optimization, is now frequently employed in the optimization of structural layout. An important point is that if enough alternative joints and members are included in the ground structure then although the optimum weight of the structure will be unique, the layout will not necessarily be unique and may become a matter for choice [1, 95]. A mathematical programming technique will normally yield just one of the choices.

Another significant factor in determining the optimum layout for a structure is the spacing of the grid points used to make up the 'ground structure'. Dorn, Gomory and Greenberg [10] appear to have been the first to have investigated the influence of the 'ground structure' grid on the optimum layout when they applied the linear programming technique to a series of grids for a bridge truss. As expected, their conclusions indicated that the 'ground structure' grid has a significant effect on both the weight and layout of the optimum structure. This work leads to another approach to optimizing structural layout in which the joint coordinates as well as the member cross-sectional properties are considered as design variables.

Mathematical programming methods for the optimization of structural layout may generally be divided into three main groups.

1. The 'Ground Structure' Approach:
   In this case member areas are allowed, if required, to reduce to zero and hence can be deleted from the structure.

2. The 'Geometric' Approach:
   Here the joint coordinates as well as member cross-sectional properties are used as design variables.

3. Hybrid Methods:
   There methods allow for topological considerations at certain points during the design process and generally keep the design variables in two separate groups or design spaces. These methods may be applied to 'ground structures' or use joint co-ordinates as design variables.

With the exception of references [82, 49], little work has been undertaken on the effect of probabilistic loading on the optimum shape of skeletal structures.

3   The Ground Structure Approach

The simplest technique to apply here is the Fully Stressed Design Stress-Ratio Method in which the structural members are resized after each elastic reanalysis using:

$$A_{inew} = A_{iold}(\sigma_i/\sigma_{ip})_{max} \tag{3}$$

in which A_i = the cross-sectional area of the $i-th$ member; σ_i, σ_{ip}, = the stress and the permissible stress in the member after the current analysis; and max indicates that the maximum value of this ratio is taken by considering all loading cases.

With this approach to a highly connected 'ground structure' many of the member areas will reduce to zero [1, 95]. For structure subject to one loading case with the stress constraints for members in tension and compression at the same level, then the resulting structural layout will be the same as that derived using the linear programming techniques applied by Hemp and Chan [4, 18, 19, 20] and will generally be statically determinate. The results for structures designed for multiple loading cases and different stress constraints are somewhat more difficult to correlate, but the resulting structures will generally be statically indeterminate. Little attention appears to have been given to this approach despite the remarks of Gallagher [16] in his 1970 review of this subject when he said *"it appears that there is a need for an algorithm that incorporates the automatic removal of members as they approach zero size so as to include consideration of subsiduary structural forms."*

The application of Linear Programming techniques for a single loading case was discussed in the last section. Pearson [54] and Chan [4] considered the problem of multiple loading cases. Pearson [54] expressed the member force in terms of the applied loads and the redundant forces; he then randomly varied the forces in the redundant members until he arrived at an optimum solution. In the case of structures subject to multiple loading cases Pearson noted that the optimum structure tends to be statically indeterminate and that each member must attain a maximum load under at least one of the loading cases since if it does not the volume is reducible. Chan [4] arrived at similar conclusions using a linear programming algorithm. The statically indeterminate forms derived using these methods are selected without considering the strcutural compatibility requirements and are therefore not elastic designs. In 1981, Lev [30] related the Stress-Ratio method to the Michell approach for a single design loading case and investigated the properties of these trusses.

To avoid the problem of indeterminate layouts and compatibility requirements, some researchers assumed (often incorrectly) that the optimum layouts will be statically determinate. Simplified work of this type includes the paper by Schmidt [75] who in 1962 correctly concluded from a study of sample structures that *"a statically indeterminate form could sometimes give a lighter structure than a statically determinate one."*

Hemp [19], Spillers and Lev [91] showed that where only two loading cases are considered, the least weight layout using plastic methods may be obtained by superimposing the least weight layouts for the single load cases $1/2(F_1 + F_2)$ and $1/2(F_1 - F_2)$, in which F_1 and F_2 represent the loading cases. Rozvany [71] extended this to the general problem where the number of load cases is greater than two. Spillers [90, 91] considered at length the problem of the combined layouts 'realizability' when an elastic design is required. He showed that 'overstress' may occur in some members due to stress redistribution when a combination of two statically determinate solutions are used in this way. Reinschmidt and Russell [66] considered the effects of different stress constraints for tension and compression members on the linear programming solution for structures subject to single and multiple loading cases. A comparison with the results for structural forms derived using the Fully Stressed Design Stress-Ratio Method and the Linear Programming formulation is to be found in reference [95]. Results indicate that the layouts derived using the Stress-

Ratio method are not always optimum. This is because the Stress-Ratio method only considers the constraints and the method does not consider an objective function of minimum weight or cost. Unfortunately, the linear programming method does not consider compatibility and if the derived layout is indeterminate then the member sizes are not necessarily sufficient to ensure compatibility with 'overstress' in at least one of the members. This problem can be avoided by resizing the layout derived by the Linear Programming Method using the Stress-Ratio Method. The essence of this idea was extended by Reinschmidt and Russell [66, 67, 73] when they formulated their iterative technique for finding the layout of trusses constructed from available steel sections. This method of deriving the layout by Linear Programming and subsequently resizing it using the Fully Stressed Design Method has shown to be particularly useful [66, 67, 73, 95].

The problem of compatibility (or 'realizability' as Spillers [90] refers to it), together with the difficulty of incorporating deflection constraints, led to the application of nonlinear programming methods to the form-finding problem. Dobbs and Felton [9] used a 'steepest descent-alternate mode' algorithm to minimize the weight of a 'ground structure' subject to stress constraints and multiple loading cases. During the process some member areas reduced to zero and these were removed and not allowed to re-enter the design problem. There is no mathematical justification for the removal of these members and no proof that they would not subsequently help reduce the weight of the structure. The members were assumed to be circular tubes when forming the compressive stress constraints and this enabled the effects of buckling to be considered without the introduction of extra design variables. If the buckling stress constraint is critical for a particular member then its area cannot reduce to zero in the design process and the member cannot be deleted. This will result in the member being included in the design when it might otherwise have been deleted. Hajela and Ashley [17] presented an algorithm which utilized a feasible and usable search algorithm, and a cumulative constraint formulation. Stress and buckling constraints for each member were compounded into a single measure of constraint violation. In the early stages of the optimization active buckling constraints were suppressed. In this way the algorithm avoided the problem of members being included in the layout due to buckling constraints.

The 'ground structure' approach is particularly applicable to the design of modular space structures constructed using standard space structure systems. However, computationally efficient procedures for the design of large structures which incorporate comprehensive design constraints have yet to be developed, without which it could be impossible for a designer to include the required members to explore all the design possibilities for a particular structure. It should be pointed out that the methods, other than plastic design, which delete members from a 'ground structure' are heuristic and give no mathematical assurance of reaching a global optimum. Considerable research is still required in this area.

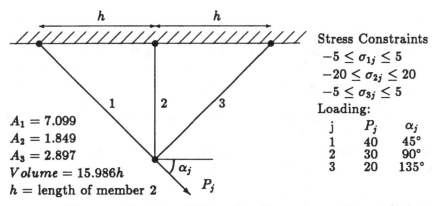

Figure 1. Optimized Three Bar Truss due to Schmit [76]

4   The Geometric Approach

With this approach the objective function becomes highly nonlinear because the member lengths vary with the changes in the joint coordinates. Continuous sizing and configuration variables were not initially treated simultaneously and early work centered on what results would be expected if joint coordinates were included among the design variables. The work of Schmit, which considered design variables when formulating the optimization problem of a three-bar truss using a 'steepest descent - alternative mode' nonlinear algorithm, is probably the earliest investigation of this type. Schmit [76] first showed that for structures of fixed topology with multiple loading cases an optimum statically indeterminate truss is not necessarily one in which each member is fully stressed in at least one condition. His optimum solution for this three bar problem, with the inclined bars fixed at 45°, is shown in Figure 1. The volume of the structure is 15.988h.

Schmit and Kicher [77] considered several configurations by designing three-bar trusses with different angles between the members thus making configuration a discrete variable. The nonlinear optimization procedure accounted for stress and deflection constraints; but the choice of design was made by comparison of the optima for each configuration and material type. No automatic method was yet available for deriving the optimum layout. The results, however, indicated that the optimum configuration or material selection often changes if design constraints or loads are changed even slightly. Schmit and Morrow [78] introduced buckling stress constraints into the problem and showed that when these constraints were active the optimum layouts are more rigid and should be constructed of lighter material. In 1963, Schmit and Mallett [79] presented a paper in which the angles between the members were treated as continuous design variables together with the member cross sectional vari-

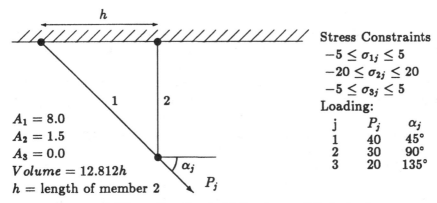

Figure 2. *Optimized Three Bar Truss (Fully Stressed Solution) due to Sved and Ginos [93]*

ables and material properties. The method presented accounted for deflection and stress constraints with multiple loading cases and used a 'steepest descent-alternative mode' algorithm in which the alternate mode direction was developed using a random direction generator in the plane of constant weight. The inclusion of configurational variables improved the optimum and showed that even under multiple loading conditions the optimum may be statically determinate. It was also shown that by starting with different feasible initial structures it is possible to derive different structures with different final design variable values that have the same displacement pattern and weight.

Sved and Ginos [94] investigated a three-bar truss subjected to three loading cases. Schmit [76] had previously studied this structure assuming a fixed topology. Schmit's fixed topology solution is shown in figure 1. Sved and Ginos showed that a global optimum could be obtained for the problem by removing one of the members and, in effect, violating the stress constraint for that member. They suggested "*if there is a single redundancy, it is necessary systematically to search all perfect structures (i.e. structures for which the stiffness is not singular) that can be obtained from the original one by omitting one member.*" These researchers did not present a suitable technique to search for these subsidiary forms, but they did illustrate the important point that an algorithm which did not test whether the removal of members was beneficial could not ensure a global optimum solution. This is because redundant constraints or members that should be removed can sometimes affect the solution of a problem due to their imposition of strain constraints on the structure. Sved and Gino's solution, to the three bar truss problem previously studied by Schmit [76], in shown in figure 2.

In 1970 Corcoran [5, 6] reconsidered Schmit's three-bar truss [76] and made it six dimensional by considering the horizontal coordinates of three of the joints and the cross-sectional areas of the bars as design variables. Two of the joints coalesced and the optimum configuration became a two bar truss. Corcoran's solution [5, 6] to Schmit's three bar truss problem [76] is shown in figure 3.

This problem considered first by Schmit [76] and shown in figures 1, 2 and 3. This solution shown in figure 2 corresponds to a singular point in the design space. This type of problem where the optimum topology is singular has been investigated in depth by Kirsch [26, 27]. Kirsch also considered the problem of sigular solutions to grillage problems [26, 25, 28]. In 1990, Kirsch [27] suggested that *"in cases of singular solutions, it might be difficult or even impossible to arrive at a true optimum by numerical search alogorithms."* It should also be noted that the inclusion of geometric variables is no guarantee that a singular optimum solution may be found.

Corcoran [5, 6] used two nonlinear algorithms; the gradient projection technique and a sequential unconstrained minimization technique. Each algorithm could account for buckling if required. His approach was a significant advance and was applied to a series of truss structures [6] showing that optimum configurations could be derived using joint coordinates as variables with substantial savings in material. In some cases a small change in topology can be shown to result in a large saving in material. This method allows for the possibility of joints moving together and the superposition of members can result in formation of a reduced topology, thus avoiding the problem of redundant constraints. The difficulty created by elastic instability constraints preventing the removal of members was also avoided with this method. The gradient projection method was found to be more efficient than the sequential unconstrained minimization technique (Powell's Method) and also avoided local optima more reliably.

Several groups of researchers have studied the problem of optimization of trusses with joint coordinates as design variables. For example, in 1968 Cornell [7] published a summary of the work undertaken at M.I.T. which included work of this same type. The sensitivities or derivatives of the stresses and deflections were determined with respect to member areas and joint coordinates and the weight of the structure minimized using a method of successive linearization. Convergence of the technique to a solution was ensured by the use of 'move limits' which restricted the changes in the design variables at each successive iteration [11]. In 1970 Pederson [55] presented an iterative technique for the design of plane trusses subject to a single loading case using a sequence of linear programming with 'move limits'. Both member areas and nodal coordinates were used as design variables. The structures were designed for only one loading case and the resulting solutions were therefore statically determinate. The method utilized a sensitivity analysis of the partial derivatives with respect to the member areas and the joint coordinates and also accounted for self weight, stress and buckling constraints. In 1972, Pederson [56, 57] extended this work to include multiple loading cases and deflection constraints. This algorithm was applied to a bridge structure and cantilever truss. A year later Pederson [58] included upper and lower limits on the design variables without increasing the size of the linear

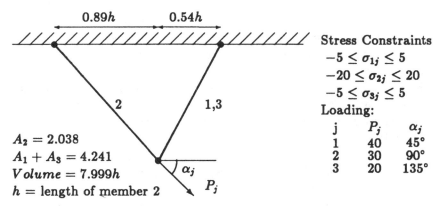

Figure 3. Optimized Three Bar Truss due to Corcoran [5, 6]
with geometric variables (x co-ordinates of the support joints)

programming problem and extended the work to space trusses. The technique was
successfully used to design the shape of a dome structure. The general conclusions
in these papers indicate that the members in a structure subject to active buckling
constraints tend to be shorter than members not thus constrained and that active
displacement constraints normally resulted in a more spread out optimal layout.
Without displacement constraints the optimal design is often fully stressed. However,
with given or fixed joint coordinates this statement sometimes does not hold true
although the difference in total mass for the optimal design and the fully stressed
design was often found to be negligible. Optimum Designs are frequently, as with
this dome, very flat or insensitive; i.e., little is gained by the last few steps of the
iteration.

Thomas and Brown [100] used a penalty function method with a sequential unconstrained minimization technique to optimize various roof systems considering a
cost objective function. Members that approached their minimum allowable cross-
sectional area were deleted except where they were needed to ensure stability. Once
a member had been deleted, the structure was re-optimized and the possibility of
removing more members investigated. A similar scheme was adopted when joints
coalesced. Graphs to relate the member areas to the dependent variables (i.e. second moments of area) were drawn using the available section data. The technique
was applied to several rigidly jointed structures using an approximate technique to
determine the effective lengths of the members. Thomas and Brown noted that when
secondary moments are not high the method gave solutions that were similar to those
derived assuming a pin-jointed structural model. Their examples indicated that the
changes in geometry can produce significant reductions in the cost of a roofing system.

Saka [74] presented a technique for the design of trusses subject to multiple loading cases where the member areas and the modal coordinates were varied and account was taken of the grouping of members with stiffness, displacement, stress and buckling constraints. In addition to the member areas and nodal coordinates, joint displacements were considered as design variables. This avoided having to analyze the structure. The constraints were linearized by direct differentiation and the simplex method used for optimization in an iterative fashion. Typical struss structures were designed to demonstrate that the technique yields results which are better than those derived by methods not using nodal coordinates as design variables. This method appeared to converge to a solution rapidly.

Topping and Robinson [69, 70, 97, 98, 99] applied sequential linear programming to the optimization of rigidly jointed timber plane and space frames using geometric and cross-sectional design variables. The dimensions of the cross-section linked section modulus and cross-sectional area. The stress constraints were formulated using the British code of practice. Deflection constraints were also included and a number of two and three dimensional examples presented. Majid and Tang [45] have also contributed to practical studies concerning the optimum shape design of space trusses using sequential linear programming and constraints formulated from British codes of practice.

Imai and Schmit [22] developed an advanced primal-dual method called the multiplier method, which is an extension of the quadratic penalty function approach. To avoid the severe nonlinearity of these truss problems, second order Taylor's series expansions for the displacement quantities (in terms of the reciprocal sizing and nodal coordinate variables) are used along a search direction. Stress, displacement and Euler buckling constraints can be imposed. The algorithm was shown to be well conditioned and has the advantage that it can be used with any starting design feasible or infeasible.

Lin et al. [33, 34] presented an algorithm based on the Kuhn-Tucker criteria for minimal weight of a structure subject to stress and dynamic constraints. They include member sizes and nodal coordinates as design variables. The dynamic constraints were a frequency prohibited band and the stress constraints were based on the fully-stressed criteria. The illustrative examples included a plane truss, a frame structure and roof space truss. Kapoor and Kumarasamy [23] have used Powell's unconstrained minimization technique to optimize the configuration of transmission line towers (trusses). Dynamic, stress and displacement constraints were considered and the design variables included; member areas, the dimensions of the tower base width and panel heights. They included that a rigorous dynamic analysis was required for optimum design and that the tower configuration was most sensitive to dynamic loads than static loads.

Little attention appears to have been given to developing methods which introduce into the structure new members or nodes that may be required to derive an optimum design. Work of this type appears to have been confined to one paper by Spillers and Friedland [15, 86, 88] who developed a heuristic scheme for modifying the connectivity of a structure so as to improve the design of truss structures subject to a single loading system. Michell's work indicates that relatively large numbers of joints and members should be included in a structure. However, other studies reviewed in this and the last section have indicated that many members and joints should be removed. An optimal design algorithm should therefore allow for the addition of joints and members in the design process as well as their removal. Comprehensive algorithms of this type have yet to be developed.

In Lev's report [32] of 1981, the fact that geometric optimization does not enjoy the same degree of progress and application as fixed geometry optimization was attributed to the following factors;

a. the increased number of decision variables

b. the different degree of nonlinearity in the numerical behaviour of these variables

c. the potential change in topology under both fixed and variable geometry optimization."

It is the second of these factors which led to the development of the hybrid methods which are described in the next section.

5   Hybrid Methods

The problem of designing both the shape and the member sizes of a structure may lead to large members of mixed design variables which, in turn, may have 'a wide range of sensitivities. In addition, as was shown by Sved and Ginos [94], the design space can become disjointed thereby making solution to a global optimum difficult. To avoid these problems some researchers have divided the variables into two design spaces. Others have allowed for topological considerations at regular intervals in the design process.

McConnel [47] has developed a method for obtaining Mitchell structures by minimising the volume of a fixed topology pin-jointed truss subject to a single load case by varying joint co-ordinates. He did not consider member instability and as expected his solutions were determinate. He utilised Powell's direct search method for the co-ordinate variation. He also suggested decomposing the problem into two sub-problems: the optimization of a highly connected 'ground structure' using linear programming (as previously suggested by Hemp [19]); and the subsequent variation of the joint co-ordinates using Powell's direct search method.

Vanderplaats and Moses [101, 102] developed a technique for truss structures subject to multiple load cases. The stress-ratio method was used to size the structure keeping the topology fixed and the steepest descent method was used to move the nodal coordinates to an optimum position. The algorithm used these two techniques one after another in an iterative fashion keeping the design spaces separate. The range of allowable member sizes were constrained and member buckling constraints were introduced. But it was not possible automatically to add or delete members or joints during the design process. They concluded that using two separate but dependent design spaces reduced the number of design variables considered at any one time. Consequently, ill-conditioning problems were seldom encountered with the method. Vanderplaats [103, 104] developed this idea further for structures which were also deflection constrained. This time the method of feasible directions, with modifications to account for initial infeasible designs, was used to optimize the structure with fixed geometry. Reciprocal design variables were used to transform the constraint functions to approximately linear form. The search direction for the geometric optimization is determined by the feasible direction method using the slope of the active constraints and the objective function. A one dimensional search was used to determine the step size. The sizes of the members were then adjusted with the geometry fixed using the feasible directions method. As before, these techniques were used iteratively until an optimum solution was determined. Once again the method was found to converge after only a few iterations resulting in a major weight reduction for the problems considered. Fu [14] developed another algorithm for the design of trusses subject to multiple loading cases using a combined gradient method and random search technique. By moving one of the nodes he investigated the properties of a truss subject to a single loading case and showed that the weight function is convex. For the simple problem that he studied the global optimum (which agrees with Michell's results) can be determined by differentiation of the weight function with respect to the node coordinates. For the general case, however, this cannot be proved and a design method consisting of two stages was presented. In the first stage, the fixed geometry was sized using the critical member forces of each loading case. The structure was sized directly if determinate, but if statically indeterminate the 'cutting plane' method using a series of simplex solutions with 'move limits' was employed. Once the sizing of the fixed geometry had been completed, the effects of varying the unloaded nodes in the structure could be considered using a search procedure to determine their optimum position. This work indicated that the surface of the weight function created by variation of the coordinates of a single node was unimodal.

Lipson and Agrawal [35] used the Complex Method to design trusses subject to stress constraints (including buckling stresses) and multiple loading cases. The member areas and node coordinates were used as design variables and were varied at the same time. Lipson and Gwin [36, 37] later concluded that when using the Complex Method it was efficient to separate the design variables and only consider either the member areas or nodal coordinates at any one time. With this later work a cost function was used and a possibility of deflection constraints included. Member deletion was permitted when the member force tended to zero during the design process and when the deletion did not render the structure unstable. Members could be linked together in groups. The problem of a non-continuous range of member sizes was considered by making the sizes discrete at each iteration rather than at the completion of the design process. The stress-ratio method with a displacement scaling technique to account for displacement constraints was used for sizing the fixed geometry problem. The Complex Method was used for the geometry optimization. As with Vanderplaat's algorithm [102], iteration between the design spaces continues until convergence. However, with this algorithm it was possible to delete members and joints during the design process. Structures were designed for both minimum weight and minimum cost, but the results showed that no general conclusions could be drawn since they were sometimes similar and at other times dissimilar depending on the type of structure, loads and unit costs.

Spillers and Friedland [85, 86, 87, 88] developed other techniques, first for statically determinate truss structures and later for statically indeterminate truss structures. They derived expressions for the gradients of the weight function with respect to the node co-ordinates and cross-sectional areas. Optimality conditions were formulated by ignoring the compatibility requirements and using the Kuhn-Tucker conditions. An iterative scheme, based on Newton's Method, was developed to ensure that these conditions were satisfied by the modified structure. The conditions for the nodal positions were linearized and the change in the coordinates computed. Again the design procedure was kept separate in two design spaces. Spillers and Kountouris [89] used the AISC allowable stress formulas to define a weight/unit length relationship with the safe load of the available sections. This effectively accounted for member buckling and the objective functions was modified accordingly. Optimality conditions were re-established for the member areas and nodal coordinates. Successive iteration of these conditions enabled the optimum structure shape and member sizes to be determined. However, this work only considered one loading case. Lev [31] suggested a heuristic and very attractive approach to the problems of geometric optimization. He showed that the solutions derived using the Spillers and Vanderplaats programs can, because of the different sequence of optimization, result in different local optimum solutions. He suggested that to optimize a structure with only one loading case, the structure should be first sized using the stress-ratio method with the geometry fixed. During this process some of the member areas will reduce to zero and a "reduced" structure will be derived. He concluded that the optimization of the geometry should only be undertaken once the topology has been reduced using this method. Where an indeterminate form is derived using the stress-ratio method each determinate subset should be investigated. The method was illustrated in the paper using an example studied earlier by Spillers [85]. Lev also suggested that the initial topology for the geometric design of a structure subject to multiple loading cases should be a layout formed from a combination of the topologies of the single loading designs. This method in effect combines the 'ground structure' approach with a subsequent iteration procedure which alternates between the member sizing variable subspace and the node co-ordinate variable subspace.

In addition to these geometric methods several hybrid methods involving the 'ground structure' approach has been developed. One of the earliest of these by Reinschmidt and Russell [66, 67, 73] has already been mentioned. Their method consisted of applying linear programming techniques followed by the Fully Stressed Design method to a 'ground structure' subject to multiple loading cases and buckling stress constraints. Successive iteration of these methods ensured that structurally inefficient members were removed by the application of the linear programming algorithm and that compatibility was restored by the Fully Stressed Design resizing. Both techniques had to be applied repeatedly to ensure that the buckling stress constraints were satisfied. They showed that the algorithm was more successful than the Fully Stressed Design Method alone in deriving optimum configurations. The method was also used with a discrete set of available steel sections by adopting (at the end of each iteration of the Linear Programming or Fully Stressed Design sizing) the minimum weight sections which satisfy the AISC specification. The iterative procedure was continued until the sections selected after resizing were unchanged.

Sheu and Schmit [81] also used a 'ground structure' and based their method on a comparison of the upper bound for the configuration derived from the feasible direction technique with the lower bound for subsets of the configuration derived using the dual-simplex algorithm with compatibility relaxed. The Linear Programming solutions for the subsets of bars were used as a guide to the optimum configuration. Farshi and Schmit [12] developed another method to avoid the problems of a disjoint feasible design space, these problems having been illustrated by Sved and Ginos [94]. In this method redundant member forces and cross-sectional areas of all the members were used as design variables. Limits on the maximum and minimum member sizes were included in the constraints. Compatibility requirements were at first ignored and the problem solved using the simplex method. This initial design was analyzed using the displacement method and strain compatibility requirements for the redundant members were formulated. These compatibility requirements were expressed in a linear form and included as constraints in a subsequent linear programming problem. The constraints are non-linear and depend on the solution of the previous linear programming problem. The technique is therefore used iteratively until the compatibility violations decrease and the linear programming solution becomes compatible. Any unnecessary members in the structure vanish during the linear programming method and technique avoids local optima and generally yields a global optimum solution. Unfortunately the speed of convergence of this method depends on the choice of redundants.

Lev [29] has considered the problem of how to design an optimum statically determinate form to support two loading cases. The initial form of the structure is selected from an overlay of the determinate solutions derived from the separate load cases. This usually results in a statically indeterminate form. The solution is made determinate by using a process similar to the simplex method to predict the order in which the members are to be removed from the structure. The method gives a series of statically indeterminate optima of decreasing redundancy and finally a statically determinate optimum. All solutions except the determinate optimum are unlikely to be elastic 'realizable'. Thus, unless a statically determinate structure is required this method would not be suitable for elastic design.

Majid and Elliott [38, 40, 41] developed a technique for the design of trusses using a 'ground structure.' A 'steepest descent-alternate mode algorithm' was used to optimize the weight of the structure when subject to multiple loading cases with deflection, stress and buckling constraints. Theorems of Structural Variation were formulated so that the structure need only be analyzed once during the whole design process. Subsequent analyses were made using the theorems together with the original analysis and a series of influence coefficients for unit loads applied at the ends of each member. The structure was first optimized keeping the topology fixed. The theorems and influence coefficients were then used to provide trade-off data to indicate the most beneficial order in which members should be deleted from the 'ground structure'. The theorems were also used to indicate if removal of a particular member would cause the structure to become unstable. The design algorithm accounted for the possibility of grouping members together. A series of local optimum designs of gradually reducing design were derived using this method. These theorems were extended by Majid and Saka to the analysis [39, 43] and optimization [42] of rigidly jointed frames. In this latter work the 'cutting plane' method incorporating move limits was used to optimize a 'ground structure'. This method was also applied to the design of steel sway frames [44] using a cost objective function and 'differential' displacement constraints to restrict the relative sway of the frame. In the case of frame structures members which decreased to zero cross-sectional during the design process were removed from the layout. Majid et. al. [44] showed that minimum weight was not a sound objective for sway frames and that cost has to be taken into consideration.

Kirsch [24] proposed a number of approximation concepts to be used with the optimization of structures with nodal coordinates included among the design variables. These concepts include:

1. Approximations based on information obtained from exact analysis of a limited number of designs.

2. Approximations based on temporarily neglecting the implicit analysis requirements.

He presented an algorithm in which the compatibility conditions were ignored until the final geometry of the structure was fixed and the number of design variables was reduced by a design variable linking procedure. Again the optimization was performed using separated design spaces.

Shiraishi and Furuta [83] used sequential linear programming to delete unnecessary members from truss structures and the Monte Carlo method for optimising truss geometry. They considered member buckling and multiple loading cases. Their case studies showed that, as expected, the initial 'ground structure' affects the final optimum topology and geometry. In addition, the studies demonstrated that different support systems have a significant effect on the optimum topology and geometry. Shiraishi, Furuta and Ikejima [83] considered the optimization of truss and arch bridges. Once again they separated the cross-sectional properties and nodal coordinate design variables into two design spaces and iterated between the two until convergence. They utilised sequential linear programming or optimality criteria for the fixed geometry optimization and the variable metric method foir the geometrical optimization. They showed that according to the initial topolgy different structural types would be derived. Their work clearly illustrated that *"topological factors may play an important role in decisions concerning the selection of structural types."*

6 Other Methods

In 1964 Porter Goff [60, 61] used dynamic programming to minimize the weight of truss structures subject to multiple loading cases. This method was based on a sequential decision process using a field of allowable members (similar to a 'ground structure'). The route of accepted members in the field is based on in a series of decisions starting from a loaded node of the structure. Porter Goff's approach is founded on an intuitive principle of optimality; viz. "At any stage in an optimal policy the sequence of remaining decisions constitutes an optimal policy from that stage irrespective of earlier decisions." The method was applied to a series of cantilever trusses subject to a single load [60] and the results were compared with those of Mitchell and Hemp. Later the same method [61] was applied to a bridging structure subject to buckling stress constraints and multiple loading cases. Palmer and Sheppard [53] also used dynamic programming to optimize the shape of symmetric pin-jointed cantilever trusses subject to a single load. They showed that this method required far less computation than the direct enumeration 'ground structure' approach. They suggested that a coarse grid could be used for an initial design and that the method could then be applied to a series of locally refined grids. The solutions to these problems are, of course, statically determinate. The work was extended to consider multiple loading cases and discrete member size. The solutions for these problems were again assumed to be statically determinate. In a later paper [80] the same authors applied these techniques to the design of the optimal layout of transmission towers.

Ho [21] using a limit analysis approach has developed a nested decomposition technique for the optimal design of multi-stage structures subject to a single loading system. With this technique the structure is divided into stages and the algorithm solves a sequence of smaller linear programming problems. This enables larger scale problems to be solved in what is termed a 'staircase' of linear programs. Configuration is considered by treating each stage of the structure as a 'ground structure'. For the single loading cases the structural configuration derived using this method will be determinate and therefore a fully stressed elastic optimum.

7 Conclusions

Virtually every mathematical programming technique has been applied to the shape optimization of skeletal structures. However, few algorithms with comprehensive design constraints exist. The convergence of these algorithms to globally optimum solutions is difficult to ensure; but for practical design situations, provided these algorithms give substantial savings in resources and or costs, then structural engineers should be encouraged to take advantage of them. Most algorithms are formulated to either remove members, or move joints and sometimes both in a design procedure which seeks to minimize weight or cost. But rigorous methods incorporating algorithms which consider the possibility of introducing new members and nodes during the optimization procedure have yet to be developed.

Kirsch [26] has recently described the toplogical design problem as *"perhaps the most challenging of the structural optimization tasks."* Considerably more progress will have to be made before the general shape optimization problem for skeletal structures can be solved.

Acknowledgment   The author acknowledges and appreciates the contributions made to this work by his research students and assistants. In particular, the contribution made by D.J. Robinson and H.F.C. Chan is gratefully acknowledged.

References

[1] Barnes, M.R., Topping, B.H.V., Wakefield, D.S., "Aspects of Form-Finding by Dynamic Relaxation," International Conference on the Behaviour of Slender Structures, London, United Kingdom, Sept., 1977.

[2] Barta, J., "On the Minimum Weight of Certain Redundant Structures," Actatechnica Academiae scientiarum hungaricae, Budapest, Hungary, Vol.18, 1957, 67-76.

[3] Chan, H.S.Y., "The Design of Michell Optimum Structures," College of Aeronautics, Cranfield, United Kingdom, Report No. 142, Dec., 1960.

[4] Chan, H.S.Y., "Optimum Structural Design and Linear Programming," College of Aeronautics, Cranfield, United Kingdom, Report No. 175, Sept., 1964.

[5] Corcoran, P.J., "Configurational Optimization of Structures," International Journal of Mechanical Sciences, Vol.12, 1970, 459-462.

[6] Corcoran, P.J., "The Design of Minimum Weight Structures of Optimum Configuration," thesis presented to The City University, at London, United Kingdom, in 1970, in fulfillment of the requirements for the degree of Doctor of Philosophy.

[7] Cornell, C.A., "Examples of Optimization in Structural Design," in An Introduction to Structural Optimization, Study No.1, Solid Mechanics Division, University of Waterloo, Waterloo, Canada, 1968.

[8] Cox, H.L., "The Design of Structures of Least Weight," Pergamon Press, London, United Kingdom, 1965.

[9] Dobbs, M.W., Felton, L.P., "Optimization of Truss Geometry," Journal of the Structural Division, ASCE, Vol. 95, No.ST10, Oct., 1969, 2105-2118.

[10] Dorn, W.S., Gomory, R.E., Greenberg, H.J., "Automatic Design of Optimal Structures," Journal de Mecanique, Vol.3, No. Mars, France, 1964.

[11] Estrada-Villegas, J.E., "Optimum Design of Planar Trusses using Linear Programming," thesis presented to Massachusetts Institute of Technology, at Cambridge, Massachusetts, in 1965, in partial fulfillment of the requirements for the degree of Master of Science.

[12] Farshi, B., Schmit, L.A., "Minimum Weight Design of Stress Limited Trusses," Journal of the Structural Division, ASCE, Vol. 100, No. ST1, Jan., 1974, 97-107.

[13] Fleron, P., "The Minimum Weight of Trusses," Bygningsstatiske Meddelelser, Vol. 35, No. 3, Dec., 1964, 81-96.

[14] Fu, Kuan-Chen, "An Application of Search Techniques in Truss Configuration Optimisation," Computers and Structures, Vol. 3, 1973, 315-328.

[15] Friedland, L.R., "Geometric Structural Behaviour," thesis presented to Columbia University, at New York, N.Y., in 1971, in partial fulfillment of the requirements for the degree of Doctor of Engineering Science.

[16] Gallagher, R.H., "Fully Stressed Design," Chapter 3 of Optimum Structural Design - Theory and Applications, R.H. Gallagher, O.C. Zienkiewicz, eds., John Wiley and Sons, Inc., New York, N.Y., 1977.

[17] Hajela, P., Ashley, H., "Hybrid Optimization of Truss Structures with Strength and Buckling Constraints," Proceedings, International Symposium on Optimum Structural Design, Tucson, Ariz., Oct. 1981.

[18] Hemp, W.S., "Studies in the Theory of Michell Structures," Proceedings International Congress of Applied Mechanics, Munich, West Germany, 1964.

[19] Hemp, W.S., "Optimum Structures," Clarendon Press, Oxford, United Kingdom, 1973.

[20] Hemp, W.S., Chan, H.S.Y., "Optimum Design of Pin-Jointed Frameworks," Aeronautical Research Council Reports and Mem. No. 3632, Her Majesty's Stationery Office, London, United Kingdom, 1970.

[21] Ho, J.K., "Optimal Design of Multi-Stage Structures: A Nested Decomposition Approach," Computers and Structures, Vol.5, 1975, pp 249-255.

[22] Imai, K., Schmit, L.A., "Configuration Optimization of Trusses," Journal of the Structural Division, ASCE, Vol. 107, No. ST5, May, 1981, 745-756.

[23] Kapoor, M.P., Kumarasamy, K., "Optimum Configuration of Transmission Towers in Dynamic Response Regime," Proceedings, International Symposium on Optimum Structural Design, Tucson, Ariz., Oct., 1981.

[24] Kirsch, U., "Synthesis of Structural Geometry Using Approximation Concepts," Computers and Structures, Vol.15, No. 3, 1982, 305-314.

[25] Kirsch, U., "Optimal Topologies of Flexural Systems", Engineering Optimization, v.11, 141-149, 1987.

[26] Kirsch, U., "Optimal Topologies of structures", Applied Mechanics Review, vol.42, n.8, 223-238, 1989

[27] Kirsch,U., "On singular topologies in optimum structural design", Structural Optimization, v.2, n.3, 133-142, 1990.

[28] Kirsch, U., Taye, S., "On Optimal Topologies of Grillage Structures", Engineering Computations, v.1, 229-243, 1986.

[29] Lev, O.E., "Optimum Choice of Determinate Trusses under Multiple Loads," Journal of the Structural Division, ASCE, Vol. 103, No. ST2, Feb., 1977, 391-403.

[30] Lev, O.E., "Topology and Optimality of Certain Trusses," Journal of the Structural Division, ASCE, Vol. 107, No. ST2, Feb., 1981, 383-393.

[31] Lev, O.E., "Sequential Geometric Optimization," Journal of the Structural Division, ASCE, Vol. 107, No. ST10, Oct., 1981, 1935-1943.

[32] Lev, O.E., ed., "Structural Optimization - Recent Developments and Applications", ASCE, New York, N.Y., 1981.

[33] Lin, J.H., Che, W.Y., Yu, Y.S., "Structural Optimization on Geometrical Configuration and Element Sizing with Statical and Dynamical Constraints," Proceedings, International Symposium on Optimum Structural Design, Tucson, Ariz., Oct., 1981.

[34] Lin, J.H., Che, W.Y., Yu, Y.S., "Structural Optimization on Geometrical Configuration and Element Sizing with Statical and Dynamical Constraints," Computers and Structures, Vol. 15, No. 5, 1982, 507-515.

[35] Lipson, S.L., Agrawal, K.M., "Weight Optimization of Plane Trusses," Journal of the Structural Division, ASCE, Vol. 100, No. ST5, May 1974, 865-879.

[36] Lipson, S.L., Gwin, L.B., "Discrete Sizing of Trusses for Optimal Geometry" Journal of the Structural Division, ASCE, Vol. 103, No. ST5, May, 1977, 1031-1046.

[37] Lipson, S.L., Gwin, L.B., "The Complex Method Applied to Optimal Truss Configuration," Computers and Structures, Vol. 7, 1977, 461-468.

[38] Majid, K.I., "Optimum Design of Structures," Newnes-Butterworths, London, United Kingdom, 1974.

[39] Majid, K.I., "Generalized Theroems of Structural Variation for Rigidly Jointed Frames," International Conference on the Behaviour of Slender Structures, London, United Kingdom, Sept. 1977.

[40] Majid, K.I., Elliott, D.W.C., "Forces and Deflections in Changing Structures," The Structural Engineer, Vol. 51, No. 3, Mar., 1973, 93-101.

[41] Majid, K.I., Elliott, D.W.C., "Topological Design of Pin Jointed Structures by Non-Linear Programming," Proceedings, Institution of Civil Engineers, Vol. 55, Pt. 2, Mar., 1973, 129-149.

[42] Majid, K.I., Saka, M.P., "Optimum Design of Rigidly Jointed Frames," Proceedings of the Symposium on Applications of Computer Methods in Engineering, Los Angeles, Aug., 1977.

[43] Majid, K.I., Saka, M.P., Celik, T., "The Theorem of Structural Variation Generalized for Rigidly Jointed Frames," Proceedings, Institution of Civil Engineers, Vol. 65, Pt. 2, Dec., 1978, 839-856.

[44] Majid, K.I., Stojanovski, P., Saka, M.P., "Minimum Cost Topological Design of Sway Frames," The Structural Engineer, Vol. 58B, No. 1, Mar., 1980, 14-20.

[45] Majid, K.I., Tang, X., "The Optimum Design of Pin-jointed Space Structures with Variable Shape", The Structural Engineer, Vol. 62B, n.2, 31-37, 1984.

[46] Maxwell, J.C., "On Reciprocal Figures, Frames and Diagrams of Forces," Scientific Papers, Cambridge University Press, United Kingdom, Vol.2, 1890, 175-177.

[47] McConnel, R.E., "Least-Weight Frameworks for Loads across Span". Journal of the Engineering Mechanics Division, American Society of Civil Engineers, v.100, n. EM5, 885-901, 1974.

[48] Michell, A.G.M., "The Limits of Economy of Materials in Frame Structures," Philosophical Magazine, Series 6, Vol.8, No. 47, 1904, 589-597.

[49] Murotsu, Y., Shao, S., "Optimum shape design of truss structures based on reliability", Stuctural Optimization, Vol. 2, n.2, 65-76, 1990.

[50] Owen, J.B.B., "The Analysis and Design of Light Structures," Edward Arnold, London, United Kingdom, 1975.

[51] Parkes, E. W., Braced Frameworks, 2nd Edn., Pergamon Press, London, United Kingdom, 1974.

[52] Parkes, E.W., "Joints in Optimum Frameworks," International Journal Solids Structures, Vol. 11, 1975, 1017-1022.

[53] Palmer, A.C., Sheppard, D.J., "Optimising the Shape of Pin-Jointed Structures," Proceedings, Institution of Civil Engineers, Vol. 47, 1970, 363-376.

[54] Pearson, C.E., "Structural Design by High-Speed Computing Machines," Conference on Electronic Computation, Structural Division, ASCE, Kansas City, Mo., Nov. 1958.

[55] Pederson, P., "On the Minimum Mass Layout of Trusses," Advisory Group for Aerospace Research and Development, Conf. Proc. No. 36, Symposium on Structural Optimization, AGARD-CP-36-70, 1970.

[56] Pederson, P., "On the Optimal Layout of Multi-Purpose Trusses," Report No. 18, The Technical University of Denmark, Dec. 1971.

[57] Pederson, P., "On the Optimal Layout of Multi-Purpose Trusses," Computers and Structures, Vol. 2, 1972, 695-712.

[58] Pederson, P., "Optimal Joint Positions for Space Trusses," Journal of the Structural Division, ASCE, Vol. 99, No. ST12, Dec. 1973, 2459-2476.

[59] Pope, G.G., "The Application of Linear Programming Techniques in the Design of Optimum Structures," Symposium on Structural Optimization, Conference Proceedings, No. 36, AGARD-CP-36-70, 1970.

[60] Porter Goff, R.F.D., "Decision Theory and the Shape of Structures," J. Roy, Aero. Soc., Vol. 70, Mar. 1974, 448-452.

[61] Porter Goff, R.F.D., "Dynamic Programming and the Shape of Structures," Proceedings International Conference, Computer Aided Design, University of Southampton, Apr. 1969.

[62] Prager, W., "A Note on Discretized Michell Structures," Computer Methods in Applied Mechanics and Engineering, Vol. 3, 1974, 349-355.

[63] Prager, W., "Optimal Layout of Cantilever Trusses," Journal of Optimization Theory and Applications, Vol. 23, No. 1, Sept., 1977, 111-117.

[64] Prager, W., Rozvany, G.I.N., "Optimisation of Structural Geometry," Dynamical Systems, Proceedings, University of Florida, International Symposium, A.R. Bednarek and L. Cesari, Eds., Academic Press, New York, N.Y., 1977.

[65] Prager, W., Rozvany, G.I.N., "Optimal Layout of Grillages," Journal of Structural Mechanics, Vol.5, No. 1, 1977, 1-18.

[66] Reinschmidt, K.F., Russell, A. D., "Linear Methods in Structural Optimisation," Research Report R 70-41, Department of Civil Eng., Cambridge, Mass., July, 1970.

[67] Reinschmidt, K.F., Russell, A.D., "Applications of Linear Programming in Structural Layout and Optimization," Computers and Structures, Vol. 4, 1974, 855-869.

[68] Richards, D.M., Chan, H.S.Y., "Developments in the Theory of Michell Optimum Structures," Advisory Group for Aerospace Research and Development, Report 543, Apr., 1;966.

[69] Robinson, D.J., "Optimization of Rigidly Jointed Frames", thesis presented for Doctor of Philosophy, University of Edinburgh, 1984.

[70] Robinson, D.J., Topping, B.H.V., "Frame Optimization and Column Design", Civil Engineering Systems, v.1, 211-223, 1984.

[71] Rozvany, G.I.N., "Optimal Design of Flexural Systems - Beams, Grillages, Slabs, Plates and Shells," Pergamon Press, Oxford, United Kingdom, 1976.

[72] Rozvany, G.I.N., "Discussion on Reference 23," Journal of the Structural Division, ASCE, Vol. 103, No. ST12, Dec., 1977, 2432-2433.

[73] Russell, A.D., Reinschmidt, K.F., Discussion on "Optimum Design of Trusses for Ultimate Loads," Lapay, W.S., Globle, G.G., Journal of the Structural Division, ASCE, Vol. 97, No. ST9, Sept., 1971, 2437-24

[74] Saka, M.P., "Shape Optimization of Trusses," Jounral of Structural Division, ASCE, Vol. 106, No. ST5, May, 1980, 1155-1174.

[75] Schmidt, L.C. "Minimum Weight Layouts of Elastic Statically Determinate, Triangulated Frames under Alternative Load Systems," J. Mech. Phys. Solids, Vol. 10, 1962, 139-149.

[76] Schmit, L.A., "Structural Design by Systematic Synthesis," Proceedings, Second National Conference on Electronic Computation, ASCE, Pittsburgh, Pa., Sept. 1960, 105-132.

[77] Schmit, L.A., Kicher, T.P., "Synthesis of Material and Configuration Selection," Journal of the Structural Division, ASCE, Vol. 88, No. ST3, June, 1962, 79-102.

[78] Schmit, L.A., Morrow, W.M., "Structural Synthesis with Buckling Constraints," Journal of the Structural Division, ASCE, Vol. 89, No. ST2, Apr. 1963, 107-126.

[79] Schmit, L.A., Mallett, R.H., "Structural-Synthesis and Design Parameter Hierarchy," Journal of the Structural Division, ASCE, Vol. 89, No.ST4, Aug. 1963, 269-299.

[80] Sheppard, D.J., Palmer, A.C., "Optimal Design of Transmission Towers by Dynamic Programming," Computers and Structures, Vol. 2, 1972, 455-468.

[81] Sheu, C.Y., Schmit, L.A., "Minimum Weight Design of Elastic Redundant Trusses under Multiple Static Loading Conditions," AIAA Journal, Vol. 10, No. 2, Feb. 1972, 155-162.

[82] Shiraishi, N., Furuta, H., "On Geometry of Truss", Memoirs of the Faculty of Engineering, Kyoto University, Vol. XLI, Part 4, 498-517, October, 1979. *(in english)*

[83] Shiraishi, N., Furuta, H., "A few Considerations in Topological and Geometrical Systems of Trussed Bridges", Theoretical and Applied Mechanics, v.28, 129-137, 1980,

[84] Shiraishi, N., Furuta, H., Ikejima, K., "Configurational Optimization of Framed Structural Systems", Theoretical and Applied Mechanics, v.29, 85-91, 1981,

[85] Spillers, W.R., "Iterative Design for Optimal Geometry," Jounral of the Structural Division, ASCE, Vol. 101, No. ST7, July 1975, 1435-1442.

[86] Spillers, W.R., Friedland, L., "On Adaptive Structural Design," Journal of the Structural Division, ASCE, Vol. 98, No. ST10, Oct. 1971, 2155-2163.

[87] Spillers, W.R., "Some Problems of Structural Design" in "Basic Questions of Design Theory," North Holland Publishing Co., Amsterdam, Holland, 1974.

[88] Spillers, W.R., "Iterative Structural Design," North Holland Publishing Co., Amsterdam, Holland, 1975.

[89] Spillers, W.R., Kountouris, G.E., "Geometric Optimisation Using Simple Code Representation," Journal of the Structural Division, ASCE, Vol. 106, No. ST5, 1980, 959-971.

[90] Spillers, W.R., Lefcochilos, E., "Elastic Realizability of Force and Displacement Systems in Structures," Quart. Appl. Math., Vol. Jan. 1980, 411-420.

[91] Spillers, W.R., Lev, O., "Design for Two Loading Conditions," International Journal of Solids and Structures, Vol. 7, 1971, 1261-1267.

[92] Spillers, W.R., "Shape Optimization of Structures", Chapter 2 of "Design Optimization", edited J.S. Gero, Academic Press, Florida, 1985.

[93] Sved, G., "The Minimum Weight of Certain Redundant Structures," Australian J. Appl. Sci., Vol. 5, 1954, 1-9.

[94] Sved, G., Ginos, Z., "Structural Optimization under Multiple Loading," Int. J. Mech. Sci., Vol. 10, 1968, 803-805.

[95] Topping, B.H.V., "Form-Finding of Modular Space Structures using Dynamic Relaxation," thesis presented to The City University, at London, United Kingdom, in 1978, in fulfillment of the requirements for the degree of Doctor of Philosophy.

[96] Topping, B.H.V., "Shape Optimization of Skeletal Structures: A Review", ACSE, Journal of Structural Engineering, v.109, n.8, 1933-1951, 1983

[97] Topping, B.H.V., Robinson, D.J., "Optimization of Timber Framed Structrures", Computers & Structures, v. 18, n.6, 1167-1177, 1984.

[98] Topping, B.H.V., Robinson, D.J., "Selecting non-linear optimization techniques for structural design", Engineering Computations, v.1, 252-262, 1984.

[99] Topping, B.H.V., Robinson, D.J., "Computer Aided Design of Timber Space Frames", CAD84, Proceedings of the 6th Int Conf on Computers in Engineering, Butterworths, 74-88, 1984.

[100] Thomas, H.R., Brown, D.M., "Optimum Least-Cost Design of a Truss Roof System," Computers and Structures, Vol. 7, 1977, 13-22.

[101] Vanderplaats, G.N., "Automated Design of Elastic Trusses for Optimum Geometry," thesis presented to Case Western Reserve Univ. at Cleveland, Ohio, in June 1971 in partial fulfillment of the requirements for the degree of Doctor or Philosophy.

[102] Vanderplaats, G.N., Moses, F., "Automated Design of Trusses for Optimum Geometry," Journal of the Structural Division, ASCE, Vol. 98, No. ST3, 1972, 671-690.

[103] Vanderplaats, G.N., "Design of Structures for Optimal Geometry," National Aeronautics and Space Administration Report, TMX-62, Aug. 1975, p. 462.

[104] Vanderplaats, G.N., "Structural Optimisation via a Design Space Hierarchy," International Journal for Numberical Methods, Vol. 10, J. Wiley, England, 1976, 713-717.

[105] Vanderplaats, G.N., "Numerical Methods for Shape Optimization - An Assessment of the State of the Art," Proceedings, International Symposium on Optimum Structural Design, Tucson, Ariz., Oct. 1981.

Chapter 18

EXACT AND APPROXIMATE STATIC
STRUCTURAL REANALYSIS

B.H.V. Topping
Heriot-Watt University, Riccarton, Edinburgh, UK

1   Introduction

The 'exact' and approximate methods of structural re-analysis are reviewed in this chapter. With the advent of less expensive computing hardware, structural engineers are employing optimization and non-linear analysis techniques in new application areas which require more computational power hence re-analysis techniques are likely to become more important over the next decade.

In structural optimization the procedures are generally iterative and require repeated structural analysis, as the structure is progressively modified. In order to avoid a fresh analysis everytime, many re-analysis techniques have been devised as reviewed by Arora [11]. He gave a concise definition of the problem when he states that the problem is:

"to find the response of a structure after modifications using the original response of the structure such that the computational time of re-analysis is less that the complete analysis time".

At the same time re-analysis techniques should allow one to efficiently compute the design sensitivity coefficients used in optimization. Re-analysis techniques are particularly important for large structures, especially in finite element structures where only a small part of the structure is progressively modified. Such re-analysis techniques should be capable of rapid implemention in existing computer codes.

Unfortunately texts specially devoted to re-analysis techniques and their computer implementation are not readily available. At most only single book chapters concerning these much neglected techniques are available. These chapters include Atrek et.al. [14] (Chapter 17), Fox [31] (Chapter 5), Kirsch [51] (Chapter 5), Majid [66] (Chapter 6), Meek [72] (Chapter 6), Pestel and Leckie [87] (Chapter 9,10) and Pipes [89] (Chapter 6).

This paper closely follows the form of review presented in reference [3]. Recent developments in computer architecture make it even more important for research to continue into the formulation and assessment of re-analysis techniques.

Re-analysis techniques may be broadly classified as either direct (or 'exact') or iterative and approximate methods. These techniques are either formulated in the force method, displacement method, or mixed methods of analysis. Judicious selection of a technique is important to ensure that it is easy to apply and is also efficient.

2   Direct (or 'Exact') Methods

These methods give an exact closed-form solutions which have the same effect of solving the modified stiffness equations by undertaking a fresh analysis. In general direct methods may be most efficient if the number of modified elements are small.

2.1   Initial Strain Techniques

The method of 'initial strain or stress' was first formulated in matrix notation by Argyris and Kelsey [5, 8] in 1956. The earlier reference [5] uses the force method of analysis because it was then felt, that it was superior to that of the stiffness method. The subsequent paper [8] however gives the formulation for both the force and stiffness methods. This re-analysis technique was used for the analysis of structures with cutouts (the total removal of elements) and for the modification of elements.

The technique was initially developed to avoid the difficulties of analysing structures with cut-outs by the force method. Instead the analysis of the complete structure was used to predict the response of the modified structure. The matrices required for the analysis of a complete structure using the force method are much more easier to form than those fora structure with cut-outs. In the force method, the original structure is analysed for the external loading and initial strains are imposed on the elements to be removed such that their stresses due to the external loadings and initial strains are reduced to zero. The magnitude of the initial strains are unknowns but may be determined from their corresponding stresses acting on the original structure due to external loadings only. The modified stresses are found in terms of the original stresses. The technique requires only one analysis of the original structure. For the stiffness method initial unknown element stresses are imposed in addition to the applied loads. This is the dual approach of the force method.

This technique had already been developed in 1945 by Best [17, 18] and was exploited by Cicala [25] and Michelson and Dijk [78]. These early attempts were suited for the repeated analysis by hand. Cicala uses special perturbation stress systems to nullify the stresses in the cut-out elements. The perturbation was then added to the stresses obtained for the continous structure. Michelson and Dijk outline the technique to account for variations in cross-sectional areas and elastic modulus. Goodey [36] proposed the same technique but based on a variational argument (the minimisation of the strain energy). This approach was purely a mathematical idea to obtain the desired modifications. The matrix form of Argyris and Kelsey [7, 8] was used by Poppleton [90, 91] for the redesign of redundant structures by the force method. Here the inverse problem was posed; the stresses are specified and the change in the elements is to be determined under a given load system. This approach is much the same as in the examples given by Best, and leads to redesign problems that requires iteration so that the stresses that are specified are satisfied.

The initial strain concept was disputed by Grzedzielski [37] who introduced fictitious thermal loads to replace the initial strains. In addition he mentioned that the method was limited for analysing structures with a purely diagonal flexibility matrix. In a series of articles by Grzedzielski, Argyris and Kelsey [7, 38] the validty of the initial strain concept was discussed. However no satisfactory agreement was obtained over this.

The initial strain or stress technique require the inverse of the original flexibility or stiffness matrix respectively. The efficiency of the technique based on operation counts was investigated by Kavlie and Powell [46, 117] for the stiffness method. Using a system of operation counts, they concluded that this re-analysis technique is inefficient compared to a fresh analysis. In addition an error in the original derivation was corrected by the authors.

2.2 Parallel Element Techniques

In 1968 Sobieszczanski [103] introduced the parallel element concept in matrix form using the force method of analysis. This is a perturbation technique originated from the earlier work of Cicala [25] and Michelson and Dijk [78]. It was indirectly suggested in the series of papers by Grzedzielski, Argyris and Kelsey [7, 37, 38] in their controversy of the initial strain concept.

The technique uses superposition of an element parallel to that of the one to be modified. It may be used for the addition (where none originally existed), deletion and modification of elements and hence more general than that of the initial strain concept. The differences to that of the initial strain concept were also mentioned. Here the original structure is updated after each modification; the results of the previous modification is used to predict the response of a new modification.

In the initial strain concept, subsequent analyses is always referred to the original structure. This technique was then formulated for the displacement method by Sobieszczanski [101, 102]. These papers include tests on efficiency and accuracy of the parallel element concept to that of the initial strain concept. The efficiency was measured in terms of operation counts and the computer time. He concluded that the technique is very efficient for a large number of modifications. The accuracy of the technique also seems not to be affected by the magnitude of the modifications. Errors that are obtained are assumed to be random and do not accumulate. At the same time a version of the technique using the displacement method was recognised to be more efficient than that of the force technique [102]. This was because of the sparsity of the stiffness matrix compared to that of the flexibility matrix.

2.3   Modified Inverse of Matrices

The two previous techniques were based on intuitive and physical reasoning rather than a mathematical approach. Many investigators have sought the relation between the change in stiffness, the inverse of the original and modified stiffness matrices. One approach is to express the modified inverse explicitly in terms of the original inverse and the changes in the stiffness.

The required relationship is based on the Sherman-Morrison identity [99, 100]. Householder [42, 43, 44], provides an equation between these three quantities in matrix form. Zielke [120] also derived a similar equation for the inverse of modified symmetric matrices. The modified inverse can be built up a row or a column at a time in any order, and it is also possible to proceed coefficient by coefficient. It was subsequently used by Sack et.al. [95] where the modified inverse was obtained one column at a time. Using operation counts the technique was shown to be efficient in comparison to Gauss elimination when the modifications are small. Kavlie and Powell [46, 117] have also presented an operation count of Sack's technique [95]. They found that it is inefficient compared to a complete re-analysis using the stiffness method.

Kirsch and Rubinstein [53, 54] also investigated various versions of the Sherman-Morrison identity. These versions involves forming the modified inverse by considering simultaneous changes of coefficients, changing the coefficients and the columns one at a time. The most efficient way involves changes of the columns.

Another variation of the Sherman-Morrison identity is given by Mohraz and Wright [79], where the size of the stiffness matrix is allowed to change. For addition of nodes the size of the modified inverse increases and for deletion it decreases. Based on operation counts their proposed method suggested savings of 20%-80% in computational effort compared to a complete inversion. Recently Hager [39] has presented a historical review of the Sherman-Morrison-Woodbury formulas.

Other similar methods due to MacNeal [65] and Goodey [36] were presented in the 1950's. These methods use the original inverse matrix to obtain the modified inverse matrix. MacNeal's method is analogous to the compensation theorem used in network theory and was somewhat simplified by Kosko [58]. Later Kosko [58] derived various techniques for inversion of matrices (including for modified matrices) by partitioning. The possibility of sparsity of a matrix common in the displacement method was not considered. This was probably due to the popularity of the force method during this time.

2.4   Modified Displacement Vector

The technique of calculating the modified inverse matrix directly is expensive. An alternative approach is to calculate the modified displacement vector using the Sherman-Morrison identity. This approach avoids the necessity of forming the inverse explicitly.

The technique of Sack et.al. [95] have several drawbacks as pointed out by Argyris et.al. [6]. The main objection was that to compute the inverse of a large and banded stiffness matrix is expensive, and that to store it may at times be uneconomical. Secondly the Sherman-Morrison identity considers modification of one column at a time. Based on this objections Argyris et.al. [6] produced yet another algorithm to be used with the displacement method. Instead of the modified inverse, the modified displacement vector is calculated by rearranging the Sherman-Morrison identity. The method made use of the sparsity and bandwidth of the stiffness matrix during the reduction process by Cholesky decomposition. This approach is particularly efficient if changes occur at the higher numbered nodes of the structure, i.e. 'near the bottom' of the stiffness matrix. This technique was shown using operation counts [6] to be more efficient than that of Sack et.al. [95].

Kirsch and Rubinstein [53, 54] also used the Sherman-Morrison identity where the generation of the modified inverse is avoided. Two versions were presented to calculate the modified displacement vector. One of these versions which they called the 'method of reduced equations' was the most efficient compared to the techniques of the previous section.

Kavlie and Powell [46, 117] also presented a similar technique to calculate the modified displacement vector. Their technique is inefficient compared to a complete re-analysis except for very small modifications. However for cases where a linear search is made in an optimization procedure, the technique becomes very competitive.

As a result of earlier work [6] Argyris and Roy [9] proposed a general method of re-analysis where the sparsity and banded nature of the stiffness matrix is taken into account. The proposed method involves changes in the size of the stiffness matrix, much in the same way as Mohraz and Wright [79]. However it is more efficient since the inverse is not required. The method is extended for updating the decomposed stiffness matrix after each modification. It may also be used in substructures where elements withnin it are modified. This technique is applied to problems of crack extension and closure by Armen [10]. Its implementation in a general purpose computer program was undertaken by Raibstein et.al. [92]. Here the efficiency was also investigated for large and complex structures that occur in the design environment. The authors showed that for an efficient re-analysis, the maximum percentage of degrees of freedom that may be modified using this method varies between 8-60%.

A development by Wang et.al. [112, 113, 114, 115] is similar to that of Argyris and Roy [9]. The modified response of a structure is expressed as a linear combination of the original response and a term depending on the pseudo loads. These pseudo loads are related to changes in the stiffness of the original system and analysed with the applied loads. A reduced set of equations are set up for the modified parts of the structure only. They called this technique the 'pseudo load method'. An application of the 'pseudo load method' with static condensation in finite element analysis is shown by Hirai et.al. [40]. The technique was used in the re-analysis of a fine mesh using the results of a coarse mesh.

2.5 Modified Decomposed Matrices

This technique has received particular attention. This is because the finite element method results in a large symmetric and banded stiffness matrix. The matrix is decomposed and the displacements obtained by back-substitution. For minor modifications, it is more efficient to update the decomposed matrix rather than form and reduced the new stiffness matrix.

Bennett [16] recognised that in the case of sparse and banded matrix, working with the complete inverse may be avoided. Instead, he proposed an algorithm that enabled the modified decomposed matrix to be obtained from the original decomposed matrix. Here the modified triangular factors of a matrix are computed from the original triangular factors. Argyris et.al. [6, 9] also considered this technique as part of their general modification method discussed in the previous section. Variants of Bennett's technique [16] have also appeared in the mathematical literature [30, 33, 34]. An earlier paper by Weiner [116] which appeared in 1948 have considered a similar problem. In this paper the solution procedure was tabulated for easy hand calculations and checking purposes. More importantly in the ensuing discussions, it was pointed out that the technique is suitable for studying varying structures of similar configurations. However its advantage in not using the inverse matrix was not recognised.

An application of Bennett's algorithm in finite element analysis was derived by Young [119]. To use the technique the local nonlinear behaviour must be expressed in a particular form. A revised version of the technique for nonsymmetric changes in an initially symmetric matrix is given by Kleiber and Lutoborski [57]. Other similar techniques of updating the decomposed matrix is given by Row et.al. [94] and Yang [118]. Row et.al. [94] developed two algorithms to decompose the matrix, using the Crout and Cholesky method. Based on operation counts and CPU (central processor unit) time, the Cholesky method was shown to be more efficient for obtaining the modified decomposed matrix.

Ertas and Fenves [28] have also introduced three different reanalysis techniques, two of which are based on the stiffness method and the third on the force method of analysis. The first two techniques were the "modified stiffness method", which is similar to those of Kosko [58] and MacNeal[65] and the "modified Gauss method", which is similar to that of Bennett's algorthm [16]. The third method was the "modified flexibility method," which is similar to Argyris and Kelsey's [7, 8] initial strain concept The efficeiency of these three techniques were compared, and the modified Gauss method was shown to be the most efficient. This was, however, for a problem where the modifications were restricted to a predefined region of interest.

The main disadvantage of the previous techniques is that for efficiency the modifications should be near the bottom of the matrix. This requires prior knowledge of the changes and possibly a rearrangement of the stiffness coeffcients in the matrix. To circumvent these difficulties a technique has been proposed by Law and Fenves [61, 62]. Their approach was to consider matrix modification problems, sparce matrix methods, substructure analysis and graph theory together. Various improvements to the algorithms by Bennett [16], Row et.al. [94] and Yang [118] were given. In addition an algorithm was suggested where the coeffcents of the original matrix need not be stored as required in the previous techniques. As the decomposed matrix is known, then to obtain the original coefficients a reverse process of the reduction is used. The use of substructures in their strategy suggests that modifications may be randomly distributed without prior knowledge of the structure. Unfortunately their algorithms rely heavily on graph theory which is generally an unfamiliar topic for engineers. Similar ideas on sparse matrix techniques for re-analysis were also given by Lam et.al. [60].

2.6   Superposition Techniques

Melosh et.al. [29, 73, 74, 75, 76, 77] introduced superposition techniques similar to the initial strain or stress concept. One technique is based on the complementary energy approach where the modifed forces are expressed as a linear combination of the original forces and self-equilibrating force vectors. The other uses the potential energy approach with the original displacements and self-straining vectors to obtain the modified displacements. The self-equilibrating forces or strains may be selected to give the exact result, but approximate procedures are also possible. The response of the modified structure is obtained by superimposing the response of the initial structure and the response due to self-equilibrating force vectors. Kavlie and Powell [46, 117] have given a count operations for this technique. For one load case a complete re-analysis is more efficient. For a structure with five load cases where the number of elements to be modifed was greater than two, the technique was shown to be inefficient.

Another superpostion technique by Majid and co-workers [24, 66, 68, 69, 70, 71] was introduced in the 1970's. This technique was based on the superpostion of unit and applied load analyses. It was called the 'theorems of structural variation' and used in the optimum design of pin and rigidly-jointed structures [66, 69, 70]. These theorems were first able to handle modification of one element at a time.

Bakry [15] has shown that simultaneous modification of two or more elements may
be undertaken. Its extension to finite element problems was done by Topping and
co-workers [1, 2, 106, 108, 109] and its simplification by Atrek [13]. Its application
in nonlinear analysis for framed structures is shown by Celik and Majid [22, 23, 67].
However the method only caters for changes in element properties such as cross-
sectional areas, thickness or moment of inertia. The extension of these theorems
to account for changes in structural geometry was achieved recently by Topping
and Chan [111]. These new theorems are hence called the 'geometric theorems of
variation' and may also include the effect of changes in element properties. The
geometric theorems have been applied to design problems [1, 110] and non-linear
analysis [4, 110].

The theorems of structural and geometric variation were reviewed in reference [107].
Although unit load analyses are required, these theorems are attractive because the
design sensitivity coefficents are obtained for use in structural optimization. However
efficiency studies (based on CPU times or operation counts) have not been carried
out.

3   Iterative and Approximate Methods

Approximate methods are derived from some form of a series expansion. An early
review on approximate methods is given by Schmit [97] in 1971. Iterative methods
apply successive corrections to the initial solution and converge to a more accurate
solution for the modified structure. In these methods accuracy and quality of the
solutions are important. Computational effort and efficiency has to be offset against
accuracy. These two factors can rarely be simultaneously satisfied, because a more
refined solution decreases the efficiency and vice versa. Therefore some compromise
must be made such that the 'run-time' is not excessive and the solution obtained is
reasonable. The optimum balance between these two opposing characteristics should
be judged with respect to the required level of accuracy of the re-analysis for a
particular problem. Usually these methods would be used to evaluate various design
alternatives quickly. For the initial and final designs, an exact analysis should be
carried out.

3.1   Iterative Techniques

In 1963 Best [17] presented a re-analysis technique using simple iteration. He called
the technique the 'equivalent load method'. The equivalent load is calculated from
$(\{F\} - [\triangle K]\{\delta\})$ where $\{F\}$ = the vector of applied loads; $[\triangle K]$ = the change in
stiffness matrix; $\{\delta\}$ = the current value of displacements at each iteration. The
reduced form of the stiffness matrix is available from the first iteration and hence
only reduction of the equivalent loads is required. The improved solution is then
obtained by back-substitution. This simple iteration technique is very similar to the

initial stiffness method used in nonlinear analysis. When $[\triangle K]$ is small, convergence is rapid but for large changes it may be slow or diverges. In Best's method the total displacements are calculated after each iteration. A slightly different form is given by Das [26] where the incremental displacements are calculated instead. Here the equivalent load is given by $-[\triangle K]\{\delta\}$ at each iteration. The numerical results obtained by Das [26] indicate that three iterations are required for most structural problems provided that changes in stiffness are kept withnin 10% of the original values. Substantial savings in computer time can be made as shown by Das [26]. The author also suggested that the error increases with more severe changes. This suggests that the simple iteration technique is only useful for small changes in stiffness where convergence is rapid and errors small.

Kavlie and Powell [46, 117] counted the number of operations per iteration required for the simple iteration technique. For small changes and a few load cases the technique is efficient compared with a fresh analysis. For large changes the technique may in some cases not converged at all. In addition they proposed that convergence may be improved for large changes if an under-relaxation factor is used.

In order to improve the convergence of the simple iteration technique, Kirsch and Rubinstein [55] introduced another technique. The change in stiffness, $[\triangle K]$ is expressed as a linear combination of two matrices. The relationship that was obtained was shown to be that of the Jacobi iteration. A further improved technique was suggested where iteration for each component of $\{\delta\}$ was performed instead of the whole vector. Each technique was compared for changes in stiffness of 100% or more. The simple iteration technique fails for such large changes although it requires fewer operations. The two proposed techniques require slightly more operations but this is more than offset by the improved convergence. The improved technique of calculating each component of $\{\delta\}$ was proved to be superior.

Phansalkar [88] considered splitting the stiffness matrix in different ways and arrived at the simple, Jacobi and Gauss-Seidel iterations. In the Gauss-Seidal iteration, components of $\{\delta\}$ for each iteration are successively used to compute the remaining components. The question of convergence and effectiveness of the various iterative schemes was also considered. One scheme that was found to be efficient is the Block Gauss-Seidel iteration. This technique involves adjustment of groups of unknowns as opposed to the point Gauss-Seidel iteration where only one unknown is adjusted at a time. It was further suggested that acceleration technqiues can be used and easily implemented with this technique to improve the efficiency.

3.2   Series Expansion Techniques

An early example of a series expansion technique was presented by Brock [21]. With this technique the modified inverse matrix is obtained by an infinite series expansion. An exact relationship is derived by summation of the series provided that the changes are small. Kosko [58] also considered an infinite series to obtain the modified inverse. A similar approach by Hoerner [41], uses only the first term of this series. The technique converges after one iteration for changes in the stiffness of less than 35%. For large changes (by a factor of 3.6 of the original stiffness) it converges after four iterations. This technique is not a serious contender because of the inefficiency of working with inverses as mentioned earlier.

Alternatively the approximate modified displacement vector can be derived from various series expansion. Kirsch and Rubinstein [55] for example, uses the binomial expansion to derive their Jacobi iteration technique. Romstad et.al. [93] considered a general power series expansion to obtain any desired accuracy for changes in the stiffness. For static analysis their power series expansion was the same as the binomial expansion. This technique was investigated by Arora and Rim [12] and found to be unsuitable even for small changes in stiffness. Zimmermann and Spence [121] also used the power series expansion to study the effect of changing one element. The change in one element affects a group of elements by a global parameter. This effect is term 'tracking sensitivity' and subsequently used as an algorithm for interactive finite element analysis.

Storaasli and Sobieszczanski [104, 105] used the first-order Taylor series expansion for re-analysis of large complex structures. This technique requires calculation of sensitivity coefficients which are then used to find a better approximation of the displacements of the modified structure. These sensitivity coefficients can be obtained cheaply by decomposing a pseudo-load term and then back-substitution using the original decomposed stiffness matrix. They investigated the efficiency and accuracy of this technique for modification of one element (changes of -100% to 500%) and more than one element (changes of -50% to 50%). The results indicate that the errors in displacement and stresses are less then 16%. The time taken was only a fraction of that for a complete analysis.

Noor and Lowder [83] investigated various re-analysis techniques. The Taylor series expansion technique is inaccurate for changes greater than 20% for one iteration. The error can be reduced after two iteration cycles. In a later paper [84] they considered the first and second-order Taylor series expansion in the mixed method of analysis. The second-order Taylor expansion was the more accurate technique but at the expense of more computations and storage requirements. If reciprocals of the design variables (which are the cross-sectional areas, thickness etc.) are used, the accuracy is improved in the first-order Taylor expansion. The use of reciprocals was also favoured by Schmit and Farshi [98] to improve the accuracy. In the first-order Taylor expansion of the displacement and mixed method of analysis, the displacements obtained were identical but the forces from the mixed method are more accurate.

Bhatia [20] considered a re-analysis technique where the stiffness matrix is first reduced by static condensation. The choice of which degrees of freedom to retain depends on the type of problem. To improve the efficiency the condensed matrix is transformed to a generalised matrix of lower order by the normal mode method of structural dynamics. The generalised matrix is then expanded by using the Taylor series about the original structure. However the formulation was not tested for any numerical example.

Kirsch [47, 48, 52] presented a Taylor series expansion for design variables in the displacement method and their reciprocals in the force method. The expansion was shown to be equivalent to a series from simple iteration. The Taylor series expansion was then used to formulate for re-analysis along a line. This particular form of re-analysis is used in optimization techniques. The changes in the design variables can be expressed in terms of a single independent variable. A polynomial and a modified nonpolynomial approximation were derived using the Taylor expansion in terms of the single independent variable. The case studies indicate that the use of reciprocals provide better results in the polynomial approximation than simply using the design variables. The nonpolynomial approximation gave results that were close to the exact solution, even when the behaviour is sensitive to changes. Since only a single independent variable is involved in the re-analysis, both techniques require less computations than the usual procedure. To improve the approximations, acceleration techniques by scaling of the design and dynamic acceleration were also introduced. Other approximations were also suggested using quadratic and cubic interpolations. This technique requires two exact analysis and better approximations are obtained but at extra computations. All these techniques were placed in the context of optimization design by Kirsch [49, 50].

3.3 Combined Series - Iterative Techniques

To overcome the disadvantages of the simple iterative and first-order Taylor series techniques, Noor and Lowder [83] suggested a combination of both would be profitable. The first approximation was obtained from the Taylor series which was then used as an estimate for the simple iterative technique. In this combined technique, one iteration cycle may significantly improve the accuracy and efficiency of the approximation. Compared to the modified reduced basis technique (see section 3.5) it leads to the same accuracy for changes less than 20%.

This combined technique was then used by Noor [80] in the mixed method of analysis for modifications of the structural geometry. By comparison with the Taylor series expansion technique, the combined technique was proved to be superior.

A similar combined technique was proposed by Kirsch and Toledano [56], for changes in the geometry of the structure. The new procedure is based on combining simple iteration and scaling of the original structure. The simple iteration part is expressed as a series expansion. To improve the quality of this approximation, scaling of the initial design was introduced. It was concluded that the new proposed technique was adequate in terms of accuracy but involves more calculations per iteration compared to other techniques.

3.4 Reduced Basis Techniques

One possible approximation technique is based on the reduced basis idea. Here the response of the modified structure is expressed as a linear combination of known independent vectors. The number of these vectors is less than the number of structural degrees of freedom. In other words the modified behaviour of the structure is approximated using a smaller degrees of freedom. Melosh and Luik [29, 73, 74, 75, 76, 77] proposed two techniques on how these vectors are chosen. The choice is as outlined in section 2.6. This approximate technique requires less operations for several reanalysis cycles [46, 117]. This is in contrast to the version of the technique for an exact analysis.

Fox and Miura [31] considered a similar technique where the modified displacements were expressed as a linear combination of previously computed displacement vectors. These displacement vectors were obtained from previous changes of the structure. The changes were the basic design vectors. The choice of these basic designs appears to be on ad hoc basis. It was also suggested that the technique is equivalent to applying a Ritz-Galerkin principle. This particular form was shown to be efficient compared to a complete re-analysis.

A similar approach was introduced by Kavanagh [45]. The displacement vector was expressed as a linear combination of the eigenvectors of the original structure. The choice of eigenvectors depended on their energy contribution; only those with significant contributions will be included in the approximation. The basis of the technique is the normal mode method of structural dynamics. The changes in the structure are introduced as a nonlinearity into the normal mode equations and solved by dynamic relaxation. The technique performs best for global changes rather than local changes.

The application of this technique to nonlinear analysis has been presented by Noor and Peters [81, 85, 86]. The independent vectors are those used in static perturbation technique.

More recently, Ding and Gallagher [27] introdueced a reduced basis formula using the force method of analysis. The formula may be derived to give the exact or approximate response of the modified structure. The redundant forces were selected as the reduced basis, since they were fewer than the total number of applied forces on the structure. The authors' case studies showed that the exact and, in particular, the approximate techniques are efficient. The authors concluded that both these techniques were effective and that even the results of the approximate method were sufficiently accurate for the purpose of redesign during an optimization process. However, the techniques are limited to the reanalysis of frame and truss structures.

3.5   Modified reduce basis technique

A modified reduced basis technique was developed by Noor and Lowder [83]. This technique is a combination of the first-order Tayor series and the reduced basis methods. The choice of the independent vectors are those of the original solution and the first-order sensitivity coefficients. Some numerical studies and results was given which showed that the modified reduced basis technique is highly accurate for a wide range of modifications. Furthermore the choice of the independent vectors is rational according to the authors. Noor and Lowder [84] concluded that this technique and the first-order Taylor series expansion in the mixed method of analysis offer the highest potential in terms of accuracy and efficiency.

The technique was further developed for use in substructuring [82]. The main difference is that the original solution is not included as the independent vector. The design variables and their reciprocals would give the same results. This version of the technique was shown to be accurate and efficient for an analysis of a large structure.

4 Final Comments

There are numerous re-analysis techniques as outlined in the review. The choice of the technique will depend on the type of problem, size of the problem, the number and magnitude of modifications, efficiency and accuracy required. There is no superior technique best suited for all problems. Comparisons between various techniques based on some 'standard tests' are limited [12, 46, 55, 56, 83, 84, 117]. In general direct methods are applicable to situations where a relatively small portion of the structure is modified. Iterative and approximate methods are more efficient, but the accuracy of the solution in some cases may not be sufficient. Experience with the various methods is the best guide to the appropriate choice and their applications.

It should be remembered that for computer implementation that there is no need to program exactly in the same way as in the derivations. The derivations are merely used for conforming to matrix formalism and should not be carried out in actual programming. This would generally lead to unnecessary large number of operations and a waste of computer time. Where possible the use of sparsity of the matrices would result in reduction in computer time and storage requirements. Therefore it is essential to use the optimum sequence of operations and data storage.

The main advantages of using a re-analysis technique are:

1. many redesign cycles in structural optimization may be done efficiently;

2. it is possible to treat nonlinearity efficiently by solving a reduced set of equations; and

3. the use of interactive analysis becomes possible for large structures.

Little work appears to have been done on the use of new computer architectures and parallel processsing for structural re-analysis. Recent research by Livesley and Modi [63, 64] considers the re-analysis problem on both serial and parallel computers. The parallel architecture investigated by the authors was a distributed transputer array and Bennett's algorithm [16] was considered in detail. Much future research effort should be directed in this area.

Acknowledgment The author acknowledges and appreciates the contributions made to this work by his research students and assistants. In particular, the contribution made by A.M. Abu Kassim, H.F.C. Chan and A.I. Khan is gratefully acknowledged.

References

[1] Abu Kassim, A.M., "The Theorems of Structural and Geometric Variation for Linear and Nonlinear Finite Element Analysis", A thesis submitted for the degree of Doctor of Philosophy, Department of Civil Engineering and Building Science, University of Edinburgh, 1985.

[2] Abu Kassim, A.M., Topping, B.H.V., "The Theorems of Structural Variation for Linear and Nonlinear Finite Element Analysis", Proceedings of the Second International Conference on Civil and Structural Engineering Computing, London, vol.2, 159-171, Civil-Comp Press, Edinburgh, 1985.

[3] Abu Kassim, A.M., Topping, B.H.V., "Static Reanalysis: A Review", Journal of Structural Engineering, American Society of Civil Engineers, v.113, n.5, 1029-1045, 1987.

[4] Abu Kassim, A.M., Topping, B.H.V., "The Theorems of Geometric Variation for Nonlinear Finite Element Analysis", Computers and Structures, v.25, n.6, 877-893, 1987.

[5] Argyris, J.H., "The matrix analysis of structures with cut outs and modifications", 9th. Int. Cong. of Applied Mechanics, Univ. of Brussels, Belgium, Vol.6, 131-140, Sept. 1956.

[6] Argyris, J.H., Bronlund, O.E., Roy, J.R., Scharpf, D.W., "A direct modification procedure for the displacement method", AIAA Journal, Vol.9, No.9, 1861-1864, Sept. 1971.

[7] Argyris, J.H., Kelsey, S., "Initial strains in the matrix force method of structural analysis", J. of the Royal Aero. Soc., Vol.64, 493-495, Aug. 1960.

[8] Argyris, J.H., Kelsey, S., "The matrix force method of structural analysis and some new applications", Great Britain Aeronautical Research Council Technical Report, R & M No.3034, Vol.93, 787-828, Feb. 1956.

[9] Argyris, J.H., Roy, J.R., "General treatment of structural modifications", J. of the Struct. Div., Proc. of the A.S.C.E., Vol.98, No.ST2, 465-492, Feb. 1972.

[10] Armen, H., "Applications of a substructuring technique to the problem of crack extension and closure", Report No.NASA-CR-132458, RE-480, Grumman Aerospace Corp., Bethpage, N.Y. Research Dept., July 1974.

[11] Arora, J.S., "Survey of structural reanalysis techniques", J. of the Struct. Div., Proc. of the A.S.C.E., Vol.102, No.ST4, 783-802, April, 1976.

[12] Arora, J.S., Rim, K., "An algorithm for fail safe structural optimization and a review of reanalysis techniques", Tech. Rept. No.11, Dept. of Mechanics and Hydraulics, College of Engineering, Univ. of Iowa, March, 1974.

[13] Atrek, E., "Theorems of structural variation : A simplification", Int. J. for Numerical Methods in Engr., Vol.21, 481-485, 1985.

[14] Atrek, E., Gallagher, R.H., Ragsdell, K.M., Zienkiewicz, O.C., Eds "NewDirections in Optimum Structural Design", John Wiley & Sons, NewYork, 1984.

[15] Bakry, M.A.E., "Optimal Design of Transmission Line Towers", A thesis submitted for the degree of Doctor of Philosophy, University of Surrey, England, 1978.

[16] Bennett, J.M., "Triangular factors of modified matrices", Numerishe Mathematik, Vol.7, 217-221, 1965.

[17] Best, G., "A method of structural weight minimization suitable for high-speed digital computers", AIAA Journal, Vol.1, No.2, 478-479, Feb. 1963.

[18] Best, G.C., "The stress-area method applied to frames", J. of the Aero. Sci., Vol.13, No.3, 151-155, March, 1946.

[19] Best, G.C., "The stress-area method of designing beams", J. of the Aero. Sci., Vol.12, No.3, 298-304, July, 1945.

[20] Bhatia, K.G., "Rapid iterative reanalysis for automated design", NASA TN D-7357, NASA, Washington D.C., Oct. 1973.

[21] Brock, J.E., "Variation of coefficients of simultaneous linear equations", Quart. of Applied Maths., Vol.11, 234-240, 1953.

[22] Celik, T., "Nonlinear moment curvature analysis by means of theorems of structural variation", Istanbul Devlet Muhendislik ve Mimarlik Akademisi Dergisi, No.6, 81-103, 1981.

[23] Celik, T., "The theorems of structural variation and their application in the elastic-plastic analysis of frames", Istanbul Devlet Muhendislik ve Mimarlik Akademisi Dergisi, No.5, 125-138, 1979.

[24] Celik, T., Saka, M.P., "The Theorems of Structural Variation in the Generalised Form", Proceedings of the Southeastern Conference on Theoretical and Applied Mechanics, University of South Carolina, 726-731, 1986.

[25] Cicala, P., "Effects of cutouts in semimonocoque structures", J. of the Aero. Sci., Vol.15, No.3, 171-179, March, 1948.

[26] Das, P.C., "Reanalysing structures with small modifications", Computer Aided Design, Vol.10, No.6, 371-374, Nov. 1978.

[27] Ding, H., Gallagher, R.H., "Approximate Force Method Reanalysis Techniques in Structural Optimization", Int. J. of Num. Methods Eng., vol.21, 1253-1267, 1985.

[28] Ertas, R., Fenves, S.J., "Automatic Analyser for Iterative Design", Civil Engineering Studies Series no 352, University of Illinois, Illinois, 1969.

[29] Fenves, S.J., Ertas, R., Discussion of "Multiple configuration analysis of structures", J. of the Struct. Div., Proc. of the A.S.C.E., Vol.95, No.ST7, 1586-1589, July, 1969.

[30] Fletcher, R., Powell, M.J.D., "On the Modification of LDT^T Factorizations" Mathematics of Computation, vol.28, no. 128, Oct, 1974, 1067-1087.

[31] Fox, R.L., "Optimization Methods for Engineering Design", Addison-Wesley Publishing Co., Reading, Massachusetts, 1971.

[32] Fox, R.L., Miura, H., "An approximate analysis technique for design calculations", AIAA Journal, Vol.9, No.1, 177-179, Jan. 1971.

[33] Gill, P.E., Golub, G.H., Murray, W., Saunders, M.A., "Methods for modifying matrix factorizations", Mathematics of Computation, Vol.28, 505-535, 1974.

[34] Gill, P.E., Murray, W., Saunders, M.A., "Methods for computing and modifying the LDV factors of a matrix", Mathematics of Computation, Vol.29, 1051-1077, 1975.

[35] Goodey, W.J., "Notes on a general method of treatment of structural discontinuities", J. of the Royal Aero. Soc., Vol.59, 695-697, Oct. 1955.

[36] Goodey, W.J., "Solution of modified linear simultaneous equations", Aircraft Engineering, Vol.31, 358-359, 364, Dec. 1959.

[37] Grzedzielski, A.L.M., "Note on some applications of the matrix force method of structural analysis", J. of the Royal Aero. Soc., Vol.64, 354-357, June, 1960.

[38] Grzedzielski, A.L.M., Argyris, J.H., Kelsey, S., Discussion and comments on "The initial strain concept", J. of the Royal Aero. Soc., Vol.65, 127-138, Feb. 1961.

[39] Hager, W.W., "Updating the Inverse of a Matrix", SIAM Review, v.31, n.2, 221-239, 1989.

[40] Hirai, I., Wang, B.P., Pilkey, W.D., "An Efficient Zooming Method for Finite Element Analysis", Int J for Num Methods in Eng, vol.20, 1671-1683, 1984.

[41] Hoerner, S.V., "Homologous deformations of tiltable telescopes", J. of the Struct. Div., Proc. of the A.S.C.E., Vol.93, No.ST5, 461-485, Oct. 1967.

[42] Householder, A.S., "Principles of Numerical Analysis", McGraw-Hill Book Company, New York, 1953.

[43] Householder, A.S., "A survey of some closed methods for inverting matrices", SIAM Journal, Vol.5, No.3, 155-169, Sept. 1957.

[44] Householder, A.S., "The Theory of Matrices in Numerical Analysis", Blaisdell Pub. Co., New York, 1964.

[45] Kavanagh, K.T., "An approximate algorithm for the reanalysis of structures by the finite element method", Computers & Structures, Vol.2, 713-722, 1972.

[46] Kavlie, D., Powell, G.H., "Efficient reanalysis of modified structures", J. of the Struct. Div., Proc. of the A.S.C.E., Vol.97, No.ST1, 377-392, Jan. 1971.

[47] Kirsch, U., "Approximate structural reanalysis based on series expansion", Computer Methods in Applied Mechanics & Engr., Vol.26, 205-223, 1981.

[48] Kirsch, U., "Approximate structural reanalysis for optimization along a line", Int. J. for Numerical Methods in Engr., Vol.18, 635-651, 1982.

[49] Kirsch, U., "On some simplified models for optimal design of structural systems", Computer Methods in Applied Mechanics & Engr., Vol.48, 155-169, 1985.

[50] Kirsch, U., "Optimal design based on approximate scaling", J. of the Struct. Div., Proc. of the A.S.C.E., Vol.108, No.ST4, 888-909, April, 1982.

[51] Kirsch, U., "Optimum Structural Design", McGraw-Hill Book Co., New York, 1981.

[52] Kirsch, U., Hofman, B., "Approximate behavior models for optimum structural design", Proc. Int. Symp. on Optimum Structural Design, Tuscon, Arizona, 7.17-7.26, 1981.

[53] Kirsch, U., Rubinstein, M.F., "Modification of structural analysis by the solution of a reduced set of equations", UCLA Paper ENG-0570, Univ. of California, Los Angeles, Calif., Dec. 1970.

[54] Kirsch, U., Rubinstein, M.F., "Reanalysis for limited structural design modifications", J. of the Engr. Mech. Div., Proc. of the A.S.C.E., Vol.98, No.EM1, 61-70, Feb. 1972.

[55] Kirsch, U., Rubinstein, M.F., "Structural reanalysis by iteration", Computers & Structures, Vol.2, 497-510, 1972.

[56] Kirsch, U., Toledano, G., "Approximate reanalysis for modifications of structural geometry", Computers & Structures, Vol.16, 269-277, 1983.

[57] Kleiber, M., Lutoborski, A., "Modified triangular factors in the incremental finite element analysis with nonsymmetric stiffness changes", Computers & Structures, Vol.9, 599-602, 1978.

[58] Kosko, E., "Effect of local modifications in redundant structures", J. of the Aero. Sci., Vol.21, No.3, 206-207, March, 1954.

[59] Kosko, E., "Matrix inversion by partitioning", Aeronautical Quarterly, Vol.8, 157-184, May, 1957.

[60] Lam, H.L., Choi, K.K., Haug, E.J., "A sparse matrix finite element technique for iterative structural optimization", Computers & Structures, Vol.16, 289-295, 1983.

[61] Law, K.H., "Sparse matrix factor modification in structural reanalysis", Int. J. for Numerical Methods in Engr., Vol.21, 37-63, 1985.

[62] Law, K.H., Fenves, S.J., "Sparse matrices, graph theory, and reanalysis", Proc. of the Int. Conf. in Civ. Eng., 1st., New York, N.Y., by ASCE, 234-249, May 12-14, 1981.

[63] Livesley, R.K., Modi, J.J., "The Re-analysis of Linear Structures on Serial and Parallel Computers", Cambridge University, Engineering Department Report, CUED/F-INFENG/TR.43, May, 1990.

[64] Livesley, R.K., Modi, J.J., "The Re-analysis of Linear Structures on Serial and Parallel Computers", Computing Systems in Engineering, vol.2, n.1, 1991.

[65] MacNeal, R.H., "Application of the compensation theorem to the modification of redundant structures", J. of the Aero. Sci., Vol.20, No.10, 726-727, Oct. 1953.

[66] Majid, K.I., "Optimum Design of Structures", Newnes Butterworth, 1974.

[67] Majid, K.I., Celik, T., "The elastic-plastic analysis of frames by the theorems of structural variation", Int. J. for Numerical Methods in Engr., Vol.21, 671-681, 1985.

[68] Majid, K.I., Elliott, D.W.C., "Forces and deflections in changing structures", The Structural Engineer, Vol.51, No.3, 93-101, 1973.

[69] Majid, K.I., Elliott, D.W.C., "Topological design of pin jointed structures by non-linear programming", Proc. of the Instn. of Civ. Eng., Vol.55, Part 2, 129-149, 1973.

[70] Majid, K.I., Saka, M.P., "Optimum shape design of rigidly jointed frames", Proc. of the Symp. on the Appl. of Comp. Meth. in Engr., Vol.1, 521-531, Univ. of Southern California, 1977.

[71] Majid, K.I., Saka, M.P., Celik, T., "The theorems of structural variation generalised for rigidly jointed frames", Proc. of the Instn. of Civ. Eng., Vol.65, Part 2, 839-856, 1978.

[72] Meek, J.L., "Matrix Structural Analysis", McGraw-Hill Book Company, New York, 1971.

[73] Melosh, R.J., "Structural analysis, fraility evaluation and redesign", Tech. Report No.TR-70-15, Vol.1, Air Force Flight Dynamics Lab., Wright-Patterson Air Force Base, Ohio, July, 1970.

[74] Melosh, R.J., Johnson, J.R., Luik, R., "Survivability analysis of structures", Proc. of the 2nd. Conf. on Matrix Methods in Structural Mechanics, Wright-Patterson Air Force Base, Ohio, AFFDL-TR-68-150, Dec. 1969.

[75] Melosh, R.J., Luik, R., "Approximate multiple configuration analysis and allocation for least weight structural design", AFFDL-TR-67-59, Wright-Patterson Air Force Base, Ohio, 1967.

[76] Melosh, R.J., Luik, R., "Multiple configuration analysis of structures", J. of the Struct. Div., Proc. of the A.S.C.E., Vol.94, No.ST11, 2581-2596, Nov. 1968.

[77] Melosh, R.J., Luik, R., Closure on "Multiple configuration analysis of structures", J. of the Struct. Div., Proc. of the A.S.C.E., Vol.96, No.ST6, 1239-1241, June 1970.

[78] Michielson, H.F., Dijk, A., "Structural modifications in redundant structures", J. of the Aero. Sci., Vol.20, No.4, 286-288, April, 1953.

[79] Mohraz, B., Wright, R.N., "Solving topologically modified structures", Computers and Structures, Vol.3, 341-353, 1973.

[80] Noor, A.K., "Multiple configuration analysis via mixed method", J. of the Struct. Div., Proc. of the A.S.C.E., Vol.100, No.ST9, 1991-1997, Sept. 1974.

[81] Noor, A.K., "Recent Advances in Reduction Methods for NonLinear Problems", Computers and Structures, Vol. 13, 31-44, 1981.

[82] Noor, A.K., Lowder, H.E., "Approximate reanalysis techniques with substruc-
      turing", J. of the Struct. Div., Proc. of the A.S.C.E., Vol.101, No.ST8, 1687-
      1698, Aug. 1975.

[83] Noor, A.K., Lowder, H.E., "Approximate techniques of structural reanalysis",
      Computers & Structures, Vol.4, 801-812, 1974.

[84] Noor, A.K., Lowder, H.E., "Structural reanalysis via a mixed method", Com-
      puters & Structures, Vol.5, 9-12, 1975.

[85] Noor, A.K., Peters, J.M., "Nonlinear analysis via global-local mixed finite el-
      ement approach", Int. J. for Numerical Methods in Engr., Vol.15, 1363-1380,
      1980.

[86] Noor, A.K., Peters, J.M., "Reduced basis technique for nonlinear analysis of
      structures", AIAA Journal, Vol.18, No.4, 455-462, April, 1980.

[87] Pestel, E.C., Leckie, F.A., "Matrix Methods in Elastomechanics", McGraw-Hill
      Bok Co, New York, N.Y., 1963.

[88] Phansalkar, S.R., "Matrix iterative methods for structural reanalysis", Com-
      puters & Structures, Vol.4, 779-800, 1974.

[89] Pipes, L.A., "Matrix Methods in Engineering", Prentice-Hall Inc., Englewood
      Cliffs, N.J., 1963.

[90] Poppleton, E.D., "Note on the design of redundant structures", UTIA Tech.
      Note, No.36, Inst. of Aerophysics, Univ. of Toronto, July, 1960.

[91] Poppleton, E.D., "The redesign of redundant structures having undesirable
      stress distributions", J. of the Aerospace Sci., Vol.28, No.5, 347-348, April,
      1961.

[92] Raibstein, A.I., Kalev, I., Pipano, A., "Efficient reanalysis of structures by a
      direct modification procedure", Fifth Nastran User's Colloquim, 1976.

[93] Romstad, K.M., Hutchinson, J.R., Runge, K.H., "Design parameter variation
      and structural response", Int. J. for Numerical Methods in Engr., Vol.5, 337-
      349, 1973.

[94] Row, D.G., Powell, G.H., Mondkar, D.P., "Solution of progressively chang-
      ing equilibrium equations for nonlinear structures", Computers & Structures,
      Vol.7, 659-665, 1977.

[95] Sack, R.L., Carpenter, W.C., Hatch, G.L., "Modification of elements in the
      displacement method", AIAA Journal, Vol.5, No.9, 1708-1710, Sept. 1967.

[96] Saka, M.P., Celik, T., "Nonlinear Analysis of Space Trusses by the Theorems of Structural Variation", Proceedings of the Second International Conference on Civil and Structural Engineering Computing, V.2, 153-158, Civil-Comp Press, Edinburgh, 1985.

[97] Schmit, L.A., "Literature review and assessment of the present position", AGARDograph No.149 on Structural Design and Applications of Mathematical Programming Techniques, 34-45, 1971.

[98] Schmit, L.A., Farshi, B., "Some approximation concepts for structural synthesis", AIAA Journal, Vol.12, No.5, 692-699, May, 1974.

[99] Sherman, J., Morrison, W.J., "Adjustment of an inverse matrix corresponding to a change in one element of a given matrix", Ann. Math. Statist., Vol.21, 124-126, 1950.

[100] Sherman, J., Morrison, W.J., "Adjustment of an inverse matrix corresponding to changes in the elements of a given column or a given row of the original matrix", Ann. Math. Statist., Vol.20, 621, 1949.

[101] Sobieszczanski, J., "Evaluation of algorithms for structrual modification", Proc. of the Conf. on Finite Element Methods in Civil Engr., Vanderbilt Univ., Nashville, Tenn., 129-153, 1969.

[102] Sobiieszczanski, J., "Matrix algorithm for structural modification based upon the parallel element concept", AIAA Journal, Vol.7, No.11, 2132-2139, Nov. 1969.

[103] Sobieszczanski, J., "Structural modification by perturbation method", J. of the Struct. Div., Proc. of the A.S.C.E., Vol.94, No.ST12, 2799-2816, Dec. 1968.

[104] Storaasli, O.O., Sobieszczanski, J., "Design oriented structural analysis", AIAA Paper 73-338, Williamsburgh, Virginia, 1973.

[105] Storaasli, O.O., Sobieszczanski, J., "On the accuracy of Taylor approximation", AIAA Journal, Vol.12, No.2, 231-233, Feb. 1974.

[106] Topping, B.H.V., "The Application of Dynamic Relaxation to the Design of Modular Space Structures", A thesis submitted for the Degree of Doctor of Philosophy, The City University, London, 1978.

[107] Topping, B.H.V., "The Theorems of Structural and Geometric Variation: A Review", Engineering Optimization, v.11, 239-250, 1987.

[108] Topping, B.H.V., "The Application of the Theorems of Structural Variation to Finite Element Problems", International Journal for Numerical Methods in Engineering, vol.19, 141-144, 1983.

[109] Topping, B.H.V., Abu Kassim, A.M., "The Use and Efficiency of the Theorems of Structural Variation for Finite Element Analysis", International Journal for Numerical Methods in Engineering, v.24, 1901-1920, 1987.

[110] Topping, B.H.V., Abu Kassim, A.M., "The Theorems of Geometric Variation for Finite Element Analysis", International Journal for Numerical Methods in Engineering, v.26, 2577-2606, 1988.

[111] Topping, B.H.V., Chan, H.F.C, "Theorems of geometric variation for engineering structures", Proceedings of Instn of Civ Engrs, part2, v.87, 469-486, 1989.

[112] Wang, B.P., Pilkey, W.D., "Efficient reanalysis of locally modified structures", Proc. 1st. Chautaqua on Finite Element Modelling, 37-61, Schaeffer Analysis, 1980.

[113] Wang, B.P., Pilkey, W.D., "Efficient reanalysis of locally modified structures", Dept. of Mech. & Aerospace Engr., Charlottesville, Virginia Univ., 1980.

[114] Wang, B.P., Pilkey, W.D., "Parameterization in finite element analysis", Proc. Int. Symp. on Optimum Structural Design, Tuscon, Arizona, 7.1-7.7, 1981.

[115] Wang, B.P., Pilkey, W.D., Palazzolo, A.R., "Reanalysis, modal synthesis and dynamic design", Chapter 8 of State-of-the-Art Surveys on Finite Element Technology, (Eds. Noor, A.K., Pilkey, W.D.), The American Society of Mechanical Engineers, New York, 1983.

[116] Weiner, B.L., "Variation of coefficients of simultaneous linear equations", Trans. A.S.C.E., Vol.113, 1349-1390, 1948.

[117] Wiberg, N.E., Discussion of "Efficient reanalysis of modified structures", by Kavle, D., Powell, G.H., J. of the Struct. Div., Proc. of the A.S.C.E., Vol.97, No.ST10, 2612-2619, Oct. 1971.

[118] Yang, W.H., "A method for updating Cholesky factorization of a band matrix", Computer Methods in Applied Mechanics & Engr., Vol.12, 281-288, 1977.

[119] Young, R.C., "Efficient nonlinear analysis of factored matrix modification", Trans. of the 4th. Int. Conf. on Struct. Mech. in Reactor Tech., (Eds. Jaeger, T.A., Boley, B.A.), San Francisco, Calif., M4/3, Aug. 1977.

[120] Zielke, G., "Inversion of modified symmetric matrices", J. Assoc. Comp. Mach., Vol.15, 402-408, 1968.

[121] Zimmerman, K.J., Spence, R., "Interactive use of finite element programs - or - How to get more out of your finite element model", Computers & Structures, Vol.12, 633-638, 1980.

Chapter 19

THE THEOREMS OF STRUCTURAL AND GEOMETRIC
VARIATION FOR ENGINEERING STRUCTURES

B.H.V. Topping
Heriot-Watt University, Riccarton, Edinburgh, UK

1 Introduction

This chapter reviews the application of the theorems of structural and geometric variation to optimization, design and non-linear analysis.

The repeated use of matrix displacement analysis for determining design sensitivities during an optimization procedure is both time consuming and costly. This is particularly the case when the problem is large and insensitive or large numbers of simultaneous equations have to be solved repeatedly. To avoid such repeated analysis a number of re-analysis algorithms [3, 5] have been developed. The objective [5] of these algorithms is:

> "to find the response of the modified structure using the original response
> of the structure such that the computational effort is less than that required
> for a fresh analysis."

One technique that is particularly attractive, since it may be interpreted in a physical sense, is that which uses the theorems of structural variation. These were originally developed for use with trusses [10, 12] and rigidly jointed frames [15]. The theorems have since been further developed [17, 19] for use with finite element problems.

More recently, these theorems have been employed to account for variations in the coordinates of structural joints [1, 22, 23]. These geometric theorems of variation have been shown to be a generalisation of the structural theorems of variation. They may be applied to both geometric and material non-linear analysis and optimization problems.

2   The Theorems of Structural Variation

These theorems may be used to predict the internal member forces and the joint displacements throughout the structure when the material or cross-sectional properties of the member are altered or when one or more members are removed altogether. It may be shown [10, 12] that the internal force in member j of a pin-jointed structure, in which the cross-sectional area of member i has been modified, is given by:

$$P_j^* = P_j + r_i f_{ji} \tag{1}$$

where: $j \neq i$; P_j^*, P_j are the internal forces in member j under the applied loads in the modified and original structure repectively; and f_{ji} is the internal force in member j of the original structure caused by unit loads applied at the ends of the member i (acting parallel to and in a direction to put member i in tension). The internal force in the modified member i is given by:

$$P_i^* = P_i + r_i(1 + \alpha_i f_{ii}) \tag{2}$$

where the change in member i is defined by $\alpha_i = dA_i/A_i$ and the new area is $A' = A_i + dA_i = (1 + \alpha_i)A_i$. For total removal of member i, $A_i' = 0$ and $dA_i = -A_i$ and hence $\alpha_i = -1$.

The scale factor r_i, called the variation factor, is determined by equilibrium conditions. These may be established by the combination of the analyses of the original structure under the applied and unit loads. Such equilibrium conditions at the ends of the modified member give the following equation:

$$r_i = -\alpha_i P_i/(1 + \alpha_i f_{ii}) \tag{3}$$

The displacements in the modified structure, δ^*, may be determined by superpostion thus:

$$\delta^* = \delta + r_i \delta_i \tag{4}$$

where δ and δ_i are the displacements of the original structure subject to the applied and unit loads respectively.

It is important to note that this procedure only allows for the modification of one member at a time. Subsequent modifications may be performed provided all the f and δ vectors are modified for changes to the $i - th$ member.

Bakry [7] has shown that simultaneous modification of two or more members may be undertaken using the theorems. Bakry's approach may be rewritten using the above notation as follows:

If n members are to be modified by $\alpha_i, \alpha_j, \alpha_k, \ldots\ldots\alpha_n$, then n equilibrium conditions, at the ends of each of the members, must be considered using the analysis for n pairs of unit loads scaled with the n factors $r_i, r_j, r_k, \ldots\ldots.r_n$. These n equilibrium conditions yield a series of simultaneous linear equations that may be expressed in the following general form and solved to determine the variation factors:

$$\alpha_i P_i = -r_i - \alpha_i \sum_{j=1}^{n} r_j f_{ij} \qquad\qquad i = 1, \ldots .n \qquad\qquad (5)$$

The internal forces in the modified members may be calculated using the expression:

$$P_i^* = (1 + \alpha_i) P_i + (1 + \alpha_i) \sum_{j=1}^{n} r_j f_{ij} \qquad\qquad (6)$$

For unaltered members the internal forces are given by:

$$P_m^* = P_m + \sum_{j=1}^{n} r_j f_{mj} \qquad\qquad (7)$$

and the joint displacements may be calculated using:

$$\delta^* = \delta + \sum_{j=1}^{n} r_j \delta_j \qquad\qquad (8)$$

Again, if members are to be further modified, the analysis under the pairs of unit loads must be adjusted using the above procedure for the alterations to the n members.

Atrek [6] derived a simplified form of equation (6) using matrix notation. In the simplified form the relationship between the the forces in the modified members and the variation factors is uncoupled. The resulting equation involves less computations and may lead to considerable economy when the number of modified members is large. A mathematical study of the theorems was undertaken by Filali [9].

The application of the these theorems to rigidly jointed frames may be formulated in a similar manner [15] and for finite element problems may be determined by analogy with pin-jointed trusses [17, 19]. A generalised formulation is discussed in reference [8].

The theorems of structural variation [10, 12] may be summarised thus: *The internal forces and nodal displacements of a modified structure may be determined from the factored analysis of the original structure subject to the applied loading and a number of unit loading cases.* The method is therefore of the influence coefficient category. It is important to note that the modifications may include the removal of members.

2.1   Applications of the Theorems of Structural Variation

These theorems have been used in the optimization of trusses [10, 13] and rigidly
jointed frames [14, 15]. An important application is in the design of structures of
variable topology. One approach developed by Majid and Elliott [10, 12] for trusses
and by Majid and Saka [14] for frames is to combine several candidate topologies
to form a 'ground structure'. Optimization procedures may then be used to determine the optimum member cross-sectional properties and to remove structurally
inefficient members by utilising the theorems to provide re-analysis information and
design sensitivities. Another approach to selecting optimium topology is to move
the coordinates of the structural joints to their most efficient positions during the
design procedure. Both these approaches to the automated design of the topology
of skeletal structures are well established [20]. In the second approach, in which geometric variables are included among the design variables, the re-analysis cannot be
accomplished using the theorems of structural variation. The re-analysis, however,
may be accomplished, however, by using the theorems of geometric variation [22].

The theorems have also been used for the nonlinear analysis of space trusses [16]
and the elasto-plastic analysis of frames [11]. The application of the theorems of
structural variation to linear and nonlinear finite element problems was discussed
in references [1, 2, 21]. It was concluded, in the second of these papers, that there
are a number of difficulties in applying the structural theorems to nonlinear finite
element problems. First, there is the difficulty of defining a single variation factor
for a particular element. If the elasticity matrix, [D] does not changing linearly
then it is not possible to define a single α for the element to specify the change
in structural properties. For example, in elasto-plastic analysis, each coefficient of
[D] will generally reduce by different amounts of the plastic part of the matrix [D].
Secondly, the structural theorems may become inefficient, requiring the calculation
of internal forces that are not generally determined during a finite element analysis.
For nonlinear finite element analysis problems the structural theorems appear to be
limited. For linear finite analysis problems such as iterative design or the analysis
of modified structures the structural theorems may be used efficiently [1, 21] for a
range of types of idealisations. Care must, however, be exercised to ensure the most
suitable element types are used for the idealisation.

Numerous different design alternatives may be considered, at little extra computational expense, by using the theorems of structural variation to provide an 'exact'
analysis.

3   The Theorems of Geometric Variation

The formulation of the theorems of geometric variation will be given here to enable
comparison with the structural theorems discussed above.

If the coordinates of a skeletal structure are to be varied, then the member lengths
and angles between the members connected to the joints will also be varied. The latter

variation is termed rotation since the clockwise angle of the member to the vertical axis changes. The former variation is termed elongation since the member length changes.

If end j of the $i-th$ member in a truss is varied the member undergoes rotation and elongation. The internal force carried by the the ith member may be determined, if first the original structure is analysed for unit loads applied at each degree of freedom of the $j-th$ and $k-th$ ends of member i. There are a total of four unit loads for each modified member as a result of joint variation. The internal force carried by member i under the action of a horizontal unit load at joint j is given by:

$$f'_{i,jh} = E_i A_i \{ (L_i^* - L_i')/L_i' \} \tag{9}$$

and

$$L_i' = \sqrt{\{(X_k - X_j)^2 + (Y_k - Y_j)^2\}} \tag{10}$$

$$L_i^* = \sqrt{\{[(X_k + \delta_{kh,jh}) - (X_j + \delta_{jh,jh})]^2 + [(Y_k + \delta_{kv,jh}) - (Y_j + \delta_{jv,jh})]^2\}} \tag{11}$$

where:

$f'_{i,jh}$ = internal force of the $i-th$ modified member resulting from a unit horizontal load;

L_i' = length of the $i-th$ member after joint variation;

L_i^* = is termed the 'stretch' length, assuming that the displacements are the same as before variation.

X_j, Y_j, X_k, Y_k, = coordinates of joint j and k respectively after joint variation;

$\delta_{jh,jh}$, $\delta_{jv,jh}$, = displacements at joint j arising from a unit horizontal load at joint j of the original structure; and

$\delta_{kh,jh}$, $\delta_{kv,jh}$, = displacements at joint k arising from a unit horizontal load at node j of the original structure.

A similar relationship also holds for the unit vertical load at end j. In addition, the member suffers a change in rotation from Θ_{ik} to Θ'_{ik} owing to the joint variation. The subscript ik denotes that the angle is measured at end k of the $i-th$ member clockwise from the positive vertical axis. The elongation and rotation will result in unbalanced forces and these are termed compensating forces. The compensating forces at joint k corresponding to the unit horizontal load at joint j are:

$$C_{kh,jh} = f_{i,jh} \sin\Theta_{ik} - f'_{i,jh} \sin\Theta'_{ik} \tag{12}$$

$$C_{kv,jh} = f_{i,jh} \cos\Theta_{ik} - f'_{i,jh} \cos\Theta'_{ik} \tag{13}$$

where:

$C_{kh,jh}$ = horizontal compensating force at joint k, arising from the modification of member i under a unit horizonatal load applied at joint j;

$C_{kv,jh}$ = vertical compensating force at joint k, arising from the modification of member i under unit horizontal load applied at joint j;

Θ_{ik} = original clockwise angle to the positive vertical axis at node k of the $i-th$ member; and

Θ'_{ik} = modified clockwise angle to the vertical at node k of the $i-th$ member.

Compensating forces at end j arising from the rotation of member i under the action of a unit horizontal load at joint j may be calculated similarly. The compensating forces for other unit loads jv, kh and kv are evaluated using similar equations to those above. The various compensating forces for a member are tabulated below: For solution, a total of 16 compensating forces must be evaluated using equations (12) and (13).

Table 1 Compensating Forces

Compensating Forces		Horiz	Vert	Horiz	Vert
At joint j	Horiz	$C_{jh,jh}$	$C_{jh,jv}$	$C_{jh,kh}$	$C_{jh,kv}$
	Vert	$C_{jv,jh}$	$C_{jv,jv}$	$C_{jv,kh}$	$C_{jv,kv}$
At joint k	Horiz	$C_{kh,jv}$	$C_{kh,jv}$	$C_{kh,kh}$	$C_{kh,kv}$
	Vert	$C_{kv,jh}$	$C_{kv,jv}$	$C_{kv,kh}$	$C_{kv,kv}$

Thus far the discussion has been based on a single modified member. If there are N members connected to joint k, all the N members will sustain elongations and rotations. Unit load analyses at the other ends of the N members are therefore also required. The net compensating forces at a typical joint k is the sum of the member contributions. This is given by:

$$C_{kh,jh} = \sum_{i=1}^{N}(f_{i,jh}sin\Theta_{ik} - f'_{i,jh}sin\Theta'_{ik}) \qquad (14)$$

$$C_{kv,jh} = \sum_{i=1}^{N}(f_{i,jh}cos\Theta_{ik} - f'_{i,jh}cos\Theta'_{ik}) \qquad (15)$$

If the applied loads are at the joints of the modified members, the equilibrium equations may be formed. This may be undertaken by using the principle of superposition of unit load analyses and applied loads. The compensating forces are evaluated from the unit load analyses and if these are scaled the equilibrium equations are:

$$
\begin{aligned}
(1 + C_{jh,jh})r_{jh} &+ C_{jh,jv}r_{jv} &+ \cdots &+ C_{jh,kv}r_{kv} &= F_{jh} \\
C_{jv,jh}r_{jh} &+ (1 + C_{jv,jv})r_{jv} &+ \cdots &+ C_{jv,kv}r_{kv} &= F_{jv} \\
\vdots & \quad\vdots & \quad\vdots & \quad\vdots \\
C_{kv,jh}r_{jh} &+ C_{kv,jv}r_{jv} &+ \cdots &+ (1 + C_{kv,kv})r_{kv} &= F_{kv}
\end{aligned}
\tag{16}
$$

where:

r_{jh}, are the horizontal scale factor for joint j etc. ; and

F_{jh}, are the applied horizontal load for joint j etc.

The subscripts j and k denote the joints of the modified members. These subscripts vary from $j = 1$ to $j = k$, where k is now the total number of joints of the N modified members. The total number of equilibrium equations is therefore equal to the total number of degrees of freedom of the joints of the N members. These joints are termed the affected joints.

The foregoing concepts may be generalised by considering the case where there are a affected joints. The total number of affected joints is the sum of the varied joints together with all joints directly connected to the joints that are varied. Similarly, the number of members affected is equal to the total number of members connected to the varied joints. At the a affected joints there are applied loadings as for example in equation (16). There may, however, be other joints, not included in a, that carry applied loads. These are termed the u unaffected joints. The equilibrium equations for each degree of freedom at the $(a + u)$ joints are formed by superposition. For a plane truss there are $2(a + u)$ equations. The equilibrium equations in matrix form are:

$$
\left[\begin{bmatrix} I & \vdots & 0 \\ \cdots & \cdots & \cdots \\ 0 & \vdots & I \end{bmatrix} + \begin{bmatrix} C_{aa} & \vdots & C_{au} \\ \cdots & \cdots & \cdots \\ C_{ua} & \vdots & C_{uu} \end{bmatrix}\right] \left\{ \begin{matrix} r_a \\ \cdots \\ r_u \end{matrix} \right\} = \left\{ \begin{matrix} F_a \\ \cdots \\ F_u \end{matrix} \right\}
\tag{17}
$$

where:

$[C_{aa}]$ = matrix of compensating forces at the a joints arising from unit loadings at the a joints;

$[C_{au}]$ = matrix of compensating forces at the a joints arising from unit loadings at the u joints;

$[C_{ua}]$ = matrix of compensating forces at the u joints arising from unit loadings at a joints;

$[C_{uu}]$ = matrix of compensating forces at the u joints due to unit loadings at the u joints;

r_a = vector of scale factors for the a joints;

r_u = vector of scale factors for the u joints;

F_a = vector of applied loads at the a joints; and

F_u = vector of applied loads at the u joints.

If there is no applied loading at the u joints equation (17) reduces to (16). The compensating forces at the u joints are zero, since they are not involved in any changes. Therefore:

$$[C_{ua}] = [0] \tag{18}$$

$$[C_{uu}] = [0] \tag{19}$$

Substitution of equations (18) and (19) into equation (17) and considering only the lower half gives:

$$\{r_u\} = \{F_u\} \tag{20}$$

Substitution of equation (20) into the upper half of equation (17) gives:

$$\{r_a\} = ([I] + [C_{aa}])^{-1}(\{F_a\} - [C_{au}]\{F_u\}) \tag{21}$$

When the scale factors have been solved, the member forces and displacements of the modified structure are obtained as follows:

Forces in the modified members may be calculated using:

$$P_i^* = \sum_{j=1}^{a+u}(r_{jh}f'_{i,jh} + r_{jv}f'_{i,jv}) \tag{22}$$

and for the unaltered members:

$$P_i^* = \sum_{j=1}^{a+u}(r_{jh}f_{i,jh} + r_{jv}f_{i,jv}) \tag{23}$$

Equations (22) and (23) represent the first theorem of geometric variation, and the second theorem is given by the modified displacements (for each component) which are:

$$\delta^* = \sum_{j=1}^{a+u} (r_{jh}\delta_{jh} + r_{jv}\delta_{jv}) \qquad (24)$$

where:

$\delta_{jh} = $ displacement arising from a unit horizontal load at joint j; and

$\delta_{jv} = $ displacement arising from a unit vertical load at node j.

This formulation of the theorems of geometric variation is a generalisation of the the theorems of structural variation which enables not only the structural properties of the members to be varied but also the geometry of the structure. The geometric theorems have the added versatility that members may be added as well as deleted from the structure. The application of the geometric theorems to rigidly-jointed frames may be readily developed [25].

3.1 The Application of the Theorems of Geometric Variation

The application of the geometric theorems to the re-analysis of trusses and rigidly jointed frames was considered in references [23] and [25] respectively. The application of the geometric theorems to the design of trusses was demonstrated in in reference [23] and to the non-linear analysis of rigidly jointed frames in reference [24]

The application of the geometric theorems to finite element problems has been considerd in references [1, 4, 22]. In references [1, 21], it was shown that the equilibrium equations for nodes of elements modified arising from the variation of the nodal coordinates may be established using the theorems. The efficiency of the formulation was investigated and it was shown that this full and asymmetric set of equilibrium equations may be assembled and solved efficiently provided the higher order elements are used and the number of modified elements is not large. A number of examples are used to illustrate the application of the geometric theorems to nonlinear analysis and design problems [1, 4, 22]. The application of the geometric theorems to non-linear finite element analysis is considered in references [1, 4]. It was noted that the geometric theorems may be easily incorporated into existing computer codes with little modification.

4 Conclusion and Discussion

The theorems of geometric variation may be used to predict joint displacements and internal member forces throughout a pin-jointed structure when:

• One or more of the joint coordinates are varied.

- One or more of the member structural properties are varied.

- One or more members are added or deleted from the structure.

- Any combination of of the above takes place simultaneously.

The geometric theorems enable geometric as well as structural variations to be made to the structure. The theorems may also be used with rigidly-jointed frames and structures idealised using finite elements.

The object of the theorems of variation is to reduce the number of equations solved in the re-analysis. Unfortunately the system of equilibrium equations (5) or (17) is full and asymmetric. This will not generally lead to inefficiency unless the number of modifications is large. For structural modifications alone the theorems of structural variation will be more efficient since fewer equilibrium equations need to be assembled and solved.

The design studies and efficiency tests presented in references [1, 20, 21, 22, 23] indicate that the theorems of structural and geometric variation may be used efficiently for the re-analysis of structures. Potential applications include: nonlinear analysis; interactive design studies; optimization and other structural sensitivity calculations.

Acknowledgment   The author acknowledges and appreciates the contributions made to this work by his research students and assistants. In particular, the contribution made by A.M. Abu Kassim, H.F.C. Chan, E.C.Y. Yuen and A.I. Khan is gratefully acknowledged.

References

[1] Abu Kassim, A.M., "The Theorems of Structural and Geometric Variation for Linear and Nonlinear Finite Element Analysis", A thesis submitted for the degree of Doctor of Philosophy, Department of Civil Engineering and Building Science, University of Edinburgh, 1985.

[2] Abu Kassim, A.M., Topping, B.H.V., "The Theorems of Structural Variation for Linear and Nonlinear Finite Element Analysis", Proceedings of the Second International Conference on Civil and Structural Engineering Computing, London, vol.2, 159-171, Civil-Comp Press, Edinburgh, 1985.

[3] Abu Kassim, A.M., Topping, B.H.V., "Static Reanalysis: A Review", Journal of Structural Engineering, American Society of Civil Engineers, v.113, n.5, 1029-1045, 1987.

[4] Abu Kassim, A.M., Topping, B.H.V., "The Theorems of Geometric Variation for Nonlinear Finite Element Analysis", Computers and Structures, v.25, n.6, 877-893, 1987.

[5] Arora, J.S., "Survey of Structural Reanalysis Techniques", Journal of the American Society of Civil Engineers, Structural Division, vol.102, n.ST4, 783-802, 1976.

[6] Atrek, E., " Theorems of Structural Variation: A simplification", Int. J. for Numerical Methods in Engineering, vol.21, 481-485, 1985.

[7] Bakry, M.A.E., "Optimal Design of Transmission Line Towers", A thesis submitted for the degree of Doctor of Philosophy, University of Surrey, 1977.

[8] Celik, T., Saka, M.P., "The Theorems of Structural Variation in the Generalised Form", Proceedings of the Southeastern Conference on Theoretical and Applied Mechanics, University of South Carolina, 726-731, 1986.

[9] Filali, A.A., "Theorems of Structural Variation - concerning the inverse of the stiffness matrix", Proceedings of the Second International Conference on Civil & Structural Engineering Computing, vol.2, 149-152, Civil-Comp Press, Edinburgh, 1985.

[10] Majid, K.I., "Optimum Design of Structures", Newnes Butterworths, London, 1974.

[11] Majid, K.I., Celik, T., "The Elastic-Plastic Analysis of Frames by the Theorems of Structural Variation", Int. J. for Numerical Methods in Engineering, vol.21, 671-681, 1985.

[12] Majid,K.I., Elliott, D.W.C., "Forces and Deflections in Changing Structures", The Structural Engineer, vol.51, n.3, 93-101, 1973.

[13] Majid,K.I., Elliott, D.W.C., "Topological Design of Pin Jointed Structures by Non-Linear Programming", Proceedings of the Institution of Civil Engineers, vol.55, part 2, 129-149, 1973.

[14] Majid, K.I., Saka,M.P., "Optimum Shape Design of Rigidly Jointed Frames", Proceedings of the Symposium on Computer Apllications in Engineering, University of Southern California, vol.1, 521-531, 1977.

[15] Majid, K.I., Saka, M.P., Celik, T., "The Theorems of Structural Variation Generalised for Rigidly Jointed Frames", Proceedings of the Institution of Civil Engineers, Part 2, vol.65, 839-856, 1978.

[16] Saka, M.P., Celik, T., "Nonlinear Analysis of Space Trusses by the Theorems of Structural Variation", Proceedings of the Second International Conference on Civil and Structural Engineering Computing, V.2, 153-158, Civil-Comp Press, Edinburgh,1985.

[17] Topping, B.H.V.,"The Application of Dynamic Relaxation to the Design of Modular Space Structures", A thesis submitted for the degree of Doctor of Philosophy, The City University, London, 1978.

[18] Topping, B.H.V., "The Theorems of Structural and Geometric Variation: A Review", Engineering Optimization, v.11, 239-250, 1987.

[19] Topping, B.H.V., "The Application of the Theorems of Structural Variation to Finite Element Problems", International Journal for Numerical Methods in Engineering, vol.19, 141-144, 1983.

[20] Topping, B.H.V., "Shape Optimisation of Skeletal Structures: A Review", Journal of Structural Engineering, American Society of Engineers, vol.119, n.8, 1933-1951, 1983.

[21] Topping, B.H.V., Abu Kassim, A.M., "The Use and Efficiency of the Theorems of Structural Variation for Finite Element Analysis", International Journal for Numerical Methods in Engineering, v.24, 1901-1920, 1987.

[22] Topping, B.H.V., Abu Kassim, A.M., "The Theorems of Geometric Variation for Finite Element Analysis", International Journal for Numerical Methods in Engineering, v.26, 2577-2606, 1988.

[23] Topping, B.H.V., Chan, H.F.C, "Theorems of geometric variation for engineering structures", Proceedings of Instn of Civ Engrs, part2, v.87, 469-486, 1989.

[24] Topping, B.H.V., Khan, A.I., "Non-linear analysis of rigidly jointed frames using the geometric theorems of variation", to be published.

[25] Topping, B.H.V., Yuen, E.C.Y., Khan, A.I., "The theorems of geometric variation for rigidly jointed frames", to be published.

Chapter 20

SHAPE OPTIMIZATION WITH FEM

G. Iancu and E. Schnack
Karlsruhe University, Karlsruhe, Germany

1 Introduction

Shape optimization with the Finite Element Method is a very powerful tool for minimizing stress concentration in machine components. The aim is to find shapes of domains so that the stress field at the critical boundary has a special characteristic. This is important because stress optimal machine components show a better fatigue behaviour if they are used in the low frequency region.

A nongradient algorithm for stress peak reduction of planar, axisymmetric and three-dimensional structures has been presented by Schnack [06-14], Schnack and Spörl [19], Spörl [21], Schnack, Spörl and Iancu [20], Schnack and Iancu [15-18], Iancu and Schnack [02-03] and Iancu [01].

Given is a homogeneous, linearly elastic, isotropic material. We consider only small deformations. The domain, see Figure 1, can be a planar or axisymmetric body

$$V \subset \mathbb{R}^n, \ n = 2,3 \tag{1}$$

As will be shown in Section 4, it is also possible to extend the present approach to three-dimensional structures. The subdomain for minimization of stress field values is defined by:

$$V^* \subset V \tag{2}$$

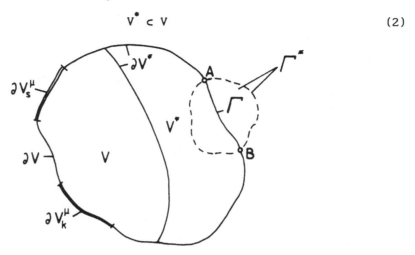

Figure 1. Problem definition

The Neumann and Dirichlet boundary conditions, respectively, are given on the boundary parts ∂V^μ_s and ∂V^μ_k:

$$\vec{t}_\mu = \tau^\mu_{km} \, n_k \, \vec{e}_m \bigg|_{\partial V^\mu_s} = \tilde{\vec{t}}_\mu, \ \partial V^\mu_s \subset \partial V \tag{3}$$

$$\vec{u}_\mu \bigg|_{\partial V^\mu_\kappa} = \tilde{\vec{u}}_\mu, \ \partial V^\mu_\kappa \subset \partial V \tag{4}$$

In this context it is important to consider quasistatic loading, where the boundary conditions change in large time intervals: $\Delta t_\mu \to \infty$, where the subscript $\mu = 1(1)M$ denotes the loading case.

The optimization of the subdomain V^* means to select the boundary $\Gamma \subset \partial V^* \cap \partial V$ which minimizes the stress concentration in this subdomain. The minimization is subjected to the constraint that the design boundary Γ has to lie within the variation domain Γ^*, see Figure 1.

2 Model equations

In the following, u_k denotes the displacement vector, e_{km} the strain tensor and τ_{km} the stress tensor. The strain tensor is defined as:

$$e_{km} = \frac{1}{2}\left(u_{k,m} + u_{m,k}\right) \tag{5}$$

The compatibility equations are:

$$i_{rs} = \varepsilon_{kpr}\varepsilon_{mqs}e_{km,pq} = 0 \tag{6}$$

with i_{rs} as the incomptability tensor. The equations of equilibrium are:

$$\tau_{km,k} = 0 \tag{7}$$

The stress-strain relation is given by:

$$\tau_{km} = 2G\left(e_{km} + \frac{\nu}{1-2\nu}\delta_{km}e_{qq}\right) \tag{8}$$

The control value of the problem is the von Mises stress $\bar{\sigma}_{\mu}$, defined as:

$$\bar{\sigma}^2 = \left(\tau_{kk}\right)^2 - \frac{3}{2}\left[\left(\tau_{kk}\right)^2 - \tau_{km}\tau_{km}\right] \tag{9}$$

The shape optimization problem is:

$$\min_{\Gamma \ V^*} (\max \bar{\sigma}_{\mu}), \ \mu = 1(1)M \tag{10}$$

$$\Gamma \subset \Gamma^* \tag{11}$$

$$\bar{\sigma}_{\mu} - \tilde{\sigma} \leq 0 \text{ in } V, \ \mu = 1(1)M \tag{12}$$

The inequality (12) means that we have an upper bound on the stress values in V, depending on the material properties. Discretizing the problem (10, 12), see Figure 2, and taking the nodal point coordinates of the FE model on the design boundary as design variables:

$$\vec{x}^T = \left[x_1, y_1, \ldots, x_i, y_i, \ldots, x_N, y_N\right] \tag{13}$$

we obtain the following objective function:

$$\min_{\Gamma \ V^*} \max\left[\bar{\sigma}_{\mu}^1(\vec{x}), \ldots, \bar{\sigma}_{\mu}^N(\vec{x})\right], \ \mu = 1(1)M \tag{14}$$

subjected to the constraints:

$$\left(x_{i}, \, y_{i}\right) \in \Gamma_{i}^{*}, \, i = 1(1)N \tag{15}$$

$$\bar{\sigma}_{\mu}^{i}(\vec{x}) - \tilde{\sigma} \leq 0, \, \mu = 1(1)M, \, i = 1(1)\bar{N} \tag{16}$$

with $\bar{N}$ number of nodal points on V.

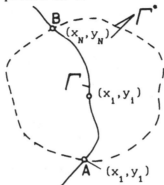

Figure 2. Discretization of the free boundary Γ

The field equations are solved by the FEM, where the accuracy control is done by the stress jump:

$$GAP = \frac{\max_{i} \left| \Delta \bar{\sigma}_{i} \right|}{\max_{i} \bar{\sigma}_{i}} 100\%, \, i = 1(1)\bar{N} \tag{17}$$

3 Boundary perturbation analysis

With the conformal mapping:

$$z = x + iy = \sqrt{w^2 - 1}, \, w = u + iv \tag{18}$$

we can construct the boundary of the notch in a half plane, see Figure 3. The torsion problem is taken here as an example. For:

$$t/\rho \gg 1 \tag{19}$$

with t as the notch depth and ρ as the radius of curvature at the notch tip, see Figure 3, it follows:

$$\bar{\sigma}_{max}/\sigma_{0} = \sqrt{t/\rho} \tag{20}$$

The result is known as Neuber's fade-away theorem, see [05]. The fade-away theorem is a kind of Saint-Venant-principle which applies to the geometry

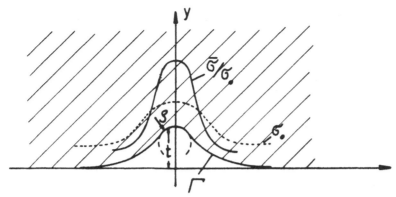

Figure 3. Perturbation by a single notch

geometry and means that at a large distance from the perturbation a fast fade-away of the stress values $\bar{\sigma}$ occurs. Moreover, Equation (20) shows a monotonical behaviour of the von Mises stress $\bar{\sigma}$ with the normalized curvature t/ρ:

$$\bar{\sigma}(t\kappa) \nearrow \quad \kappa = \frac{1}{\rho} \tag{21}$$

With the next mapping:

$$\cos z = \lambda \cos w \tag{22}$$

we can describe the boundary of a multiple notch, see Figure 4a. The normalized stress distribution is:

$$\bar{\sigma}_{max}/\sigma_0 = \sqrt{\frac{t}{\rho} \underbrace{\frac{b}{\pi t} \tan h \frac{\pi t}{b}}_{\gamma}} \tag{23}$$

This means a relieving effect due to the multiple notch boundary, described by the parameter γ, see Figure 4b. The reaction law was stated first by Thum and Oschatz [22]. The theory of relieving notches can be shortly formulated as:

$$\sigma_{min}^{-\Gamma} \uparrow \rightarrow \sigma_{max}^{-\Gamma} \downarrow \tag{24}$$

Using the statements (21) and (24) we can give a rough description of the solution idea. The controlling parameter is the curvature of the design boundary. In a first step, we reduce the maximum von Mises stress value of all loading cases on Γ locally by using the fade-away theorem. Additionally, by increasing the mimimum von Mises stress value on Γ, the objective

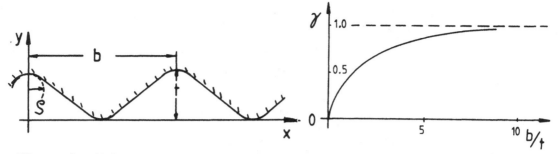

Figure 4a. Multiple-notched
surface

Figure 4b. Relief factor γ as function
function of $\frac{b}{t}$

function will be further decreased. Then we must check the geometrical constraints:

$$1. \; \bar{\sigma}_j^{1^*} \; \downarrow, \; \bar{\sigma}_j^{1^*} = \max_{i=1(1)N} \left(\max_{\mu=1(1)M} \bar{\sigma}_{(j)\mu}^1 \right) \tag{25}$$

$$2. \; \bar{\sigma}_j^{\tilde{1}} \; \uparrow, \; \bar{\sigma}_j^{\tilde{1}} = \min_{i=1(1)N} \left(\max_{\mu=1(1)M} \bar{\sigma}_{(j)\mu}^1 \right) \tag{26}$$

$$3. \; \left(x_{(j)i}, \; y_{(j)i} \right) \in \Gamma^*, \quad i = 1(1)N \tag{27}$$

We must also pay attention to the physical constraints:

$$4. \; \bar{\sigma}_{(j)\mu}^1 - \tilde{\sigma} \le 0, \quad i = 1(1)\bar{N}, \; \mu = 1(1)M \tag{28}$$

Note that the shifting of the nodal points with maximum or minimum von Mises stress introduces jagged boundaries, so that a smoothing process for
the residual nodes on Γ becomes necessary.

The local change of the curvature, see Figure 5, is done for the piecewise
linear finite element approximation used here by shifting the node i to i′
according to:

$$x_j^1 = x_{j-1}^1 + u_j^1 \; \beta_{j-1}^1 \left[y_{j-1}^1 + \gamma_{j-1}^1 \left(y_{j-1}^{i-1} - y_{j-1}^1 \right) - y_{j-1}^{i+1} \right] \tag{29}$$

$$y_j^1 = y_{j-1}^1 + u_j^1 \; \beta_{j-1}^1 \left\{ -\left[x_{j-1}^1 + \gamma_{j-1}^1 \left(x_{j-1}^{i-1} - x_{j-1}^1 \right) - x_{j-1}^{i+1} \right] \right\} \tag{30}$$

with

$$\gamma^i_{j-1} = \sqrt{\frac{\left(x^{i+1}_{j-1}-x^i_{j-1}\right)^2 + \left(y^{i+1}_{j-1}-y^i_{j-1}\right)^2}{\left(x^{i-1}_{j-1}-x^i_{j-1}\right)^2 + \left(y^{i-1}_{j-1}-y^i_{j-1}\right)^2}} \tag{31}$$

$$\beta^i_{j-1} = \left\{ \left[x^i_{j-1}+\gamma^i_{j-1}\left(x^{i-1}_{j-1}-x^i_{j-1}\right)-x^{i+1}_{j-1}\right]^2 + \left[y^i_{j-1}+\gamma^i_{j-1}\left(y^{i-1}_{j-1}-y^i_{j-1}\right)-y^{i+1}_{j-1}\right]^2 \right\}^{-1/2} \tag{32}$$

whereby $i = 2(1)N-1$.

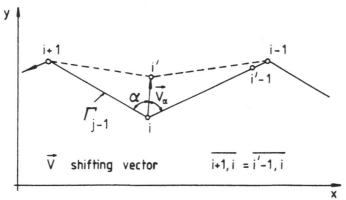

Figure 5. Local change of the curvature

Besides the direction of shifting which is determined by Equations (29–32), we need the sign and magnitude of shifting u^i_j at each nodal point i on Γ_{j-1}. As shown in Figure 6, this will be determined from the stress distribution and the sign of local curvature κ^i_{j-1} of Γ_{j-1}:

$$\kappa^i_{j-1} = -\frac{\det D}{|\det D|} \tag{33}$$

where

$$\det D = \begin{vmatrix} x^{i-1}_{j-1} & x^i_{j-1} & x^{i+1}_{j-1} \\ y^{i-1}_{j-1} & y^i_{j-1} & y^{i+1}_{j-1} \\ 1 & 1 & 1 \end{vmatrix} \tag{34}$$

The iterative change of shape is formulated with the transition function $\vec{f}_j\left(\vec{x}_{j-1}, \vec{u}_j\right)$:

$$\vec{x}_j = \vec{f}_j(\vec{x}_{j-1}, \vec{u}_j) \tag{35}$$

or in matrix notation:

$$
\begin{bmatrix} x_j^1 \\ \cdot \\ \cdot \\ \cdot \\ x_j^i \\ \cdot \\ \cdot \\ \cdot \\ x_j^N \\ \hline y_j^1 \\ \cdot \\ \cdot \\ \cdot \\ y_j^i \\ \cdot \\ \cdot \\ \cdot \\ y_j^N \end{bmatrix}
=
\begin{bmatrix} x_{j-1}^1 \\ \cdot \\ \cdot \\ \cdot \\ x_{j-1}^i \\ \cdot \\ \cdot \\ \cdot \\ x_{j-1}^N \\ \hline y_{j-1}^1 \\ \cdot \\ \cdot \\ \cdot \\ y_{j-1}^i \\ \cdot \\ \cdot \\ \cdot \\ y_{j-1}^N \end{bmatrix}
+
\begin{bmatrix} u_j^1 \\ \cdot \\ \cdot \\ \cdot \\ u_j^i \beta_{j-1}^i \left[y_{j-1}^i + \gamma_{j-1}^i \left(y_{j-1}^{i-1} - y_{j-1}^i \right) - y_{j-1}^{i+1} \right] \\ \cdot \\ \cdot \\ \cdot \\ u_j^N \\ \hline u_j^1 \\ \cdot \\ \cdot \\ \cdot \\ u_j^i \beta_{j-1}^i \left\{ - \left[x_{j-1}^i + \gamma_{j-1}^i \left(x_{j-1}^{i-1} - x_{j-1}^i \right) - x_{j-1}^{i+1} \right] \right\} \\ \cdot \\ \cdot \\ \cdot \\ u_j^N \end{bmatrix}
$$

whereby $i = 2(1)N-1$. $\tag{36}$

The control vector is defined by:

$$\vec{u}_j^T = \left[u_j^1, \ldots, u_j^i, \ldots, u_j^n \right] \tag{37}$$

In the following, the shape optimization problem (14-16) is formulated as a discrete dynamic program. In the j-th stage of the dynamic programming process a cost function g_j is defined:

$$g_j = \bar{\sigma}_j^{max} - \bar{\sigma}_{j-1}^{max}, \quad j = 1(1)1 \tag{38}$$

$$\text{with} \begin{cases} \bar{\sigma}_j^{max} = \max \bar{\sigma}_\mu^i, & i = 1(1)N, \ \mu = 1(1)M \\[2mm] \bar{\sigma}_0^{max} = \max \bar{\sigma}_\mu^i(\vec{x}_0), & i = 1(1)N, \ \mu = 1(1)M \end{cases}$$

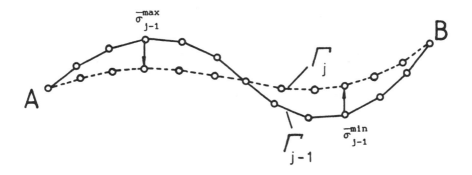

Figure 6. Control of the curvature

The objective function is:

$$\min \sum_{j=1}^{1} g_j(\vec{x}_{j-1}, \vec{u}_j) \tag{39}$$

$$\vec{x}_j = \vec{f}_j(\vec{x}_{j-1}, \vec{u}_j) \in \Xi_j, \ \Xi_j \subset \mathbb{R}^m \tag{40}$$

with $\vec{x}_j$ state vector and Ξ_j state space. The control space Ω_j is defined by:

$$\vec{u}_j \in \Omega_j(\vec{x}_{j-1}), \ \Omega_j \subset \mathbb{R}^r \tag{41}$$

We must consider the geometrical constraints as illustrated in Figure 7:

$$x_j^i \in [a,b] \tag{42}$$

and

$$y_j^i \in [c,d] \tag{43}$$

This yields:

$$\Xi_j = [a,b]^N \times [c,d]^N \subset \mathbb{R}^{2N} \tag{44}$$

$$u_j^i \in \left[\max\left(\frac{a-x_{j-1}^i}{D} ; \frac{c-y_{j-1}^i}{E} \right), \min\left(\frac{b-x_{j-1}^i}{D}, \frac{d-y_{j-1}^i}{E} \right) \right] \tag{45}$$

$$[\cdot] = \Omega_j^i(\vec{x}_{j-1}) \subset \mathbb{R} \qquad j = 1(1)1, \ i = 2(1)N-1 \tag{46}$$

with D and E given by:

$$D = \beta_{j-1}^i \left[y_{j-1}^i + \gamma_{j-1}^{i-1}(y_{j-1}^{i-1} - y_{j-1}^i) - y_{j-1}^{i+1} \right] \tag{47}$$

$$E = \beta_{j-1}^i \left\{ -\left[x_{j-1}^i + \gamma_{j-1}^i(x_{j-1}^{i-1} - x_{j-1}^i) - x_{j-1}^{i+1} \right] \right\} \tag{48}$$

Because nodes 1 and N are fixed, we have:

$$\Omega_j^1 = \Omega_j^N = \{0\} \tag{49}$$

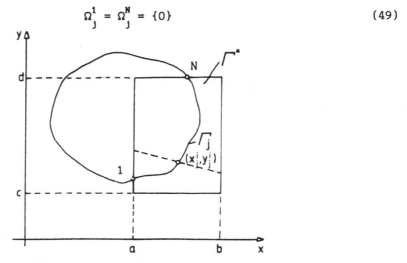

Figure 7. Feasible state space

As a result:

$$\vec{u}_j \in \Omega_j(\vec{x}_{j-1}) = \prod_{i=1}^{N} \Omega_j^i(\vec{x}_{j-1}) \subset \mathbb{R}^N \tag{50}$$

With the theorems of Tychonov and Weierstrass we can show the existence of the solution. The solution of the dynamic program with the nongradient transition function from Equation (36) is of feasible direction type, see Zoutendijk [23], as shown in the following:

$$\min_{Z} W(\vec{x}^*) \tag{51}$$

$$Z: \ g_m(\vec{x}^*) \leq 0 \qquad m \in I \tag{52}$$

where I means the index set of constraints.

With the instruction:

$$\vec{x}_j^* = \vec{x}_{j-1}^* + \alpha_j \ \vec{s}_j^* \ , \tag{53}$$

the descent condition for $\vec{s}_j^*$ is fulfilled if:

$$\left[\nabla W(\vec{x}_{j-1}^*)\right]^T \vec{s}_j^* < 0 \tag{54}$$

Moreover, we must prove the feasibility of $\vec{s}_j^*$, see Figure 8, for:

$$\alpha_j \in (0, \alpha'), \ \alpha' > 0 \tag{55}$$

$$\left[\nabla g_m(\vec{x}_{j-1}^*)\right] \ \vec{s}_j^* < 0, \ m \in I_0 \tag{56}$$

$$I_0 = \left\{ m \in I: \ g_m(\vec{x}_{j-1}^*) = 0 \right\} \tag{57}$$

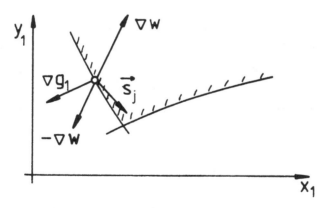

Figure 8. Transformation of the problem

Because the original problem is generally not differentiable with respect to the design variables, it is necessary to transform them as shown by Gill et al. [04]. We introduce a new variable x_{2N+1} and make an extension of the design vector:

$$\vec{x}^{*T} = \left[x_1, \ldots, x_N, \ y_1, \ y_N, \ x_{2N+1}\right] \tag{58}$$

$$\text{with } x_{2N+1} = \text{Max } \bar{\sigma}^{-1}_{\mu}, \ \mu = 1(1)M, \ i = 1(1)N$$

and obtain the new problem:

$$\min_{\vec{x}^*} x_{2N+1} \tag{59}$$

$$\left.\begin{array}{ll}
\bar{\sigma}^{i}_{\mu} - x_{2N+1} \leq 0 & i = 1(1)N, \ \mu = 1(1)M \\[2mm]
\sigma^{-i}_{\mu} - \tilde{\sigma} \leq 0 & i = 1(1)\bar{N}, \ \mu = 1(1)M \\[2mm]
A\vec{x} - \vec{b} \leq 0 &
\end{array}\right\} \ g_m \leq 0 \tag{60}$$

$$\left.\begin{array}{ll}
K_{\mu} r_{\mu} = \vec{R}_{\mu} & \mu = 1(1)M \\[2mm]
\vec{d}^{\mu}_e = T_e \vec{r}_{\mu} & \mu = 1(1)M \\[2mm]
\bar{\sigma}^{i}_{\mu} = \dfrac{1}{n_1} \displaystyle\sum_{e=1}^{n_1} D \, B^{i}_e \, \vec{d}^{\mu}_e & i = 1(1)\bar{N}, \ \mu = 1(1)M
\end{array}\right\} \ h_k = 0 \tag{61}$$

The search direction s^*_j can be given finally by considering the smoothing process on Γ_{j-1}. From the nodal position with maximum or minimum von Mises stress defined by the Equations (25) and (26), the magnitude of shifting decreases pointwise according to a law of an arithmetical series. There-fore, for the nodal point with maximum von Mises stress i^* we can write:

$$i^* \in \{1,\ldots,N\} \text{ on } \Gamma_j \tag{62}$$

$$i \in \{1,\ldots,N\} \setminus \{i^*\} \tag{63}$$

$$u^i_j = \left\{\begin{array}{ll}
0 & i = 1 \\[3mm]
u^{i^*}_j \, \xi^i_{j-1}\left(1 - \dfrac{i^*-i}{i^*-1}\right) & i < i^* \\[4mm]
& \quad\quad\quad i = 2(1)N-1 \\[2mm]
u^{i^*}_j \, \xi^i_{j-1}\left(1 - \dfrac{i-i^*}{N-i^*}\right) & i > i^* \\[4mm]
0 & i = N
\end{array}\right. \tag{64}$$

The sign parameter ξ^i_{j-1} is given by:

$$\xi^1_{j-1} = \begin{cases} -1 \text{ for } \left(i = i^* \wedge \kappa^1_{j-1} = -1\right) \vee \left[i = \tilde{i} \wedge \left(\kappa^1_{j-1} = 1 \vee \kappa^1_{j-1} = 0\right)\right] \\ +1 \text{ for } \left[i = i^* \wedge \left(\kappa^1_{j-1} = 1 \vee \kappa^1_{j-1} = 0\right)\right] \vee \left(i = \tilde{i} \wedge \kappa^1_{j-1} = -1\right) \quad (65) \\ 0 \text{ for } \left(A\vec{x} - \vec{b} > 0\right) \end{cases}$$

As a result we have for the geometrical part of the direction of search:

$$\begin{bmatrix} x^1_j \\ \cdot \\ \cdot \\ \cdot \\ x^1_j{}^* \\ \cdot \\ \cdot \\ \cdot \\ x^1_j \\ \cdot \\ \cdot \\ \cdot \\ x^N_j \\ \hline y^1_j \\ \cdot \\ \cdot \\ \cdot \\ y^1_j{}^* \\ \cdot \\ \cdot \\ \cdot \\ y^1_j \\ \cdot \\ \cdot \\ \cdot \\ y^N_j \end{bmatrix} = \begin{bmatrix} x^1_{j-1} \\ \cdot \\ \cdot \\ \cdot \\ x^1_{j-1}{}^* \\ \cdot \\ \cdot \\ \cdot \\ x^1_{j-1} \\ \cdot \\ \cdot \\ \cdot \\ x^N_{j-1} \\ \hline y^1_{j-1} \\ \cdot \\ \cdot \\ \cdot \\ y^1_{j-1}{}^* \\ \cdot \\ \cdot \\ \cdot \\ y^1_{j-1} \\ \cdot \\ \cdot \\ \cdot \\ y^N_{j-1} \end{bmatrix} + u^1{}^* \begin{bmatrix} 0 \\ \cdot \\ \cdot \\ \cdot \\ \xi^1_{j-1}{}^*\beta^1_{j-1}{}^*\left[y^1_{j-1}{}^* + \gamma^1_{j-1}{}^*\left(y^1_{j-1}{}^{*-1} - y^1_{j-1}{}^*\right) - y^1_{j-1}{}^{*+1}\right] \\ \cdot \\ \cdot \\ \cdot \\ \xi^1_{j-1}\beta^1_{j-1}\left(1 - \frac{i-1}{N-1}{}^*\right)\left[y^1_{j-1} + \gamma^1_{j-1}\left(y^{i-1}_{j-1} - y^1_{j-1}\right) - y^{i+1}_{j-1}\right] \\ \cdot \\ \cdot \\ \cdot \\ 0 \\ \hline 0 \\ \cdot \\ \cdot \\ \cdot \\ \xi^1_{j-1}{}^*\beta^1_{j-1}{}^*\left\{-\left[x^1_{j-1}{}^* + \gamma^1_{j-1}{}^*\left(x^1_{j-1}{}^{*-1} - x^1_{j-1}{}^*\right) - y^1_{j-1}{}^{*+1}\right]\right\} \\ \cdot \\ \cdot \\ \cdot \\ \xi_{j-1}\beta^1_{j-1}\left(1 - \frac{i-1}{N-1}{}^*\right)\left\{-\left[x^1_{j-1} + \gamma^1_{j-1}\left(x^{i-1}_{j-1} - x^1_{j-1}\right) - y^{i+1}_{j-1}\right]\right\} \\ \cdot \\ \cdot \\ \cdot \\ 0 \end{bmatrix}$$

$$(66)$$

Or in a short form:

$$\vec{x}_j = \vec{x}_{j-1} + \alpha_j \vec{s}_j \qquad (67)$$

$$\text{where } \alpha_j \equiv u^{1^*}$$

Using the physical component of the direction of search defined by:

$$s_j^{2N+1} = \frac{1}{\alpha j}\left[\overline{\sigma}_{max}\left(\vec{x}_{j-1} + \alpha_j \vec{s}_j\right) - \overline{\sigma}_{max}\left(\vec{x}_{j-1}\right)\right], \tag{68}$$

yields the iteration rule for the extended design vector:

$$\left[\begin{array}{c} \vec{x}_j \\ \\ x_j^{2N+1} \end{array}\right] = \left[\begin{array}{c} \vec{x}_{j-1} \\ \\ x_{j-1}^{2N+1} \end{array}\right] + \alpha_j \left[\begin{array}{c} \vec{s}_j \\ \\ s_j^{2N+1} \end{array}\right]; \quad \vec{x}_j^* = \vec{x}_{j-1}^* + \alpha_j \vec{s}_j^* \tag{69}$$

with $\vec{s}_j$ defined by Equations (66) and (67). One can show that the descent, as well as the feasibility conditions for the direction of search $\vec{s}_j^*$ hold only if the supplementary stress constraints, see first of the inequalities (60), are stable. The step length α_j is determined by an interpolation procedure, see [21].

4 Text example

The presented procedure has been used in the references cited in the Introduction to solve several mechanical engineering shape optimization problems for planar, axisymmetric and three-dimensional structural components. In this section the three-dimensional model of the part of the engine-connecting rod connected to the piston pin is considered. We take into consideration only the tesile load due to inertia. The objective is to minimize the relative maximum von Mises stress which occurs on the boundary part Γ_6, see Figure 9, by changing this boundary (between the curve segments AB and A'B'). The variation domain is represented in Figure 9 by dot lines. Note that the shape of the boundary Γ_5 which touches the piston pin is kept unchanged. Because of symmetry only a quarter of the connecting rod must be modelled as shown in Figures 9 and 10. The FE-Modell for the starting geometry, see Figure 10, presents 462 nodal points and 1511 elements. The load is modelled by a cos-shaped internal pressure which acts on the boundary Γ_5. In order to eliminate rigid body motions the boundary Γ_4 is maintained fixed.

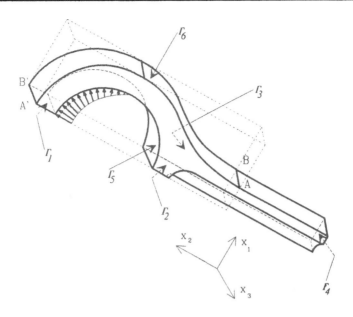

Figure 9. Connecting rod with boundaries, variation domain and loading

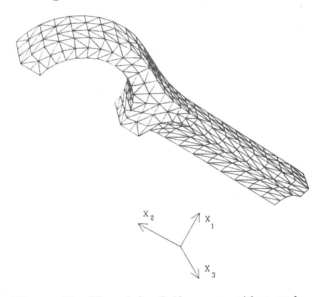

Figure 10. FE-model of the connecting rod

The distribution of the principal stresses and of the von Mises stress for the starting geometry on the boundary Γ_6 are given in Figures 11 and 12, respectively. In the solution algorithm for three-dimensional surfaces we keep the boundary curves BB' and AA' in the planes given at the beginning

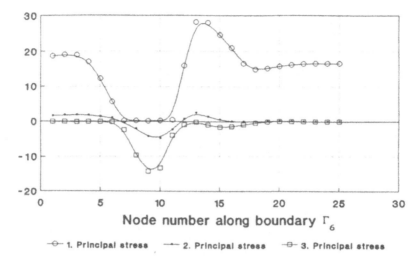

Figure 11. Distribution of principal stresses on the boundary Γ_6 of the starting geometry

of the procedure by the starting geometry. The local shifting on the design boundary is done in the direction of the average outward normal at a nodal point, whewereas the sign of shifting is determined from the optimality condition (constant von Mises stress on Γ_6). The magnitude of shifting is determined as shown in section 3 by a smoothing process which works polygonwise on the boundary. This affects all nodal points except those with maximum or minimum von Mises stress.

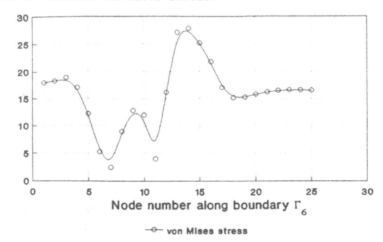

Figure 12. Distribution of the von Mises stress on the boundary Γ_6 of the starting geometry

In the implementation of this nongradient optimization procedure, we can use very efficiently the iterative CG-method for structural reanalysis. The geometry of the connecting-rod after 29 iteration steps is given in Figure 13. The almost constant distribution of the principal stresses and of the von Mises effective stress on the outer boundary Γ_6 of the optimized connecting rod is given in Figures 14 and 15. Each iteration requires (on a S400/10 machine) about 29s computational time.

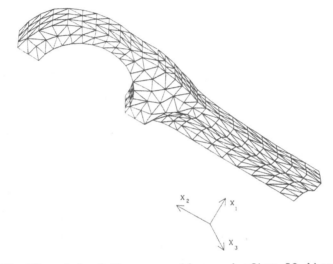

Figure 13. FE model of the connecting rod after 29 iterations

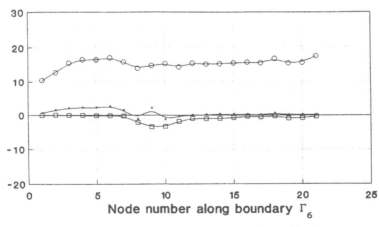

-⊖- 1. Principal stress   —— 2. Principal stress   -⊟- 3. Principal stress

Figure 14. Distribution of principal stresses on the boundary Γ_6
of the optimized connecting rod

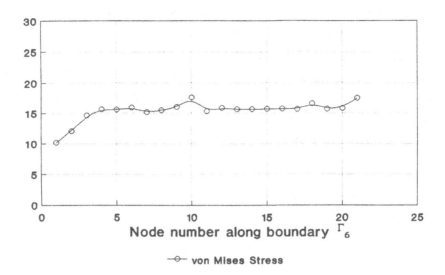

Figure 15. Distribution of the von Mises stresses on the boundary Γ_6
of the optimized connecting-rod

5 References

[01] Iancu, G.: Optimierung von Spannungskonzentrationen bei dreidimensionalen elastischen Strukturen. Doctoral Thesis, Karlsruhe University,
1991.

[02] Iancu, G., E. Schnack: Knowledge-Based Shape Optimization. First Conference on Computer Aided Optimum Design of Structures (CAOD) - OPTI
89, 20 - 23 June 1989 in Southampton/UK. In: Computer Aided Optimum
Design of Structures: Recent Advances. Eds.: C.A. Brebbia and S. Hernandez. Springer-Verlag, Berlin, Heidelberg, New York 1989, pp. 71-83.

[03] Iancu, G., E. Schnack: Shape Optimization Scheme for Large Scale
Structures. Proceedings of the Second World Congress on Computational
Mechanics, 27 - 31 August 1990, Stuttgart/Germany.
to appear

[04] Gill, P.E., W. Murray, M.A. Saunders, M.H. Wright: Aspects of Mathematical Modelling Related to Optimization. Appl. Math. Modelling 5
(1981), pp. 71-83.

[05] Neuber, H.: Kerbspannungslehre. Grundlagen für genaue Festigkeitsberechnung, 3. Auflage. Springer-Verlag, Berlin, Heidelberg, 1985.

[06] Schnack, E.: Ein Iterationsverfahren zur Optimierung von Spannungskonzentrationen. Habilitationsschrift, Univ. Kaiserslautern, 1977.

[07] Schnack, E.: Ein Iterationsverfahren zur Optimierung von Kerboberflächen. VDI-Forschungsheft, Nr. 589, Düsseldorf, VDI-Verlag 1978.

[08] Schnack, E.: An Optimization Procedure for Stress Concentrations by the Finite Element Technique. IJNME 14, No. 1 (1979), pp. 115-124.

[09] Schnack, E.: Optimierung von Spannungskonzentrationen bei Viellastbeanspruchung. ZAMM 60 (1980), T151-T152.

[10] Schnack, E.: Optimal Designing of Notched Structures without Gradient Computation. Proceedings of the 3rd IFAC-Symposium, Toulouse/France, 29 June - 2nd July 1982. In: Control of Distributed Parameter Systems. Eds: J.P. Barbary and L. Le Letty. Pergamon Press, Oxford, New York, Toronto, Sydney, Paris, Frankfurt 1982, pp. 365-369.

[11] Schnack, E.: Computer Simulation of an Experimental Method for Notch-Shape-Optimization. Proceedings of the Int. Symp. of IMACS, 9 - 11 May 1983 in Nantes. In: Simulation in Engineering Sciences, Tome 2. Eds: J. Burger and Y. Janny. Elsevier Science Publishers B.V. (North Holland), Amsterdam 1985, pp. 269-275.

[12] Schnack, E.: Local Effects of Geometry Variation in the Analysis of Structures. Studies in Applied Mechanics 12: Local Effects in the Analysis of Structures. Ed.: P. Ladevèze. Elsevier Science Publisher, Amsterdam 1985, pp. 325-342.

[13] Schnack, E.: Free Boundary Value Problems in Elastostatics. Proceedings of the 4th Int. Symp. on Numerical Methods in Engineering, Atlanta, Georgia/ USA, 24 - 28 March 1986. In: Innovative Numerical Methods in Engineering. Eds.: R.P. Shaw, J. Periaux, A. Chaudouet, J. Wu, C. Marino, C.A. Brebbia. Computational Mechanics Publications Southampton, Springer-Verlag, Berlin, Heidelberg, New York, Tokyo 1986, pp. 435-440.

[14] Schnack, E.: A Method of Feasible Direction with FEM for Shape Optimi-

zation. Invited lecture: Proceedings of the IUTAM-Symp. on Structural Optimization, Melbourne, 9 - 13 February 1988. In: Structural Optimization. Eds.: G.I.N. Rozvany, B.L: Karihaloo. Kluwer Academic Publishers, Dordrecht, Boston, London 1988, pp. 299-306.

[15] Schnack, E., G. Iancu: Control of the von Mises Stress with Dynamic Programming. GAMM-Seminar on Discretization Methods and Structural Optimization - Procedures and Applications, 5 - 7 October 1988, University of Siegen. In: Proceedings of the GAMM-Seminar, Vol. 43. Eds.: H.A. Eschenauer and G. Thierauf. Springer-Verlag, Berlin, Heidelberg 1989, pp. 154-161.

[16] Schnack, E., G. Iancu: Shape Design of Elastostatics Structures Based on Local Perturbation Analysis. Structural Optimization 1 (1989), pp. 117-125.

[17] Schnack, E., G. Iancu: Non-Linear Programming Applicable for the Control of Elastic Structures. In: Preprints of the 5th IFAC Symposium on Control of Distributed Parameter Systems, 26 - 29 June 1989 in Perpignan/France. Eds.: A. El Jai and M. Amouroux. Institut de Science et de Génie des Matériaux et Procédés (CNRS), Groupe d'Automatique, Université de Perpignan 1989, pp. 163-168.

[18] Schnack, E., I. Iancu: Shape Design of Elastostatic Structures Based on Local Perturbation Analysis. Structural Optimization 1 (1989), pp. 117-125.

[19] Schnack, E., U. Spörl: A Mechanical Dynamic Programming Algorithm for Structure Optimization. IJNME 23, No. 11 (1986), pp. 1985-2004.

[20] Schnack, E., U. Spörl, G. Iancu: Gradientless Shape Optimization with FEM. VDI Forschungsheft 647/88 (1988), pp. 1-44.

[21] Spörl, U.: Spannungsoptimale Auslegung elastischer Strukturen. Doctoral Thesis, Karlsruhe University, 1985.

[22] Thum, A., H. Oschatz: Steigerung der Dauerfestigkeit bei Rundstäben mit Querbohrungen. Forsch. Ing. Wes. (1932), S. 87.

[23] Zoutendijk, G.: Methods of Feasible Directions. Elsevier Publ. Company. Amsterdam, Princeton, 1960.

Chapter 21

SENSITIVITY ANALYSIS WITH BEM

E. Schnack and G. Iancu
Karlsruhe University, Karlsruhe, Germany

1 Introduction

Shape optimization with the Boundary Element Method, see [01,02,04,08,09]
has great advantages compared to the Finite Element Method based shape optimization. For linear problems, we have to discretize only the surface of
the body. This means that the dimension of the problem is reduced by one
and a re-zoning process for the internal nodal points is not necessary. Additionally, we have a higher sensitivity of displacement and stress fields
to geometrical disturbances. These advantagas make the BE method more attractive for shape optimization than the FE method, especially when three-
dimensional poblems are involved, see [5].

A typical problem in mechanical engineering is to minimize the volume f (z)
of a structural component:

$$\min f(z); \ z \in \mathbb{R}^n \tag{1}$$

subjected to the constraints:

$$g_j(z) \leq 0, \ j = 1(1)n; \ z \in \mathbb{R}^n \tag{2}$$

$g_j(z)$ can be, for example, the difference of stress-norms, whereby the design variables are summarized in z.

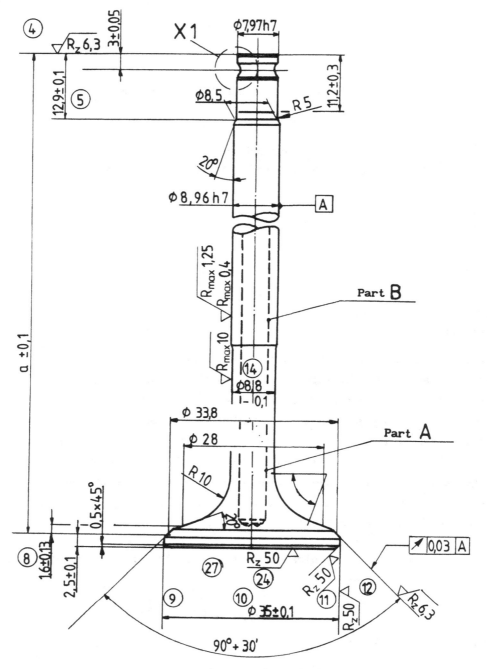

Figure 1. Valve construction

A typical shape optimization problem in mechanical engineering is valve construction, see Figure 1. The problem here is the volume minimization subjected to stress constraints. The analysis with the Boundary Element program system DBETSY, see [02-04,08,09], shows that generally for geometries of boundary curves that give a lower volume of the structural component, the stress peaks are larger. In the valve construction shown in Figure 1 one can see the critical boundary which is the transition domain between the cylindrical shaft and the taper shank of the valve. Figure 2 shows two loading cases of the valve: (a) bending and (b) compression. This valve construction leads to a notch problem with stress peaks in the transition domain. Figure 3 shows the stress distribution along the boundary of the valve in the bending case, where x means tension, and □ compression. This is also shown in Figure 4 for the compression loading case of the valve, where x and □ stand as above for tension and compression, respectively. The sensitivity of stress to shape variation has been analyzed in the bending case in Figures 5.1-6 for some critical notch configurations. Figures 5.1-6 show that notches with sharp corners lead to high stress peaks. Stress peaks are unavoidable, but we see that boundary curves with a large radius of the curvature or approximated by parabolic boundary elements lead to lower stress peaks and, therefore, the temporal beginning of crack initiation should be delayed. In Figures 6.1-6 the same problem is demonstrated for pressure loading. Once more one can see, that sharp notch radii lead to high stress preaks and the best once more is to work with a large radius (R10) or with parabolic approximation which assures a greater flexibility of the boundary curvature. Moreover, by working with two different radii, see Figure 6.5., we can reduce stress peaks.

Using the parabolic elements of the DBETSY program system we have the possiblity to reduce stress peaks by variation of the curvature along the notch boundary.

2 Basic equations

Given is a homogenous, isotropic material with Young's modulus E, Poisson's ratio ν and the shear modulus G. The thermal-expansion coefficient is given by α and the temperature difference by Θ. First, we have the equilibrium equation:

$$\tau_{ij,i} + b_j = 0 \tag{3}$$

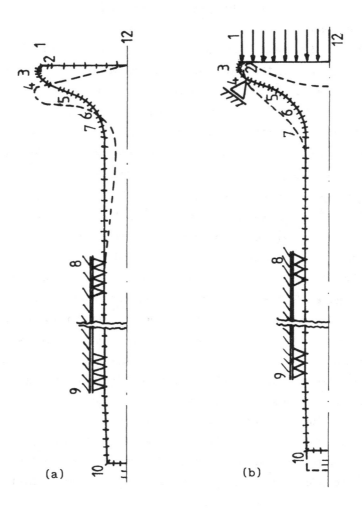

(a)   (b)

Fig. 2. Two loading cases

The definition of the strain tensor:

$$\varepsilon_{ij} = \varepsilon_{ij}^{el} + \varepsilon_{ij}^{th} \tag{4}$$

and the thermal strain tensor:

$$\varepsilon_{ij}^{th} = \alpha \ominus \delta_{ij} \tag{5}$$

Hooke's law is given by:

$$\varepsilon_{ij}^{el} = \frac{1}{2G} \left[\tau_{ij} - \frac{\nu}{1 + \nu} \delta_{ij} \tau_{kk} \right] \tag{6}$$

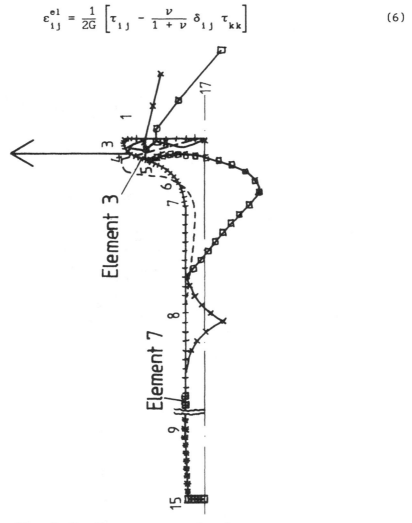

Fig. 3. Bending-case - x: tension

- □: compression

We can derive from these basic equations another partial differential equation for the displacement:

$$u_{j,ii} + \frac{1}{1 - 2\nu} u_{k,kj} + \frac{1}{G} b_j^* = 0 \tag{7}$$

with:

$$b_j^* = b_j - 2G \frac{1 + \nu}{1 - 2\nu} \alpha \Theta_{,j} = b_j - \gamma \Theta_{,j} \tag{8}$$

where:

$$\gamma = \frac{\alpha E}{1 - 2\nu} \tag{9}$$

The tractions are given by:

$$t_i = \left\{ 2G\left[\frac{1}{2}\left(u_{i,j} + u_{j,i}\right) + \frac{\nu}{1 - 2\nu} \delta_{ij} u_{k,k}\right] - \gamma \delta_{ij} \Theta \right\} n_j \tag{10}$$

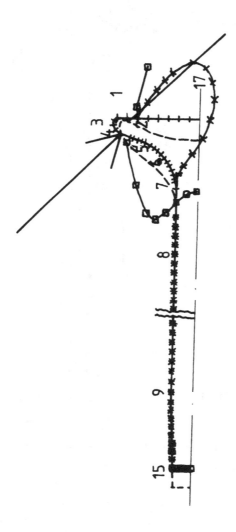

Fig. 4. Pressure-loading - x: tension

- □: compression

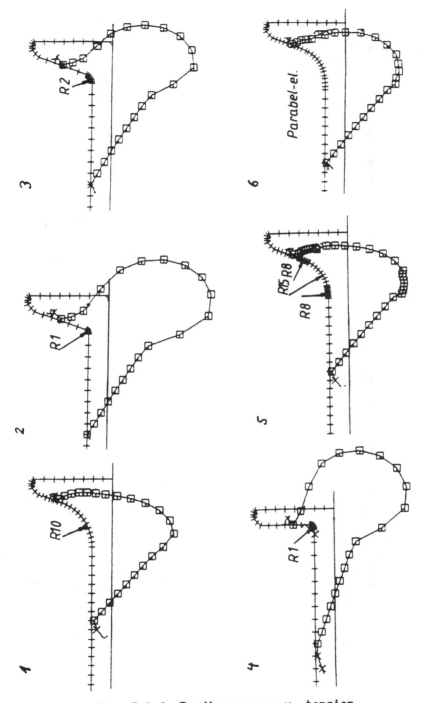

Fig. 5.1-6. Bending-case - x: tension
- □: compression

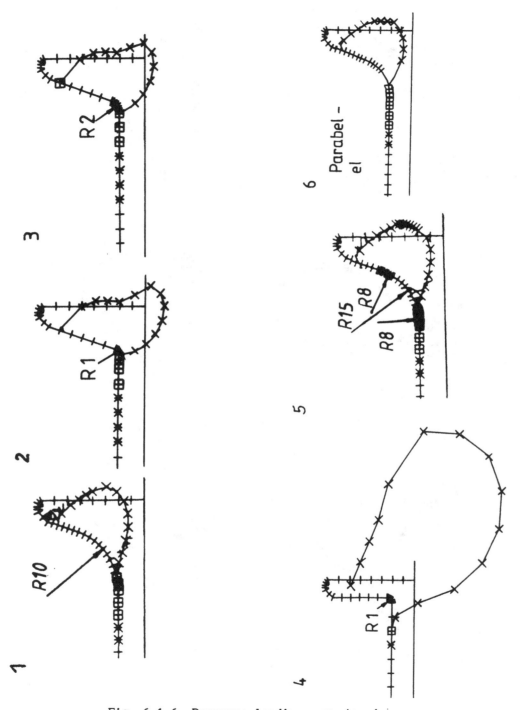

Fig. 6.1-6. Pressure-loading - x: tension
              - □: compression

Using the weighted residual formulation with U_j as the weighting function:

$$\int_\Omega \left[G\, u_{j,11} + \frac{G}{1-2\nu}\, u_{k,kj} + b_j^* \right] U_j\, d\Omega = 0 \tag{11}$$

we obtain after integration by parts:

$$- \int_\Omega G \left[U_{j,11} + \frac{1}{1-2\nu}\, U_{k,kj} \right] u_j\, d\Omega = \int_\Omega \left[b_j\, U_j + \gamma\, \Theta\, U_{j,j} \right] d\Omega$$

$$+ \int_\Gamma G \left[u_{j,1}\, U_j\, n_1 + \left(\frac{1}{1-2\nu}\, u_{k,k} - \gamma\, \Theta \right) U_j\, n_j \right] d\Gamma$$

$$- \int_\Gamma G \left[U_{j,1}\, u_j\, n_1 + \frac{1}{1-2\nu}\, U_{k,k}\, u_j\, n_j \right] d\Gamma \tag{12}$$

The boundary integrals on the right-hand side of Equation (12) are defined by:

$$\int_\Gamma G[1] = \int_\Gamma G \left[(u_{i,j} + u_{j,i}) + \frac{2\nu}{1-2\nu}\, \delta_{ij}\, u_{k,k} - \delta_{ij}\, \gamma\, \Theta \right] n_i\, U_k\, d\Gamma$$

$$= \int_\Gamma t_j\, U_j\, d\Gamma \tag{13}$$

$$\int_\Gamma G[2] = \int_\Gamma G \left[(U_{i,j} + U_{j,i}) + \frac{2\nu}{1-2\nu}\, \delta_{ij}\, U_{k,k} \right] n_i\, u_j\, d\Gamma$$

$$= \int_\Gamma T_j\, u_j\, d\Gamma \tag{14}$$

If we choose for U_j and T_j, respectively, KELVIN's solution, with the GAUCHY principal value definition we obtain the following boundary integral equation:

$$C_{ij}(x)\, u_j(x) + \int_\Gamma T_{ij}(\xi,x)\, u_j(\xi)\, d\Gamma$$

$$= \int_\Gamma U_{ij}(\xi,x)\, t_j(\xi)\, d\Gamma + \int_\Omega \left[b_j(\xi)\, U_{ij}(\xi,x) + \gamma\, \Theta(\xi)\, U_{ij,j}(\xi,x) \right] d\Omega \tag{15}$$

The fundamental solutions are:

$$U_{ij}(\xi,x) = \frac{1}{8\pi G}\left[-(3-\nu)\ \ln r\ \delta_{ij} + (1+\nu)\ r,_{i}\ r,_{j}\right] \tag{16}$$

$$T_{ij}(\xi,x) = \frac{1}{4\pi r}\left\{-\left[(1-\nu)\delta_{ij} + 2(1+\nu)r,_{i}\ r,_{j}\right]\frac{\partial r}{\partial n} + (1-\nu)\left(r,_{i}\ n_{j} - r,_{j}\ n_{i}\right)\right\}$$

$$\tag{17}$$

where U_{ij} and T_{ij} have the following order:

$$U_{ij} = O(\ln r) \tag{18}$$

$$T_{ij} = \tilde{O}\left(\frac{1}{r}\right) \tag{19}$$

In the next step we define the boundary approximation for two-dimensional strctures with the following shape functions:

$$x_{i}(\zeta) = \sum_{\mu=1}^{m} \psi^{\mu}(\zeta)x_{i}^{\mu} \tag{20}$$

with:

$$x_{i}(\Gamma) \in C^{\circ};\ i = 1,2;\ \zeta \in [-1,1] \tag{21}$$

We have the following definition for $d\Gamma$:

$$d\Gamma = ld\zeta = \sqrt{\left(\frac{\partial x_{1}(\zeta)}{\partial \zeta}\right)^{2} + \left(\frac{\partial x_{2}(\zeta)}{\partial \zeta}\right)^{2}}\ d\xi \tag{22}$$

The outward normal vector on Γ is defined by:

$$\vec{n}(\zeta) = \frac{1}{l}\left[\begin{array}{c} \dfrac{\partial x_{2}(\zeta)}{\partial \zeta} \\[2mm] -\dfrac{\partial x_{1}(\zeta)}{\partial \zeta} \end{array}\right] = \frac{1}{l}\left[\begin{array}{c} x'_{2} \\[2mm] -x'_{1} \end{array}\right] \tag{23}$$

In the DBETSY code, we have 2D linear, quadratic and circle segment elements, which are shown in Figures 7, 8 and 9. The shape functions for the linear element, see Figure 7, are:

$$\psi^{1}(\zeta) = \frac{1}{2}\ (1-\zeta) \tag{24}$$

$$\psi^2(\zeta) = \frac{1}{2}(1+\zeta) \tag{25}$$

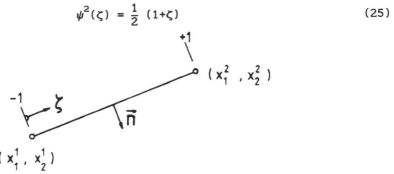

Figure 7. Linear DBETSY-2D element

For the quadratic element, see Figure 8, the shape functions are:

$$\psi^1(\zeta) = \frac{1}{2}\zeta(\zeta-1) \tag{26}$$

$$\psi^2(\zeta) = 1 - \zeta^2 \tag{27}$$

$$\psi^3(\zeta) = \frac{1}{2}\zeta(1+\zeta) \tag{28}$$

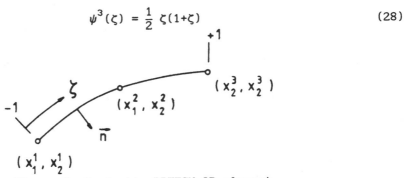

Figure 8. Quadratic DBETSY-2D element

In the case of a circle segment element, see Figure 9, the shape functions can be defined by using the angles β_1 and β_3 as:

$$\beta(\zeta) = \beta_1\psi^1(\zeta) + \beta_3\psi^2(\zeta) \tag{29}$$

with:

$$\psi^1(\zeta) = \frac{1}{2}(1-\zeta) \tag{30}$$

$$\psi^2(\zeta) = \frac{1}{2}(1+\zeta) \tag{31}$$

Now we can define the trial functions for the physical values such as displacement and traction:

$$u_j(\zeta) = \sum_{\mu=1}^{m} \phi^\mu(\zeta) \ u_j^\mu \qquad (32)$$

$$t_j(\zeta) = \sum_{\mu=1}^{m} \phi^\mu(\zeta) \ t_j^\mu \qquad (33)$$

with $j = 1,2$ and $\zeta \in [-1, +1]$

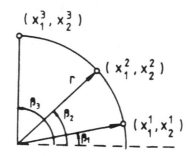

Figure 9. Circle segment DBETSY-2D element

For linear interpolation (m = 2) we have the following shape functions:

$$\phi^1(\zeta) = \frac{1}{2}(1-\zeta) \qquad (34)$$

$$\phi^2(\zeta) = \frac{1}{2}(1+\zeta) \qquad (35)$$

For quadratic interpolation (m = 3) we have:

$$\phi^1(\zeta) = \frac{1}{2} \ \zeta(\zeta-1) \qquad (36)$$

$$\phi^2(\zeta) = 1 - \zeta^2 \qquad (37)$$

$$\phi^3(\zeta) = \frac{1}{2} \ \zeta(1+\zeta) \qquad (38)$$

In the next step we define the algebraic system of integral equations, where the whole boundary is defined by:

$$\Gamma = \bigcup_{\lambda=1}^{l} \ \Gamma^\lambda \qquad (39)$$

The approximation of the integral term has the following form:

$$\int_{\Gamma} T_{ij}(\xi,x)\, u_j(\xi)\, d\Gamma(\xi)$$

$$\approx \sum_{\lambda=1}^{l} \sum_{\mu=1}^{m} u_j^{\lambda,\mu} \int_{\Gamma^\lambda} T_{ij}^\lambda \left(\xi(\zeta),\, x\right)\, \phi^\mu(\zeta)\, d\Gamma^\lambda(\zeta) \qquad (40)$$

where $d\Gamma^\lambda$ is defined by:

$$d\Gamma^\lambda(\zeta) = J^\lambda(\zeta)\, d\zeta \qquad (41)$$

where $J^\lambda(\zeta)$ is the Jacobian of transformation.

For the source points $\gamma = 1(1)g$ we have:

$$c_{ij}^\gamma\, u_j^\gamma + \sum_{\lambda=1}^{l} \sum_{\mu=1}^{m} \left[\int_{-1}^{+1} T_{ij}^\lambda \left(\xi(\zeta),\, x^\gamma\right)\, \phi^\mu(\zeta)\, J^\lambda(\zeta)\, d\zeta \right] u_j^{\lambda,\mu} =$$

$$\sum_{\lambda=1}^{l} \sum_{\mu=1}^{m} \left[\int_{-1}^{+1} U_{ij}^\lambda \left(\xi(\zeta),\, x^\gamma\right)\, \phi^\mu(\zeta)\, J^\lambda(\zeta)\, d\zeta \right] t_j^{\lambda,\mu} \qquad (42)$$

The numerical integration can be carried out as follows. If we have only regular terms as integrants

$$p^\gamma \notin \Gamma^\lambda \qquad (43)$$

we can use the GAUSS quadrature. For this case, we have the following integral equation:

$$\int_{-1}^{+1} f(\zeta)\, d\zeta \approx \sum_{j=1}^{n} h_j\, f(\psi_j) \qquad (44)$$

$$\text{for } n \in [3,8]$$

For elements with the weak singularity:

$$p^\gamma \in \Gamma^\lambda \qquad (45)$$

we have the following formula:

$$\int_0^1 f(\zeta) \ln \zeta \, d\zeta = \sum_{j=1}^m h_j \, f(\eta_j) \tag{46}$$

with $\eta_j \in [0, 1]$

For strong singularity, we insert the rigid body motions:

$$u_j^\gamma = \delta_{j1}^\gamma \tag{47}$$

with $l = 1,2$

With this Equation (42) becomes:

$$c_{11}^\gamma + \int_{-1}^{+1} T_{11}^\lambda \left(\xi(\zeta), x^\gamma \right) \phi^\mu \, J^\lambda \, d\zeta = -\sum_{\lambda=1}^l \sum_{\mu=1}^m \int_{-1}^{+1} T_{11}^\lambda \, \phi^\lambda \, J^\lambda \, d\zeta \tag{48}$$

$$\left(p^\gamma \notin \Gamma^\lambda \right)$$

According to Equation (48) we obtain the integration result by taking the sum of the integrals with regular integrants. This yields the following system of linear equations:

$$Au = Bt \tag{49}$$

Considering the DIRICHLET and NEUMANN boundary conditions, we obtain the final system:

$$Dx = C \tag{50}$$

3 Optimization procedure

First, we must define the objective function, which is here the volume of the machinary part, see [03]. For this plane structure, we must at first define the area of the domain Ω. With the GAUSS theorem:

$$\int_\Omega a,_i \, d\Omega = \int_\Gamma a \, \omega(n, x_i) \, d\Gamma \tag{51}$$

we obtain, for example, if $a = x$ and $i = 1$:

$$\int_{\Omega} d\Omega = \Omega = \int_{\Gamma} x_1 \ \omega(n, x_1) \ d\Gamma \tag{52}$$

If we now approximate the boundary Γ piecewise by straight, circle and parabolic lines, for the right-hand side of Equation (52) we obtain:

$$\int_{\Gamma} x_1 \ \omega(n, \ x_1) \ d\Gamma \approx \sum_{\lambda=1}^{1} \int_{-1}^{+1} x_1^{\lambda} \ (\zeta) n_1^{\lambda} \ (\zeta) \ J^{\lambda}(\zeta) \ d\zeta \tag{53}$$

with

$$x_1^{\lambda}(\zeta) = \sum_{\mu=1}^{m} \psi^{\mu}(\zeta) \ x_1^{\mu(\lambda)} \tag{54}$$

and

$$\Omega = \sum_{\lambda=1}^{1} \int_{-1}^{+1} x_1^{\lambda}(\zeta) \ \frac{\partial x_2^{\lambda}}{\partial \xi}(\zeta) \ d\zeta = \sum_{\lambda=1}^{1} F^{\lambda} \tag{55}$$

For straight lines we obtain the following functions for an element:

$$a = x_1 \qquad F = \frac{1}{2}\left(x_1^1 + x_1^2\right)\left(x_2^2 - x_2^1\right) \tag{56}$$

$$a = x_2 \qquad F = -\frac{1}{2}\left(x_2^1 + x_2^2\right)\left(x_2^2 - x_1^1\right) \tag{57}$$

For quadratic elements we obtain the following expressions:

$$a = x_1 \qquad F = \frac{1}{3}\left(x_1^3 - x_1^1\right)\left(x_2^1 - 2x_2^2 + x_2^3\right) + \frac{1}{6}\left(x_1^1 + 4x_1^2 + x_1^3\right)\left(x_2^3 - x_2^1\right) \tag{58}$$

$$a = x_2 \qquad F = -\frac{1}{3}\left(x_1^1 - 2x_1^2 + x_1^3\right)\left(x_2^3 - x_2^1\right) - \frac{1}{6}\left(x_2^1 + 4x_2^2 + x_2^3\right)\left(x_1^3 - x_1^1\right) \tag{59}$$

For circle lines we have the following procedure:

$$a = x_1 \tag{60}$$

$$F = -rm_1\left(\sin\beta_3 - \sin\beta_3\right) + r^2 \left[\frac{\beta_3 - \beta_1}{2} + \frac{1}{4}\left(\sin 2\beta_3 - \sin 2\beta_1\right)\right] \tag{61}$$

$$a = x_2 \tag{62}$$

$$F = -rm_2 \left(\cos\beta_3 - \cos\beta_1 \right) + r^2 \left[\frac{\beta_3 - \beta_1}{2} - \frac{1}{4} \left(\sin 2\beta_3 - \sin 2\beta_1 \right) \right] \tag{63}$$

Physical constraints are formulated in the case when the surface (Γ_s) is traction free, see Figure 10. Therefore, we must compute the tangential stress:

$$\sigma_2 = \frac{\nu}{1 + \nu} t_1 + \frac{1 + 2\nu}{(1 + \nu)^2} E u_{2,2} \tag{64}$$

which can be put in the stress constraint:

$$|\sigma_2| \le \sigma_0 \tag{65}$$

With this we can formulate the stress constraint by means of a displacement derivative as:

$$|u_{2,2}| \le \frac{\sigma_0}{E} \frac{(1 + \nu)^2}{1 + 2\nu} \tag{66}$$

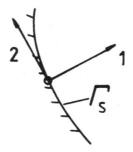

Figure 10. Restrictions

The next step is to define the design vector. To use nodal point coordinates is not successful because this leads to roughening of the structure, see [07]. It is better to work with spline parameters for the definition of the design vector, see [10]. Here, we are working with the parabolic approximation, see Figure 11. The angle β is given by:

$$\beta = \sphericalangle(n, x_1) \tag{67}$$

where

$$x_1^2 = \frac{2\left(x_2^1 - x_2^3\right)\phi_x + x_1^1\left(3\phi_2 - \phi_1\right) + x_1^3\left(\phi_3 - 3\phi_1\right)}{4\left(\phi_3 - \phi_1\right)} \tag{68}$$

$$x_2^2 = \frac{2\left(x_1^1 - x_1^3\right)\phi_y + x_2^1\left(3\phi_1 - \phi_3\right) + x_2^3\left(\phi_1 - 3\phi_3\right)}{4\left(\phi_1 - \phi_3\right)} \tag{69}$$

with the following auxiliary functions:

$$\phi_x = \sin\beta_1 \, \sin\beta_3 \tag{70}$$

$$\phi_y = \cos\beta_1 \, \cos\beta_3 \tag{71}$$

$$\phi_1 = \sin\beta_1 \, \cos\beta_3 \tag{72}$$

$$\phi_3 = \sin\beta_3 \, \cos\beta_1 \tag{73}$$

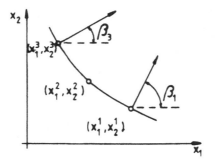

Figure 11. Design vector - parabolic approximation

The smoothness of the boundary can be controlled by changing the position of node 2. In the following, the shape will be defined by shifting nodes 1 and 3 in the direction normal to the boundary.

4 Sensitivity analysis

We need the derivatives with respect to design a variable p in a short form:

$$\frac{\partial}{\partial p} = \partial_p \tag{74}$$

The solution of the shape optimization problem requires the partial deriva-
tives ∂_p of the displacement field and $\partial_p u$ and $\partial_p (d\Omega)$ of the domain, see
[06,03]. The sensitivity of the integral Equation (42) in a local coordina-
te system is:

$$
\partial_p C_{ij}^{\gamma} u_j^{\gamma} + C_{ij}^{\gamma} \partial_p u_j^{\gamma} + \sum_{\lambda=1}^{1} \sum_{\mu=m}^{m} \left\{ \left[\int_{-1}^{+1} \partial_p \hat{T}_{ij}^{\lambda} \left(\xi(\zeta), x^{\gamma} \right) \phi^{\mu} J^{\lambda} d\zeta + \right. \right.
$$

$$
\left. \int_{1}^{+1} \hat{T}_{ij}^{\lambda} \left(\xi(\zeta), x^{\gamma} \right) \phi^{\mu} \partial_p J^{\lambda} d\zeta \right] u_j^{\lambda,\mu} + \int_{-1}^{+1} \hat{T}_{ij}^{\lambda} \left(\xi(\zeta), x^{\gamma} \right) \phi^{\mu} J^{\lambda} d\zeta \, \partial_p u_j^{\lambda,\mu} \right\} =
$$

$$
\sum_{\lambda=1}^{1} \sum_{\mu=1}^{m} \left\{ \left[\int_{-1}^{+1} \partial_p \hat{U}_{ij} \left(\xi(\zeta), x^{\gamma} \right) \phi^{\mu} J^{\lambda} d\zeta + \int_{-1}^{+1} \hat{U}_{ij} \left(\xi(\zeta), x^{\gamma} \right) \phi^{\mu} \partial_p J^{\lambda} d\zeta \right] t_j^{\lambda,\mu} + \right.
$$

$$
\left. \int_{-1}^{+1} \hat{U}_{ij} \left(\xi(\zeta), x^{\gamma} \right) \phi^{\mu} \partial_p J^{\lambda} d\zeta \right] t_j^{\lambda,\mu} + \int_{-1}^{+1} \hat{U}_{ij} \left(\xi(\zeta), x^{\gamma} \right) \phi^{\mu} J^{\lambda} d\zeta \, \partial_p t_j^{\lambda,\mu} \right\} \quad (75)
$$

where

$$
\hat{T}_{ij} (\xi, x) = L_{ik}(x) T_{kl} (\xi, x) L_{lj}^{-1}(\xi) \quad (76)
$$

with

$$
L_{ik} (x) = \begin{bmatrix} \cos\beta(x) & \sin\beta(x) \\ -\sin\beta(x) & \cos\beta(x) \end{bmatrix} \quad (77)
$$

For $\hat{U}_{ij}$, a definition analogous to Equation (76) is valid.

At boundaries with given displacements and tractions we have:

$$
\partial_p u \Big|_{\Gamma_u} = 0 \quad (78)
$$

and

$$\left. \partial_p t \right|_{\Gamma_t} = 0 \tag{79}$$

respectively.

In the operator description we can write the integral equation:

$$Au - Bt = 0 \tag{80}$$

The derivative from this integral equation has the following form:

$$\partial_p A \cdot u + A \cdot \partial_p u - \partial_p B \cdot t - B \cdot \partial_p t = 0 \tag{81}$$

for the whole system with DIRICHLET and NEUMANN conditions, and we, therefore, obtain:

$$-D \cdot \partial_p x = \partial_p A \cdot u - \partial_p B \cdot t \tag{82}$$

From this follows:

$$\partial_p x = -D^{-1} \left[\partial_p A \cdot u - \partial_p B \cdot t \right] \tag{83}$$

Note that D, u and t are known from structural analysis. For $\partial_p \Omega$ we obtain:

$$\partial_p \Omega = \sum_{\lambda=1}^{l} \left[\int_{-1}^{+1} \partial_p x_1^{\lambda}(\zeta) \frac{\partial x_2}{\partial \zeta}(\zeta) + x_1^{\lambda}(\zeta) \partial_p \frac{\partial x_2}{\partial \zeta}(\zeta) \right] d\zeta \tag{84}$$

We now need the sensitivity of singular integrals:

$$\partial_p U_{kj}(\xi, x) \tag{85}$$

The Kelvin solution is defined by:

$$U_{ij}(\xi, x) = \frac{1}{8\Pi G} \left[-(3 - \nu)\delta_{ij} \ln r + (1 + \nu) \frac{r_{,k} r_{,j}}{r^2} \right] \tag{86}$$

with

$$r = |\vec{r}| = \sqrt{(\xi_1 - x_1)^2 + (\xi_2 - x_2)} \tag{87}$$

$$\vec{r} = (r_1, r_2), \quad r_1 = \xi_1 - x_1, \quad r_2 = \xi_2 - x_2 \tag{88}$$

and in the element:

$$\xi_i(\zeta) = \sum_{\mu=1}^{m} \psi^{\mu}(\zeta)x_i^{\mu} \tag{89}$$

The derivatives of the position vector are:

$$\frac{\partial r}{\partial \xi_i} = r,_i = \frac{r_i}{r} \tag{90}$$

The sensitivity of r is:

$$\partial_p r = \frac{\partial r}{\partial r_1} \partial_p r_1 + \frac{\partial r}{\partial r_2} \partial_p r_2 = \frac{r_1}{r} \partial_p r_1 + \frac{r_2}{r} \partial_p r_2 \tag{91}$$

Therefore, we obtain:

$$\partial_p r = \frac{\left(\vec{r} \, \partial_p \vec{r}\right)}{r} \tag{92}$$

For $\partial_p r^2$ we have:

$$\partial_p r^2 = 2 \, \vec{r} \, \partial_p \vec{r} \tag{93}$$

Additionally, we have:

$$\partial_p \ln r = \frac{\vec{r} \, \partial_p \vec{r}}{r^2} \tag{94}$$

$$\partial_p r^4 = 4 \, r^2 \, \vec{r} \, \partial_p \vec{r} \tag{95}$$

It follows for $\partial_p U_{ij}$:

$$\partial_p U_{ij}(\xi, x) = \frac{1}{8\pi G} \left[-(3 - \nu)\delta_{ij} \frac{\vec{r} \, \partial_p \vec{r}}{r^2} \right.$$

$$\left. + (1 + \nu) \frac{r_j \partial_p r_i + r_i \partial_p r_j}{r^2} - 2(1 + \nu) \frac{r_i r_j \vec{r} \, \partial_p \vec{r}}{r^4} \right] \tag{96}$$

We now need the limit values for the following terms:

$$\frac{\vec{r} \, \partial_p \vec{r}}{r^2}, \frac{r_j \partial_p r_i}{r^2} \text{ and } \frac{r_i r_j}{r^2} \tag{97}$$

With Hosital's rule, we obtain:

$$G_1 = \lim_{\xi \to x} \frac{\vec{r} \, \partial_p \vec{r}}{r^2}$$

$$= \lim_{\xi \to x} \frac{(\xi_1 - x_1)(\partial_p \xi_1 - \partial_p x_1) + (\xi_2 - x_2)(\partial_p \xi_2 - \partial_p x_2)}{(\xi_1 - x_1)^2 + (\xi_2 - x_2)^2}$$

$$= \frac{x_1' \, \partial_p x_1' + x_2' \, \partial_p x_2'}{x_1'^2 + x_2'^2} \tag{98}$$

with $x_i' \to \xi_i'$

The final result is:

$$\lim_{\xi \to x} \partial_p U_{ij}(\xi, x) = \frac{1}{8\pi G}\left[-(3-\nu)\partial_{ij}G_1 + (1+\nu)\left(G_{21j} + G_{2j1} - 2G_3 G_1\right)\right] \tag{99}$$

Now we need ∂_p of the Jacobian:

$$J^\lambda(\zeta) = |\vec{N}| = \sqrt{\left[\frac{\partial \xi_1(\zeta)}{\partial \zeta}\right]^2 + \left[\frac{\partial \xi_2(\zeta)}{\partial \zeta}\right]^2} \tag{100}$$

leading to:

$$\partial_p J^\lambda = \frac{1}{2|\vec{N}|}\left[2\frac{\partial \xi_1}{\partial \zeta}\partial_p\left(\frac{\partial \xi_1}{\partial \zeta}\right) + 2\frac{\partial \xi_2}{\partial \zeta}\partial_p\left(\frac{\partial \xi_2}{\partial \zeta}\right)\right] = \frac{\vec{N}\partial_p \vec{N}}{|\vec{N}|} = \vec{n}\partial_p \vec{N} \tag{101}$$

We also need the derivatives of the rotation matrix, see Figure 12.

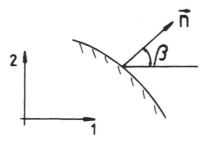

Figure 12. D_p of the rotation matrix

We have:

$$\partial_p \left[L_{ik}^\lambda(\zeta) \right]^{-1} \begin{bmatrix} \partial_p \cos\beta(\zeta) & -\partial_p \sin\beta(\zeta) \\ \\ \partial_p \sin\beta(\zeta) & \partial_p \cos\beta(\zeta) \end{bmatrix} \tag{102}$$

Additionally, we need ∂_p of the shape function, see Figure 13.

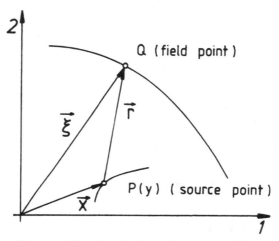

Figure 13. D_p of the shape function

In Figure 13, the position vector $\vec{r}$ is defined by the difference of position vectors $\vec{\xi}$ and $\vec{x}$, and in Figure 14 we have the definition of the design variable.

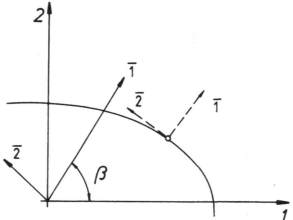

Figure 14. Definition of design variable

Therefore, we can write the following relationships:

$$
\begin{bmatrix} \xi_1^1 \\ \xi_2^1 \end{bmatrix} = \begin{bmatrix} c & -s \\ s & c \end{bmatrix} \begin{bmatrix} \bar{\xi}_2^1 \\ \bar{\xi}_2^1 \end{bmatrix} \tag{103}
$$

where

$$
c = \cos\beta, \quad s = \sin\beta \tag{104}
$$

The design variable is defined by the magnitude of shifting in the direction normal to the boundary $\bar{\xi}_1^1$. Therefore, we have:

$$
\partial_p \xi_1^1 = \frac{\partial \xi_1}{\partial_p} = \frac{\partial \xi_1}{\partial \bar{\xi}_1^1} = \cos\beta_1 \tag{105}
$$

and

$$
\partial_p \xi_2^1 = \sin\beta_1 \tag{106}
$$

The same rules are to be used for the derivatives of the objective function.

4 Numerical tests

The numerical tests were done with the software system DBETSY-2D, see [03]. We now compare the results for sensitivities obtained with the analytical and finite difference methods, respectively. An engine valve serves as an example. The change of design variable for the finite difference method is:

$$
\Delta_p = 0.01 \text{ mm} \tag{107}
$$

For the displacement $\bar{u}_1$ we have:

$$
\Delta \bar{u}_1 = \frac{\bar{u}_1\left(\Delta_p = 0.01 \text{ mm}\right) - \bar{u}_1\left(\Delta_p = 0\right)}{0.01 \text{ mm}} \tag{108}
$$

In Figures 15 and 16 the sensitivity of displacements $\bar{u}_1$ along the boundary for compression and bending, respectively, is shown. In Tables 1 and 2 the comparison between results obtained with the analytical and the finite difference methods for the sensitivity of the displacement in the two loading cases is shown. For compression a good agreement between both methods is obvious, whereas for bending, which is a more complex loading case, the difference is quite large.

scaling: displacement deriva-
        tive: $\longmapsto$ 10^{-3} units
scaling: geometry:
        $\longmapsto$ 1   unit

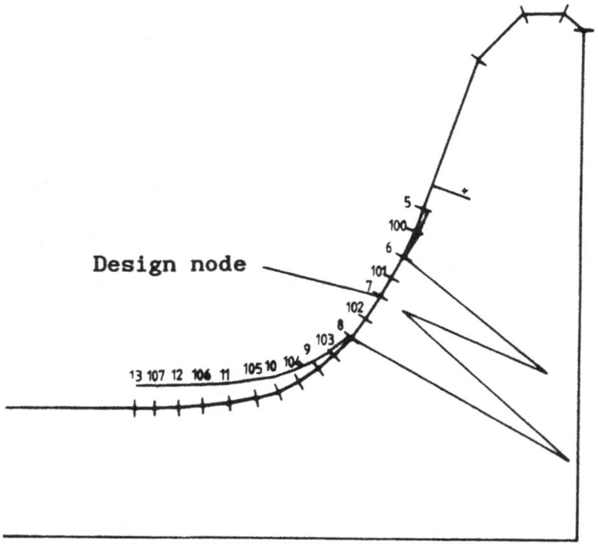

Figure 15. Sensitivity of displacement $\bar{u}_1$ in the compression loading case

scaling: displacement deriva-
        tive: $\longmapsto$ 10^{-3} units
scaling: geometry:
        $\longmapsto$ 1   unit

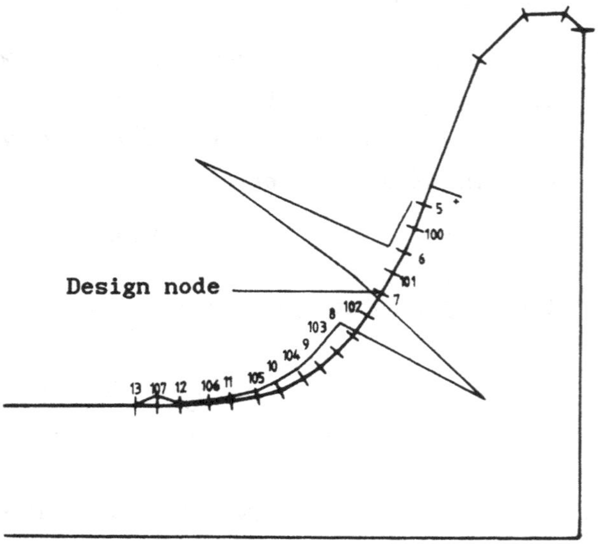

Figure 16. Sensitivity of displacement $\bar{u}_1$ in the bending loading case

Number of nodes	$\partial_p \bar{u}_1 \cdot 10^{-4}$	$\dfrac{\bar{u}_1(0,01)-\bar{u}_1}{0,01} \cdot 10^{-4}$
5	-4.247	-4.26
100	-4.732	-4.84
6	-4.989	-5.01
101	-75.20	-75.6
7	-10.96	-11.0
102	47.11	46.8
8	-5.801	-5.78
103	-5.361	-5.38
9	-4.634	-4.56
104	-3.721	-3.73
10	-2.881	-3.02
105	-1.957	-1.97
11	-1.414	-1.44
106	-0.8444	-0.819
12	-0.6104	-0.626
107	-3.696	-3.84
13	0.0062	0.014

Number of nodes	$\partial_p \bar{u}_1 \cdot 10^{-4}$	$\dfrac{\bar{u}_1(0,01)-\bar{u}_1}{0,01} \cdot 10^{-4}$
5	1.290	0.719
100	1.232	0.748
6	0.566	0.054
101	60.802	54.5
7	8.675	8.84
102	81.899	85.02
8	-0.737	-0.611
103	-1.901	-1.86
9	-3.004	-2.79
104	-4.159	-3.90
10	-5.146	-4.87
105	-5.952	-5.81
11	-6.534	-6.41
106	-6.9144	-6.84
12	-7.186	-7.16
107	-7.357	-7.42
13	-7.466	-7.52

Table 1. Displacement sensitivities for compression

Table 2. Displacement sensitivities for bending

3 References

[01] Banerjee, P.K., R. Butterfield: Boundary-Element-Methods in Engineering Science. MacGraw Hill Company, London 1981.

[02] Bausinger, R., G. Kuhn, W. Bauer, G. Seeger, W. Möhrmann: Die Boundary-Element-Methode, Theorie und industrielle Anwendung. Kontakt & Studium, Bd. 227. Ed. J. Bartz. Expert Verlag, Ehningen 1987.

[03] Bohnert, K. : 2D-Shape Optimization with BEM. Master Thesis under guidance of W. Möhrmann and E. Schnack, Karlsruhe University, 1989.

[04] Butenschön, H.J., W. Möhrmann, W. Bauer: Advanced Stress Analysis by a Commercial BEM-code. In: Industrial Applications of BEM - Developments in BEM, Vol. 5. Ed.: P.K. Banerjee, R.B. Wilson. Elsevier Applied Science, London, New York 1989, 231-261.

[05] Chaudouet-Miranda, A., F. El Yafi: 3D Optimum Design Using BEM Technique. Proceedings of BEM IX-Conference, 31 August - 4 September 1987,

Stuttgart/FRG. In: Boundary Elements IX, Vol. 2: Stress Analysis Applications. Eds. C.A. Brebbia, W.L. Wendland, G. Kuhn. Computational Mechanics Publications, Springer-Verlag, Berlin, Heidelberg, New York 1987, 449-462.

[06] Defourny, M.: Optimization Techniques and Boundary Element Method. Proceedings of BEM X-Conference, 6 - 9 September 1988, Southampton/UK. In: Boundary Elements X. Ed. C.A. Brebbia. Computational Mechanics Publications, Springer-Verlag, Berlin, Heidelberg, New York 1988, pp. 479-490.

[07] Miamoto, Y., S. Iwasaki, H. Sugimoto: On Study of Shape Optimization of 2-Dimensional Elastic Bodies by BEM. Proceedings of BEM VIII-Conference, 22 - 25 September 1986, Tokyo/Japan. In: Boundary Elements VIII. Eds. M. Tanaka, C.A. Brebbia. Computational Mechanics Publications, Springer-Verlag, Berlin, Heidelberg, New York 1986, pp. 403-412.

[08] Möhrmann, W.: DBETSY - Die Boundary-Element-Methode in der industriellen Berechnung. VDI-Berichte 537 (1984), pp. 627-650.

[09] Möhrmann, W., W. Bauer: DBETSY - Industrial Application of the BEM. Proc. of BEM IX-Conference, 31 August - 4 September 1987, Stuttgart/FRG. In: Boundary Elements IX, Vol. 1: Mathematical and Computational Aspects. Eds. C.A. Brebbia, W.L. Wendland, G. Kuhn. Computational Mechanics Publications, Springer-Verlag, Berlin, Heidelberg 1987, pp. 593-607.

[10] Sundgren, E., S.-Y. Wu: Shape Optimization Using BEM with Substructuring. Int. J. Num. Meth. in Engng. 26 (1988), pp. 1913-1924.

Chapter 22

2D- AND 3D-SHAPE OPTIMIZATION
WITH FEM AND BEM

E. Schnack and G. Iancu
Karlsruhe University, Karlsruhe, Germany

1 Introduction

The finite element formulation has been used by many researchers for shape
optimization. Nonlinear programming with sensitivities obtained by implicitly differentiating the discretized equations has been used by Zienkiewicz and Campbell [47], Francavilla, Ramakrishnan and Zienkiewicz [07], Ramakrishnan and Francavilla [24] and Kristensen and Madsen [19] to solve
this problem in two dimensions. The papers of Pedersen and Laursen [23],
Zhang and Beckers [45] and Trompette and Marcelin [42] treat shape optimization of axisymmetric structures in a similar manner. Aspects associated
with three-dimensional structures are discussed in this context by Botkin,
Yang and Benett [03], Imam [15], and Kodiyalam and Vanderplaats [18]. A detailed description of the computation of structural response using the FE
based discrete approach and numerical problems associated with this have
been presented by Haftka [09], Wang, Sun and Gallagher [43] and Haftka and
Barthelemy [10].

Variational equations such as those derived in papers by Choi and Haug
[04], Chun and Haug [05,06], Zolesio [46] and in the book by Haugh, Choi
and Komkov [11] take advantage of the variational character of the FE for-

mulation. A comparison between the discrete and continuum approaches can be
found, for instance, in [45].

Apart from the classical gradient methods of mathematical programming,
there exists the possibility of developing a nongradient strategy of the
feasible direction type for the minimization of stress concentrations. Research has shown that this is possible for two-dimensional, axisymmetric
and general three-dimensional problems. For previous work and the actual
state of research on this topic, see Schnack [27-35], Schnack and Iancu
[36-38], Schnack and Spörl [39], Schnack, Spörl and Iancu [40], Iancu and
Schnack [13-14], Iancu [12] and Spörl [41].

Attention has also been paid to shape optimization by the boundary element
approach. Mota Soares et al. [21,22], Rodriguez and Mota Soares [25] and
Rodriguez [26] use a variational approach to optimize the shape of shafts.
A paper by Kane and Saigal [16] is devoted to sensititivity analysis by
differentiating discretized BE-equations for two-dimensional problems.
Shape sensitivity using analytical differentiation of the boundary integral
equation has been formulated for three-dimensional linearly elastic structures by Barone and Yang [01] and by Zhang and Mukherjee [44] for the plane
case.

Works by Benett and Botkin [02], Kikuchi et al. [17] and Leal [20] deal
with adaptive meshing in context of shape optimization.

2 Analysis

Several engineering problems are formulated for weight and cost minimization of structures involving beam, truss and plate elements with constraints on the design variables, displacements, stresses or eigenvalues.
These are called sizing problems because the design variables, cross-
sectional dimensions, moments of inertia, moments of resistance as well as
physical constraints, are defined on a given domain. In the linear case one
can derive analytical expressions for partial derivatives of the problem
functions without any difficulties.

Because shape optimization problems with an objective function or constraints depending on the state of the system do not fall into this category, they require a more complex treatment. An example is shape optimization
with the objective minimization of the stress concentration of the rotor
mast of a helicopter shown in Figure 1. An axial section of this axisymmetric structural component together with the boundary conditions, design
boundary and variation domain are given in Figure 2. Here, Γ denotes the
design boundary and Γ^* the variation domain. In Figure 3 a FE mesh, typical
for the starting design is shown. In the case of axisymmetric loading this
problem has also been treated in [45]. Herein we compare two optimal solutions for nonaxisymmetric loading. They have been obtained in [42] by use
of the augmented Lagrangian multiplier together with the DFP method of
nonlinear programming and in [14] by a nongradient method.

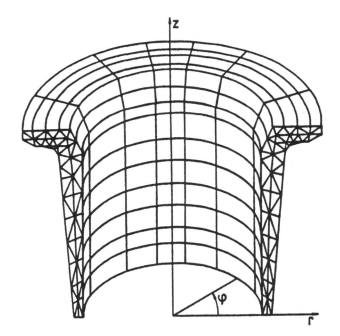

Figure 1. Axisymmetric structure - rotor mast

The optimal solution with the nongradient strategy is shown in Figure 4(a),
while the optimal design with the DFP-Method from [42] is demonstrated in
Figure 4(b). A comparison of the stress distributions of the starting design and of the solutions from [42] and [14] is shown in Fig. 5.

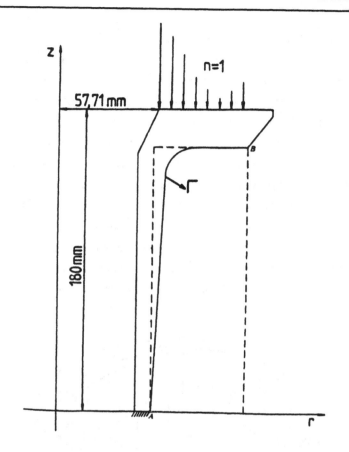

Figure 2. Axial section of the rotor mast

It can be seen that in the critical area, the stress peak is rapidly reduced by the nongradient strategy. The design variable of such a problem is
the shape. This will be described in the following by the vector of design
variables b. The vector b appears in the functions which describe the physical state of structure, such as the stress components B in two ways: explicitly and implicitly through the displacement vector u(b):

$$B(b, u(b)) \qquad b \in \mathbb{R}^n \qquad (1)$$

The minimization of stress concentrations can be written as follows:

$$\min \quad f(B(b, u(b))) \qquad b \in \mathbb{R}^n \qquad (2)$$

$$g_i(b) \leq 0 \tag{3}$$

$$h_j(B(b, u(b))) \leq 0 \tag{4}$$

with $i = 1(1)k$, $j = 1(1)\ell$.

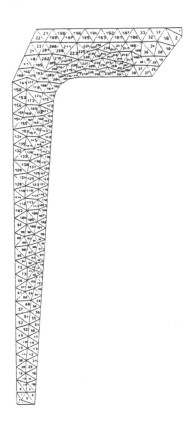

Figure 3. Starting Design

The function f is defined as the maximum von Mises stress value $\bar{\sigma}_\mu$ of all M loading cases in a subdomain of the boundary value problem Ω^*:

$$f: \text{Max } \bar{\sigma} \qquad \text{in } \Omega^* \subset \Omega \text{ and for } \mu = 1(1)M \tag{5}$$

The geometrical and physical constraints denoted by g and h, respectively, mean:

$$g: \qquad \Gamma \subset \Gamma^* \tag{6}$$

$$h: \qquad \bar{\sigma}_\mu \leq \sigma \text{ in } \Omega \tag{7}$$

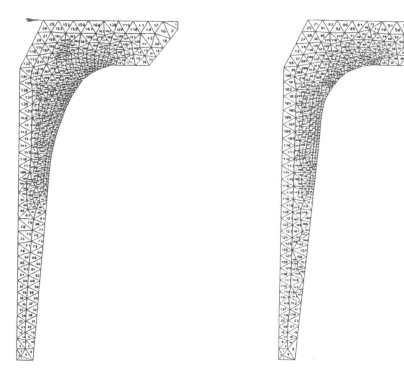

Figure 4(a): Optimal design with   Fig. 4(b): Optimal design with
the nongradient method from [14]   the DFP-method from [42]

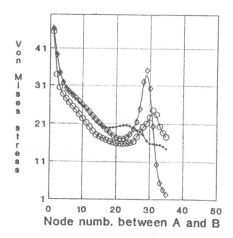

Figure 5. Stress distributions on Γ for:- ◇ starting design
                    - ◦ DFP-method from [42]
                    - ─ nongradient method from [14]

In the inequality (7) σ denotes an upper stress bound. Because the shape optimization problem (5)-(7) is, in general, nondifferentiable, we must transform it, if we want to use mathematical programming procedures for the solution. This can be done by many methods. In all these cases we must compute the sensitivities of functions of type B. This is done in the following using the FE formulation of the boundary value problem for a fixed shape. The unknown derivatives of the stress components with respect to b can be calculated as:

$$C(b) = B(b, u(b)) \qquad (8)$$

$$\frac{dC}{db} = \frac{\partial B}{\partial b} + \frac{\partial B}{\partial u} \frac{\partial u}{\partial b} \qquad (9)$$

The equilibrium conditions are given by:

$$Ku = F \qquad (10)$$

The differentiation of Equation (10) leads to:

$$K \frac{\partial u}{\partial b} = \left(\frac{\partial F}{\partial b} - \frac{\partial K}{\partial b} u\right) \qquad (11)$$

$$\frac{dC}{db} = \frac{\partial B}{\partial b} + \underbrace{\frac{\partial B}{\partial u} K^{-1} \left(\frac{\partial F}{\partial b} - \frac{\partial K}{\partial b} u\right)}_{(2)} \qquad (12)$$

Term (2) of Equation (12) can be computed either directly or by introducing the adjoint variable τ defined by:

$$K\tau = \left(\frac{\partial B}{\partial u}\right)^T \qquad (13)$$

$$\frac{dC}{db} = \frac{\partial B}{\partial b} + \tau^T \left(\frac{\partial F}{\partial b} - \frac{\partial K}{\partial b} u\right) \qquad (14)$$

The computation of dC/db using Equations (8)-(12) can now follow analytically as shown in [07], [24], [43] and [12]. This approach is very efficient, excluding any other additional errors besides the already existing discretization error. Another alternative is to use approximations such as the overall finite-differences approach:

$$\frac{\partial B}{\partial b} \cong \frac{\Delta B}{\Delta b} = \frac{B(b+\Delta b, \ u(b)) \ - \ B(b, \ u(b))}{\Delta b} \qquad (15)$$

or the semi-analytical method:

$$\frac{\partial K}{\partial b} \cong \frac{\Delta K}{\Delta b} \quad \text{and} \quad \frac{\partial F}{\partial b} \cong \frac{\Delta F}{\Delta b} \qquad (16)$$

The convergence behaviour of the displacement derivatives from Equation (11), computed by the analytical method, has been shown in [12] for a three dimensional structural model. Reference [10] deals with the same subject in the critical case of a two-material beam problem with moving interface using the overall finite-differences, semi-analytical and the adjoint variational domain method.

As shown in the Introduction, another possibility is the variational formulation of sensitivity analysis, i.e. the derivation of formulae for the first variation of functionals defined on variable domains.

The difficulties with the FE based sensitivity analysis are:

- lower shape sensitivity (when the derivatives of the solution of the Lamé equations on the boundary are required),

- for more accurate shape sensitivity a higher degree of mesh refinement is necessary,

- mesh re-zoning, at least in a partial domain of BVP, must take place,

- change in the accuracy of the solution with the shape of elements; also after the perturbation of the internal nodes.

These disadvantages can be overcome using the BE formulation which presents the following advantages:

- high shape sensitivity,

- internal nodal points are not existent (the change in accuracy of the structural model depends only on the shape of the boundary elements),

- no re-zoning process in Ω is necessary,

- the dimension of the problem is reduced by one.

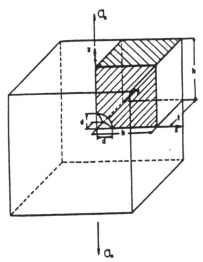

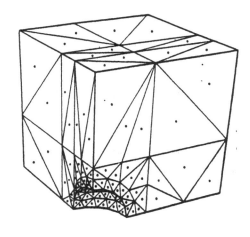

Figure 6a. Internal cavity   Figure 6b. Discretization

The computational efficiency of BEM and FEM is compared in Table 1 for the
problem of an internal cavity in an elastic cube under tension, see Figures
6a and 6b:

	No. of elements	Dim. of Sys.Matrix	CPU-time (sec)	Max (σ/σ_0)
analytical value				2.04
FEM	261 TET10	1575 x 618	280	1.99
BEM	154	462 x 462	62	1.94

Table 1. CPU-times, comparison between BEM and FEM

It is shown that with nearly the same accuracy, the computational time for
the BE approach is 4.5 lower than with the FE formulation. However, BEM has
some disadvantages, for example,

- supplementary computations are necessary in order to obtain the field
  values in the domain (this can be more expensive with BEM than with FEM),

- nodal points having weak singularities must be specially considered (because of the oscillation of the solution).

In the BE based discrete sensitivity analysis of linearly elastic structures, the possibility of implicit differentiation of discretized BE equa-

tions also exists:

$$C_{ij}(\xi)u_j(\xi) + \int_{\Gamma} T_{ij}(\xi,x)u_j(x)ds = \int_{\Gamma} U_{ij}(\xi,x)t_j(x)ds \qquad (17)$$

which gives in a discretized form:

$$Ay = F \qquad (18)$$

The unknown sensitivity of the vector y is obtained after solving the system of linear equations:

$$Ay^* = r^*$$

$$\text{with } (\ )^* = \frac{\partial}{\partial b}(\ ) \qquad (19)$$

and:

$$r^* = F^* - A^*y \qquad (20)$$

The problem is now to find the partial derivatives of the matrices A and F with respect to b. It is to be noted that A^* and F^* contain derivatives of the fundamental solutions U_{ij}, T_{ij} and of the Jacobian J of transformation with respect to b. Using the vector r^* we can compute the stress sensitivity. It is now very important that the 'new' integrals are also singular. However, the order of the singularity is the same as in the original problem. This means that we can work with same quadrature formula as for A and F. Another method which has been used for BEM in [01] is the analytical differentiation before discretization. Hypersingular kernels result in the computation of stress sensitivity at the boundary. We can also work with the boundary element formulation given by Gosh et al. [08] and Zhang and Mukherjee [44] in the linearly elastic case. In the integral equation we now obtain new kernels, so that T_{ij} is replaced by W_{ij}, having a logarithmic singularity from the type $\ln(r)$. The differentiated integral equation for multiple connected domains in this case leads to:

$$0 = \int_{\Gamma}\left[U_{ij}(\xi,x)t_j^*(x) - W_{ij}(\xi,x)\left(\frac{\partial u_j(x)}{\partial s}\right)^*\right]ds +$$

$$\int_{\Gamma}\left[U_{ij}^*(\xi,x)t_j(x) - W_{ij}^*(\xi,x)\left(\frac{\partial u_j(x)}{\partial s}\right)\right]ds +$$

$$\int \left[U_{ij}(\xi,x)t_j(x) - W_{ij}(\xi,x)\left(\frac{\partial u_j(x)}{\partial s} \right) \right] ds^* \tag{21}$$

$$\text{with } (\ )^* = \frac{\partial}{\partial b}(\ ); \quad ds^* = \frac{\partial x_k^*}{\partial x_k} - n_i n_j \frac{\partial x_i^*}{\partial x_k}$$

For plane strain, the kernel W_{ij} is given by:

$$W_{ij} = \frac{1}{4\pi(1-\nu)} \ 2(1-\nu)\Phi \ \delta_{ij} + \varepsilon_{jk}\frac{\partial r}{\partial x_i}\frac{\partial r}{\partial x_j} + (1-2\nu)\varepsilon_{ij}\ln r \tag{22}$$

It is now very important that the kernels of Equation (21) are regular.

The sensitivity coefficients can be used in connection with a nonlinear programming method, such as SQP. After writing the stationarity condition for the Lagrange function, we obtain a nonlinear system of equations that can be solved by using the Newton method. Thus, we have obtained the optimality conditions for the quadratical subproblem:

$$\min \frac{1}{2}\delta^T H_i \delta + \delta^T \nabla f_i \tag{23}$$

$$N_i^T \delta + g_i \leq 0 \qquad b \in \mathbb{R}^n, \ \delta = b_{i+1} - b_i \tag{24}$$

$$g_i^T = \left[g_1(b_i), \ldots, g_k(b_i) \right] \tag{25}$$

$$N_i = \left[\nabla g_1(b_i), \ldots, \nabla g_k(b_i) \right] \tag{26}$$

$$(h_{lm}) = \frac{\partial^2 L(b_i, \lambda_i)}{\partial b_l \partial b_m} \qquad l, m = 1(1)n \tag{27}$$

$$L = f(b) + \sum_{j=1}^{k} \lambda_j \ g_j(b) \tag{28}$$

which has to be solved for the optimal search direction vector δ in each iteration step.

Additionally, we can predict pointwise the stress response to shape variation at a traction free boundary using the monotonicity theorem. The nongradient procedure for the minimization of stress concentration has the following form:

$$b^T = \left[x_1, \ldots, x_{NB}, \ y_1, \ldots, y_{NB}, \ z_1, \ldots, z_{NB} \right] \tag{29}$$

$$\min \max \left(\bar{\sigma}_1^{\mu}(b), \ldots, \bar{\sigma}_{NB}^{\mu}(b) \right) \tag{30}$$

$$\bar{\sigma}_i^{\mu}(b) - \sigma \leq 0 \tag{31}$$

$$Ab - B \leq 0 \tag{32}$$

$$\text{with } \mu = 1(1)M \text{ and } i = 1(1)N$$

where NB is the number of nodal points on the design boundary.

The transition function f_j which describes the iteration rule for changing the nodal point coordinates on the optimizing boundary is defined on a physical basis.

• A geometrical perturbation on the design boundary Γ produces a rapid fade away of the von Mises stress $\bar{\sigma}$ in the neighbourhood of the perturbation.

• From the monotonicity relation of the two principal stresses with respect to the corresponding normal curvatures in the principal stress directions, we can derive a relation for the control of the von Mises stress on Γ. We have pointwise:

$$\delta\bar{\sigma}^{\mu} = K_w^s \delta h_n \tag{33}$$

where K_w^s is the weighted surface curvature which depends generally on the stress state and the geometrical data on the surface, and δh_n is the perturbation in the normal direction of the boundary Γ.

• By increasing the minimum effective stress, the maximum effective stress in the direct neighbourhood can also be reduced:

$$b_j = f_j(b_{j-1}, v_j) \quad \text{explicit (for tetrahedron elements)} \tag{34}$$

$$x_j^i = x_{j-1}^i + v_j^i \cdot \frac{1}{N_i} \sum_{k=1}^{N_i} \bar{a}_{j-1}^{i,k} \tag{35}$$

$$y^i_j = y^i_{j-1} + v^i_j \cdot \frac{1}{N_i} \sum_{k=1}^{N_i} \bar{b}^{i,k}_{j-1} \tag{36}$$

$$z^i_j = z^i_{j-1} + v^i_j \cdot \frac{1}{N_i} \sum_{k=1}^{N_i} \bar{c}^{i,k}_{j-1} \tag{37}$$

for i = 1(1)NB

where $\bar{a}^{i,k}_{j-1}$, $\bar{b}^{i,k}_{j-1}$ and $\bar{c}^{i,k}_{j-1}$ are coefficients of the Hessian form for each of the N_i surface triangles connected to node i. Because the magnitude of shifting v_j is controlled by an arithmetical smoothing algorithm, we have at the nodal point i:

$$v^i_j = v^{i*}_j \xi^i_{j-1} \left(1 - \frac{s^i}{M_s}\right), \quad s^i = 1(1)M_s \tag{38}$$

where M_s = number of smoothing zones

v^{i*}_j = magnitude of shifting at a point with maximum or minimum stress

Making the approximation that each of two principal stresses is a linear function of the normal curvature, for the sign of shiftig ξ^i_{j-1} we have:

$$\xi^i_{j-1} = \text{sgn}\,(\sigma^i_{1,j-1} + \sigma^i_{2,j-1})\,\text{sgn}(\bar{\sigma}^i_{j-1} - \overset{\approx}{\sigma}_{j-1}) \tag{39}$$

for i = 1(1)NB

where σ^i_1 and σ^i_2 denote the two principal stresses at the nodal point i and $\overset{\approx}{\sigma}$ the average von Mises stress on the design boundary.

As a result, we have a discrete, dynamic optimization problem with the following cost function' g_j:

$$g_j: = (\bar{\sigma})^{max}_j - (\bar{\sigma})^{max}_{j-1} \tag{40}$$

$$(\bar{\sigma})^{max}_j = \text{Max}\,((\bar{\sigma}^\mu_1)_j, \dots, (\bar{\sigma}^\mu_{NB})_j) \quad \mu = 1(1)M \tag{41}$$

Problem:

$$\min \sum_{j=1}^{l} g_j\left(b_{j-1}, v_j\right) \tag{42}$$

$$b \in \Xi$$

$$v_j \in \Omega_j\left(b_{j-1}\right) \tag{43}$$

where $\Omega_j(b_{j-1})$ is the feasible control space

$$\left(x_j^i, y_j^i, z_j^i\right) \in \bar{\Omega}_j^* \subset \mathbb{R}^n \tag{44}$$

where $n = 3$, $i = 1(1)NB$, $j = 1(1)\ell$

$$\bar{\Omega}^* \text{ closed set: } \bar{\Omega}^* = \Omega \cup \Gamma_- \cup \Gamma_+$$

State space Ξ: $\Xi \subset \mathbb{R}^{3NB}$

where Ξ is compact, as a product of compact sets from geometrical constraints:

$$v_j^i \in \left[-d^i\left(\Gamma_{d, j-1}, \Gamma_-\right), \; d^i\left(\Gamma_{d, j-1}, \Gamma_+\right)\right] = \Omega_j^i\left(b_{j-1}\right) \tag{45}$$

For the decision space it follows:

$$\Omega_j\left(b_{j-1}\right) = \prod_{i=1}^{NB} \Omega_j^i\left(b_{j-1}\right) \subset \mathbb{R}^{NB} \tag{46}$$

with the theorems of Tychonoff (Ω_j^i is compact) and Weierstrass and making the supposition that the objective function from the dynamic optimization problem is continuous, we have the existence of the solution.

3 Examples

In the following, we give some results obtained with the nongradient method. The optimal geometries for an axisymmetric structure (shaft) subjected to the loading cases bending, torsion and tension are given in Fig. 7.

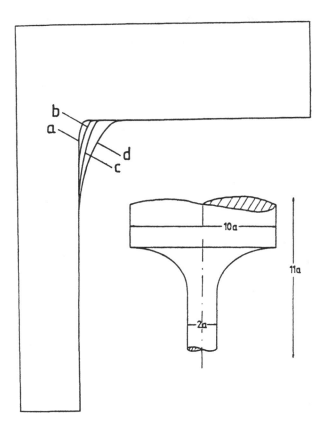

Figure 7. a - bending;   b - torsion
          c - tension/axisym.; d - tension/plane

These are compared with the optimal profile of a structure modelled with plane stress elements under tension. Details of these examples can be found in [41].

- a = axisymmetric structure subjected to bending

- b = curvature for the axisymmetric rod loaded by torsion

- c = curve for tension of the axisymmetric rod

- d = shape of the plane problem loaded by tension

In Figures 8, 9, 10 one can see a plate under tension with the variation domain Γ^*. The optimal shape for this problem has been computed in [30,41].

We have here a double loading case (see also the definition of the objective function in statement (5)). Figure 10 shows the starting profile (b) and the optimal profile (a).

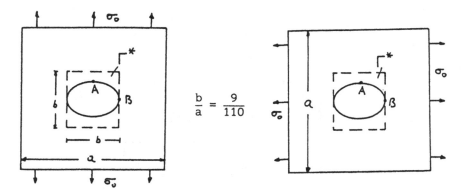

$$\frac{b}{a} = \frac{9}{110}$$

Figure 8. Plate under vertical   Figure 9. Plate under horizontal
        tension (LC = 1)   tension (LC = 2)

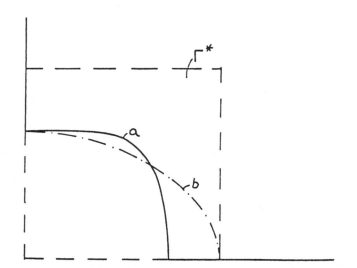

Figure 10. Plate under tension - a = optimal solution
                              - b = starting profile

The stress distribution after 12 and 24 iteration steps is given in Figures 12 and 13, and the stress distribution of the optimal profile for such a problem after 36 iteration steps in Figure 14. The letters LC in Figures 8 to 14 mean "Loading Case".

Figure 11. Multiple loading
case - Stress
distribution at
starting boundary
    +—+— LC = 1
    •—•— LC = 2

Figure 12. Multiple loading
case - Stress
distribution at
design boundary
after 12 itera-
tion steps:
    +—+— LC = 1
    •—•— LC = 2

Figure 13. Multiple loading
case - Stress
distribution at
design boundary
after 24 itera-
tion steps:
    +—+— LC = 1
    •—•— LC = 2

Figure 14. Multiple loading
case - Stress
distribution at
optimal boundary:
(36 iterations)
    +—+— LC = 1
    •—•— LC = 2

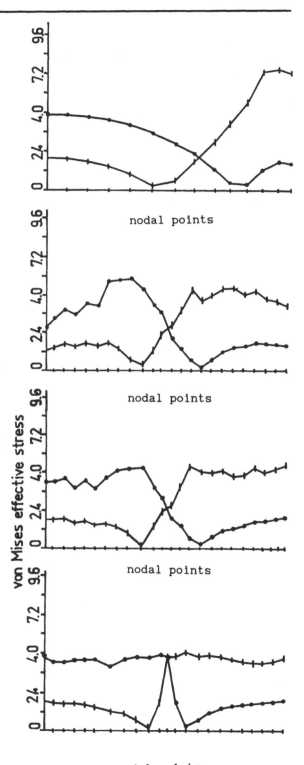

In Figure 15 a double notch problem is represented, i.e. we have a mutual influence of notch effects (see [12]). Figure 16 shows the finite element mesh for the optimal design. In Figure 17 the stress distribution of the starting profile (with circles) and the stress distribution for the optimal notch shapes is compared.

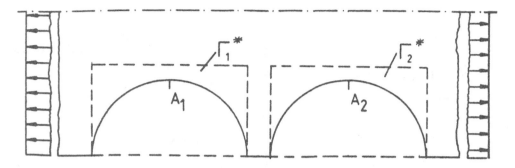

Figure 15. Double notch problem with interference of notch effects

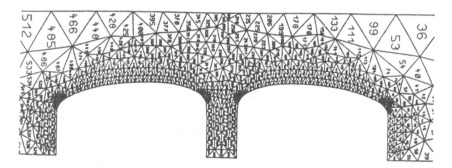

Figure 16. FE model for the optimal design of a double notch problem

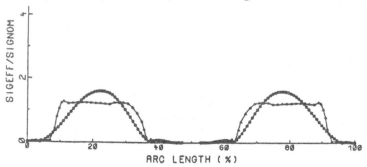

Figure 17. Stress distribution - ⊙ starting profile
                                 - Δ optimal design

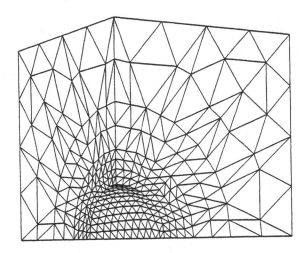

Figure 18. FE-discretization of body with spherical hole

If we want to extend an optimization procedure to the three-dimensional case, we must first be able to efficiently solve large systems of equations. This is done for the nongradient optimization using a vectorized CG algorithm which works with a full packed system matrix. In Table 2 some data obtained with this linear solver on a VP400-EX vector machine are shown. Details of these computations can be found in reference [12]. The fourth line in Table 2 corresponds to the FE model for the cavity problem shown in Figure 18.

An example of the application of the nongradient procedure for three-dimensional structures is the stress optimal shape of a cavity in a large elastic domain (see [12]). The starting geometry for this problem (see Figure 19) was a spherical hole:

$$a : b = a : c = 1, \tag{47}$$

with the maximum stress $3.36\ \sigma_{0x}$.

The loading case is:

$$\sigma_{0x} : \sigma_{0y} : \sigma_{0z} = 1 : 2 : 2. \tag{48}$$

After 24 iteration steps we obtain an axisymmetric ellipsoid with:

$$b : a = c : a = 2.160 \tag{49}$$

No. of Nodes	No. of Elements	Dimension of Matrix	Bandwidth of Matrix	Sparsity (%)	No. of MVM	CG (sec)
729	384	2187	690	4.07	147	0.97
919	521	2757	684	4.53	174	1.41
2805	1511	8415	579	5.45	615	11.04
4913	3072	14739	2709	1.29	319	12.26
5557	3529	16671	2589	1.35	334	14.28
6409	4168	19277	2679	1.37	376	33.10

Table 2. Solution of the system of linear equations

The maximum stress value is reduced here to 2.70 σ_{0x}. This shape is shown in Figure 20. The proof of optimality for the cavity problem can be given analytically with the maximum principle. The result is:

$$b : a = c : a = 2.614 \tag{50}$$

and a constant von Mises stress on Γ of 2.5 σ_{0x}.

The examples of the previous section demonstrate the high performance of the nongradient strategies in optimizing notch problems of elasticity.

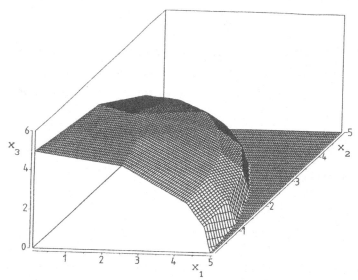

Figure 19. Shape of the starting surface

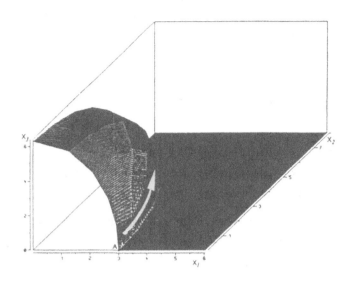

Fig. 20. Optimal shape of cavity in a large elastic domain

4 References

[01] Barone, M.R., R.J. Yang: A Boundary Element Approach for Recovery of
     Shape Sensitivities in Three-Dimensional Elastic Solids. Computer
     Methods in Applied Mechanics and Engng. 74 (1989), pp. 69-82.

[02] Benett, J.A., M.E. Botkin: Shape optimization of Two-Dimensonal Struc-
     tures with Geometric Problem Description and Adaptive Mesh Refinement,
     AIAA, 1983.

[03] Botkin, M.E., R.J. Yang and J.A. Benett: Shape Optimization of Three-
     Dimensional Stamped and and Solid Automotive Components. Paper presen-
     ted at the International Symposium on Optimum Shape, General Motors
     Research Labs, Warren, Michigan, 1985.

[04] Choi, K.K., E.J. Haug: Shape Design Sensitivity Analysis of Elastic
     Structures. J. of Structural Mech. 11, No. 2 (1983), pp. 231-269.

[05] Chun, Y.W., E.J. Haug: Two-Dimensional Shape Optimal Design. IJNME 13
     (1978), pp. 311-336.

[06] Chun, Y.W., E.J. Haug: Shape Optimization of a Solid of Revolution. J.
     of Engng. Mech. 109, No. 1 (1983), pp. 30-46.

[07] Francavilla, A., C.v. Ramakrishnan, O.C. Zienkiewicz: Optimization of Shape to Minimize Stress Concentration. J. of Strain Analysis 10/2 (1975), pp. 63-69.

[08] Ghosh, N., H. Rajiyah, S. Ghosh, S. Mukherjee: A New Boundary Element Method Formulation for Linear Elasticity. J. of Appl. Mech. 53 (1986), pp. 69-76.

[09] Haftka, R.T.: Finite Elements in Optimal Structural Design. In: Computer Aided Optimal Design: Structural and Mechanical Systems. Ed.: C.A. Mota Soares. Springer-Verlag, Berlin, New York 1986, pp. 271-297.

[10] Haftka, R.T., B. Barthelemy: On the Accuracy of Shape Sensitivity. In: Computer Aided Optimum Design of Structues: Recent Advances. Eds.: C.A. Brebbia, S. Hernandez. Springer-Verlag, Berlin, Heidelberg, New York 1989, pp. 327-336.

[11] Haug, E.J., R.K. Choi, V. Komkov: Design Sensitivity Analysis of Structural Systems. Ed.: W.F. Ames. Mathematical Science and Engineering. Academic Press, Orlando, San Diego, New York, 1986.

[12] Iancu, G.: Optimierung von Spannungskonzentrationen bei dreidimensionalen elastischen Strukturen. Doctoral Thesis, Karlsruhe University, 1991

[13] Iancu, G., E. Schnack: Knowledge-Based Shape Optimization. First Conference on Computer Aided Optimum Design of Structures (CAOD) - OPTI 89, 20 - 23 June 1989 in Southampton/UK. In: Computer Aided Optimum Design of Structures: Recent Advances. Eds.: C.A. Brebbia and S. Hernandez. Springer-Verlag, Berlin, Heidelberg, New York 1989, pp. 71-83.

[14] Iancu, G., E. Schnack: Shape Optimization Scheme for Large Scale Structures. Proceedings of the Second World Congress on Computational Mechanics, 27 - 31 August 1990, Stuttgart/Germany. to appear

[15] Imam, M.H.: Three-Dimensional Shape Optimization. IJNME 18 (1982), pp. 661-673.

[16] Kane, J., S. Saigal: Design-Sensitivity Analysis of Solids Using BEM. J. of Engng. Mech. 114, No. 10 (1988), pp. 1703-1722.

[17] Kikuchi, N., K.Y. Chung, T. Torigaki and J.E. Taylor: Adaptive Finite Element Methods for Shape Optimization of Linearly Elastic Structures. Comp. Meth. in Appl. Mech. and Engng. 57 (1986), pp. 67-89.

[18] Kodiyalam, S. and G.N. Vanderplaats: Shape Optimization of Three-Dimensional Continuum Structures via Force Approximation Techniques. AIAA Journal, 27 No. 9 (1989), pp. 1256-1263.

[19] Kristensen, E.S. N.F. Madsen: On the Optimum Shape of Fillets in Plates Subjected to Multiple In-Plane Loading Cases. IJNME 10 (1976), pp. 1007-1019.

[20] Leal, R.P.: Boundary Elements in Bidimensional Elasticity. Master Sc. Thesis, Technical University of Lisbon, 1985.

[21] Mota Soares, C.A, H.C. Rodrigues, L.M. Oliveira Faria, E.J. Haug: Optimization of the Geometry of Shafts Using Boundary Elements. ASME J. of Mechanisms, Transmissions and Automation in Design 106 (1984), pp. 199-203.

[22] Mota Soares, C.A., H.C. Rodriguez, L.M. Oliviera Faria, E.J. Haug: Boundary Elements in Shape Optimal design of Shafts. In: Optimization in Computer Aided Design. Ed. J.S. Gero. North-Holland 1985, pp. 155-175.

[23] Pedersen, P. and L.L. Laursen: Design for Minimum Stress Concentration by Finite Element Elements and Linear Programming. J. of Struct. Mech. 10/4 (1982-83), pp. 375-391.

[24] Ramakrishnan C.V., A. Francavilla: Structural Shape Optimization Using Penalty Funktions. J. of Struct. Mech. 3/4 (1974-1975), pp. 1974-1975.

[25] Rodriguez, H.C., C.A. Mota Soares: Shape Optimization of Shafts. Third National Congress of Theoretical and Applied Mechanics, Lisbon 1983.

[26] Rodriguez, H.C.: Shape Optimization of Shafts Using Boundary Elements. Master Sc. Thesis, Technical University of Lisbon, 1984.

[27] Schnack, E.: Ein Iterationsverfahren zur Optimierung von Spannungskonzentrationen. Habilitationsschrift, Univ. Kaiserslautern, 1977.

[28] Schnack, E.: Ein Iterationsverfahren zur Optimierung von Kerbober-

flächen. VDI-Forschungsheft, Nr. 589, Düsseldorf, VDI-Verlag 1978.

[29] Schnack, E.: An Optimization Procedure for Stress Concentrations by the Finite Element Technique. IJNME 14, No. 1 (1979), pp. 115-124.

[30] Schnack, E.: Optimierung von Spannungskonzentrationen bei Viellastbeanspruchung. ZAMM 60 (1980), T151-T152.

[31] Schnack, E.: Optimal Designing of Notched Structures without Gradient Computation. Proceedings of the 3rd IFAC-Symposium, Toulouse/France, 29 June - 2nd July 1982. In: Control of Distributed Parameter Systems. Eds: J.P. Barbary and L. Le Letty. Pergamon Press, Oxford, New York, Toronto, Sydney, Paris, Frankfurt 1982, pp. 365-369.

[32] Schnack, E.: Computer Simulation of an Experimental Method for Notch-Shape-Optimization. Proceedings of the Int. Symp. of IMACS, 9 - 11 May 1983 in Nantes. In: Simulation in Engineering Sciences, Tome 2. Eds: J. Burger and Y. Janny. Elsevier Science Publishers B.V. (North Holland), Amsterdam 1985, pp. 269-275.

[33] Schnack, E.: Local Effects of Geometry Variation in the Analysis of Structures. Studies in Applied Mechanics 12: Local Effects in the Analysis of Structures. Ed.: P. Ladevèze. Elsevier Science Publisher, Amsterdam 1985, pp. 325-342.

[34] Schnack, E.: Free Boundary Value Problems in Elastostatics. Proceedings of the 4th Int. Symp. on Numerical Methods in Engineering, Atlanta, Georgia/ USA, 24 - 28 March 1986. In: Innovative Numerical Methods in Engineering. Eds.: R.P. Shaw, J. Periaux, A. Chaudouet, J. Wu, C. Marino, C.A. Brebbia. Computational Mechanics Publications Southampton, Springer-Verlag, Berlin, Heidelberg, New York, Tokyo 1986, pp. 435-440.

[35] Schnack, E.: A Method of Feasible Direction with FEM for Shape Optimization. Invited lecture: Proceedings of the IUTAM-Symp. on Structural Optimization, Melbourne, 9 - 13 February 1988. In: Structural Optimization. Eds.: G.I.N. Rozvany, B.L: Karihaloo. Kluwer Academic Publishers, Dordrecht, Boston, London 1988, pp. 299-306.

[36] Schnack, E., G. Iancu: Control of the von Mises Stress with Dynamic Programming. GAMM-Seminar on Discretization Methods and Structural Op-

timization - Procedures and Applications, 5 - 7 October 1988, University of Siegen. In: Proceedings of the GAMM-Seminar, Vol. 43. Eds.: H.A. Eschenauer and G. Thierauf. Springer-Verlag, Berlin, Heidelberg 1989, pp. 154-161.

[37] Schnack, E., G. Iancu: Shape Design of Elastostatics Structures Based on Local Perturbation Analysis. Structural Optimization 1 (1989), pp. 117-125.

[38] Schnack, E., G. Iancu: Non-Linear Programming Applicable for the Control of Elastic Structures. In: Preprints of the 5th IFAC Symposium on Control of Distributed Parameter Systems, 26 - 29 June 1989 in Perpignan/France. Eds.: A. El Jai and M. Amouroux. Institut de Science et de Génie des Matériaux et Procédés (CNRS), Groupe d'Automatique, Université de Perpignan 1989, pp. 163-168.

[39] Schnack, E., U. Spörl: A Mechanical Dynamic Programming Algorithm for Structure Optimization. IJNME 23, No. 11 (1986), pp. 1985-2004.

[40] Schnack, E., U. Spörl, G. Iancu: Gradientless Shape Optimization with FEM. VDI Forschungsheft 647/88 (1988), pp. 1-44.

[41] Spörl, U.: Spannungsoptimale Auslegung elastischer Strukturen. Doctoral Thesis, Karlsruhe University, 1985.

[42] Trompette, Ph. and J.L. Marcelin: On the Choice of the Objectives in Shape Optimization. In: Computer Aided Optimal Design: Structural and Mechanical Systems. Ed. C.A. Mota Soares, Springer-Verlag, Berlin, New-York 1986, pp. 247-261.

[43] Wang, S.-Y., Y. Sun, R.H. Gallagher: Sensitivity Analysis in Shape Optimization of Continuum Structures. Computer & Structures 20, No. 5 (1985), pp. 855-867.

[44] Zhang, Q., S. Mukherjee: Design Sensitivity Coefficients for Linear Elasticity Problems by Boundary Element Methods. Proceedings of the IUTAM/IACM Symposium on Discretized Methods in Structural Mechanics, 5 - 9 June 1989, Vienna/Austria. Eds.: G. Kuhn and H. Mang. Springer-Verlag, Berlin, Heidelberg 1990, pp. 283-289.

[45] Zhang, W.H., P. Beckers: Comparison of Different Sensitivity Analysis Approaches for Structural Shape Optimization. In: Computer Aided Optimum Design of Structures: Recent Advances. Eds. C.A. Brebbia and S. Hernandez. Springer-Verlag, Berlin, Heidelberg, New-York 1989, pp.346-356.

[46] Zolesio, J.-P.: The Material Derivative (or speed) Method for Shape Optimization of Distributed Parameter Structures. Eds.: Haug E.J., J. Cea. Sijthoff and Noordhoff, Alphen aan den Rhijn (1981), pp. 1089-1151.

[47] Zienkiewicz, O.C. and J.S. Campbell: Shape Optimization and Sequential Linear Programming. In: Optimum Structural Design. Eds. R.H. Gallagher and O.C. Zienkiewicz. John Wiley & Sons, London, New-York, Sydney, Toronto, 1973.

Chapter 23

DOMAIN COMPOSITION

E. Schnack
Karlsruhe University, Karlsruhe, Germany

1 Introduction

A theory of decomposition has already been developed by Sobolev, Babuska
and Morgenstern and P.L. Lions. The idea is to decompose the whole structure into subregions. The standard methods involve substructuring techniques, for example, in finite element or boundary element methods. Here we
introduce different approaches for the partial differential equation in
each substructure. The basic idea of this paper has already been handled by
the author in earlier papers [06-07].

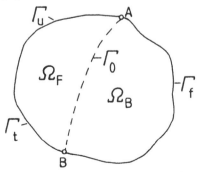

Fig. 1. Free boundary Γ_f

2 Model problem

Figure 1 shows a domain which is split into two parts, Ω_F and Ω_B. The solution in Ω_F is approximated by finite elements and that in Ω_B by boundary elements. The interface between Ω_F and Ω_B is denoted by Γ_0. In the case of shape optimization, for example, the portion Γ_f is traction-free. The fish plate construction shown in Figure 2 provides a typical technical application. The objective is to compute the free boundary so that, for example, the maximum stress is a minimum. The region around the fillet (subdomain Ω_B) is approximated by boundary elements, whereas the remainder (Ω_F) is discretized with finite elements. The reasons for proceeding this way are:

1. High gradient approximation is far better in BEM than in FEM;

2. BEM is more sensitive to boundary variation than FEM;

3. mesh grading in FEM is more expensive than in BEM (the problem is reduced by one dimension).

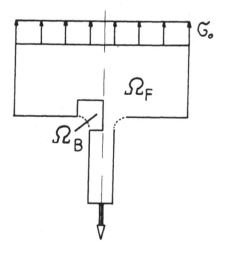

Fig. 2. Technical application

The model problem is:

$$\Omega \subset \mathbb{R}^i \qquad (1)$$

with $i = 2,3$

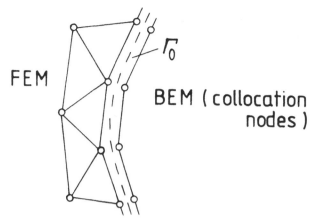

Figure 3. Definition of the characteristics of the interface Γ_0

We define the characteristics of the interface Γ_0, see Figure 3. In classical coupling procedures, the finite element nodes have the same position as
the BEM-(collocation)nodes. However, in practice we have the situation that
for gradients of the stress field only a rough description can be obtained
with FEM, whereas with BEM a high degree of accuracy can be achieved. This
leads us to the remark that the positions of FEM nodes must be different
from those of BEM nodes. This is accomplished by a *m i x e d   v a r i a -
t i o n a l   f o r m u l a t i o n* for coupling of the field equations.
The assumptions are:

- minimization of the stresss norm:

$$\min_{\Gamma_f} \quad \max_{\Gamma_f} \quad \| \tau \|, \tag{2}$$

- no body forces,
- no thermal effects,
- small deformations,
- Hooke's law.

First, consider the Laplace problem:

$$\Delta u = 0 \text{ in } \Omega \tag{3}$$

$$u = \bar{u} \text{ on } \Gamma_u \tag{4}$$

$$\frac{\partial u}{\partial n} = \bar{t} \text{ on } \Gamma_t \tag{5}$$

where $\bar{u}$ and $\bar{t}$ are the corresponding prescribed Dirichlet and Neumann data on Γ_u and Γ_t, respectively. The boundary, Γ, of Ω has two parts, namely:

$$\Gamma_u \text{ and } \Gamma_t : = \Gamma \setminus \Gamma_u \tag{6}$$

We introduce:

$$\Gamma_f \subset \Gamma_t, \text{ with } \bar{t} = 0 \text{ on } \Gamma_t \text{ always} \tag{7}$$

The stress field is defined by:

$$\sigma = \nabla u \tag{8}$$

For the tractions we have:

$$\frac{\partial u}{\partial n} = \sigma \cdot n \bigg|_{\Gamma} \tag{9}$$

The strain energy of Ω has the following form:

$$U_\Omega = \frac{1}{2} \int_\Omega |\nabla u|^2 \, d\Omega_x = \frac{1}{2} \int_\Omega \sigma \, \nabla u \, d\Omega_x \tag{10}$$

For the Lamé problem, we can proceed in the same way:

$$\Delta^* u : = \mu \, \Delta u + (\lambda + \mu) \text{ grad div } u = 0 \text{ in } \Omega, \tag{11}$$

where λ, μ are the Lamé constants.

We introduce the traction operator:

$$T[u] : = \lambda(\text{div } u)n + 2\mu \frac{\partial}{\partial n} u + \mu \, n \times \text{rot } u = \bar{t} \text{ on } \Gamma_t \tag{12}$$

We have the strain displacement relationship:

$$e_{km} = \frac{1}{2}\left(u_{k,m} + u_{m,k}\right) \tag{13}$$

Introducing Hook's law, we obtain:

$$\tau_{km} = D_{kmpq} e_{pq} \tag{14}$$

Additionally, the strain energy is given by:

$$U_\Omega = \frac{1}{2} \int_\Omega \tau_{km} e_{km} d\Omega_x \tag{15}$$

In the next step, we need the BEM equations. We start with the Betti theorem:

$$u(x) = -\int_{\Gamma_u} T(y,x)\ \bar{u}(y)ds_y + \int_{\Gamma_u} U(y,x)\ T[u](y)ds_y -$$

$$-\int_{\Gamma_t} T(y,x)\ u(y)ds_y + \int_{\Gamma_t} U(y,x)\ \bar{t}(y)ds_y$$

$$\text{where } x \in \Omega \tag{16}$$

where

$$U(y,x): = \frac{\lambda+3\mu}{4(i-1)\pi(\lambda+2\mu)\mu}\left[\gamma(y,x)I + \frac{\lambda+\mu}{\lambda+3\mu}\ \frac{1}{|x-y|^i}\ (x-y)(x-y)^T\right] \tag{17}$$

is the fundamental solution of Δ^* and

$$\gamma(y,x): = \begin{cases} -\log|x-y| & i = 2 \\ & \quad\quad\text{for} \\ \dfrac{1}{|x-y|} & i = 3 \end{cases} \tag{18}$$

is the fundamental solution of the Laplace problem.

T is given by:

$$T(y,x): = \left(T_y(u(y,x))\right)^T \tag{19}$$

The subscript y denotes differentiation with respect to the components of y. By taking the limit as the point x tends to the boundary, Γ, we obtain:

$$\frac{1}{2}\ u(x) = -\int_{\Gamma_u} T(y,x)\ \bar{u}(y)ds_y + \int_{\Gamma_u} U(y,x)\ T[u](y)ds_y \tag{20}$$

$$-\int_{\Gamma_t} T(y,x)\ u(y)ds_y + \int_{\Gamma_t} U(y,x)\ \bar{t}(y)ds_y$$

$$\text{with } x \in \Gamma$$

It also follows, by denoting T[u] by t that:

$$\frac{1}{2} \, t(x) \; = \; - \, T_x \int_{\Gamma_u} T(y,x) \; \bar{u}(y) ds_y \; + \int_{\Gamma_u} \bigl(T(x,y)\bigr)^T \, t(y) ds_y$$

$$- \, T_x \int_{\Gamma_t} T(y,x) \; u(y) ds_y \; + \int_{\Gamma_t} \bigl(T(x,y)\bigr)^T \, \bar{t}(y) ds_y$$

$$\text{with } T_x U(y,x) = \bigl(T(x,y)\bigr)^T \text{ and } x \in \Gamma \qquad (21)$$

In the following we define the boundary integral operators. We have a regular operator:

$$Vt(x): \; = \int_\Gamma U(y,x) \; t(y) ds_y \qquad (22)$$

two singular operators:

$$Ku(x): \; = \int_\Gamma T(y,x) \; u(y) ds_y \qquad (23)$$

$$K't(x): \; = \int_\Gamma (T(x,y))^T \; t(y) ds_y \qquad (24)$$

and a hypersingular integral operator:

$$Wu(x): \; = -T_x \int_\Gamma T(y,x) \; u(y) ds_y \qquad (25)$$

Now, we define the transmission problem. We introduce the interface Γ_0 and define traction $\bar{t} = 0$ on Γ_f, see Figure 4. Working with the condensation coupling procedure, we can use either a positive-definite or a symmetric formulation, but not both, see [10]. This motivates us to consider the idea of boundary energy functionals.

The Laplace problem, specified by Equations (3), (4) and (5), may be reformulated in terms of three sets of equations. The first equation set is:

$$\Delta u_F = 0 \text{ in } \Omega_F \qquad (26)$$

$$u_F\bigg|_{\Gamma_u} = \bar{u} \tag{27}$$

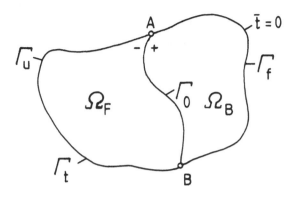

Figure 4. Transmission problem

$$\frac{\partial u_F}{\partial n}\bigg|_{\Gamma_t} = t \tag{28}$$

$$\frac{\partial u_F^-}{\partial n} = \sigma \text{ on } \Gamma_0 \tag{29}$$

The second equation set is:

$$\Delta u_B = 0 \text{ in } \Omega_B \tag{30}$$

$$\frac{\partial u_B}{\partial n}\bigg|_{\Gamma_f} = 0 \tag{31}$$

$$u_B^+ = \phi \text{ on } \Gamma_0 \tag{32}$$

The transition conditions are expressed by the third set of equations:

$$\sigma = \frac{\partial u_B^+}{\partial n} \tag{33}$$

$$\phi = u_F^- \tag{34}$$

It is important to note the identification of σ with $\dfrac{\partial u_B^+}{\partial n}$ and of ϕ with u_F^-

The reason for associating σ with $\dfrac{\partial u_B^+}{\partial n}$ is the higher accuracy for gradients with BEM. Once more, in analogy with the case of the Laplace problem, we have associated σ with $T[u_B^+]$ because of the higher accuracy for gradients with BEM. We can proceed in the same way in elasticity:

$$\Delta^* u_F = 0 \text{ in } \Omega_F \tag{35}$$

$$u_F\Big|_{\Gamma_u} = \bar{u} \tag{36}$$

$$T[u_F]\Big|_{\Gamma_t} = \bar{t} \tag{37}$$

$$T[u_F^-] = \sigma \text{ on } \Gamma_0 \tag{38}$$

Additionally we have:

$$\Delta^* u_B = 0 \text{ in } \Omega_B \tag{39}$$

$$T[u_B]\Big|_{\Gamma_f} = 0 \tag{40}$$

$$u_B^+ = \phi \text{ on } \Gamma_0 \tag{41}$$

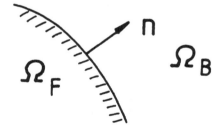

Figure 5. Definition of normal unit vector for the transmission condition

The transmission condition is:

$$\sigma = T[u_B^+] \tag{42}$$
$$\phi = u_F^- \tag{43}$$

Once more, we have here defined σ by u_B^+ because of the better accuracy for gradients with BEM.

The boundary energy for the subdomain Ω_B has, in the case of the Laplace problem, the form:

$$u_B = \frac{1}{2} \int_{\Gamma_0} u_B^+ \frac{\partial u_B^+}{\partial n} \, ds_y \tag{44}$$

and, in the case of elasticity:

$$u_B = \frac{1}{2} \int_{\Gamma_0} u_B^+ \, T[u_B^+] ds_y \tag{45}$$

3 Variational principle

The problem is to find the extremum of the total potential energy for Ω_F and Ω_B over a class of admissible functions. Without loss of generality, we assume:

$$\bar{u} = 0 \tag{46}$$

To relax the continuity requirements for the displacement transmission condition, we introduce *o n e* Lagrange multiplier λ. For the Laplace problem we obtain:

$$\Pi(u_F, u_B, \lambda): = \frac{1}{2} \int_{\Omega_F} |\nabla u_F|^2 \, d\Omega_x + \frac{1}{2} \int_{\Gamma_0} u_B^+ \frac{\partial u_B^+}{\partial n} \, ds_y$$

$$+ \int_{\Gamma_0 \cup \Gamma_f} \lambda(u_F^- - u_B^+) \, ds_y - \int_{\Gamma_t} \bar{t} \, u_F \, ds_y \tag{47}$$

For elasticity we have:

$$\Pi(u_F, u_B, \lambda): = \frac{1}{2} \int_{\Omega} a(u_F, u_F) \, d\Omega_x + \frac{1}{2} \int_{\Gamma_0} u_B^+ \, T[u_B^+] \, ds_y$$

$$+ \int_{\Gamma_0 \cup \Gamma_f} \lambda(u_F^- - u_B^+) \, ds_y - \int_{\Gamma_t} \bar{t} \, u_F \, ds_y \tag{48}$$

It is important to note that the Neumann condition of Γ_f is satisfied (for both the Laplace and the elasticity problems) a-priori. The appropriate function space for the trial and test functions is denoted in V_{ad}. The following notation is used:

$$\text{trial functions:} \quad \left(u_F, \; u_B, \; \lambda\right) \in V_{ad} \tag{49}$$

$$\text{test functions:} \quad \left(v_F, \; v_B, \; \mu\right) \in V_{ad} \tag{50}$$

The first variation of Π leads to three equations:

$$a\left(u_F, \; v_F\right) + \langle \lambda, \; v_F^- \rangle = \int_{\Gamma_t} \bar{t} \; v_F \; ds_y \tag{51}$$

$$\langle \mu, \; u_F^- - u_B^+ \rangle = 0 \tag{52}$$

$$\langle T[u_B^+] - \lambda, \; v_B^+ \rangle = 0 \tag{53}$$

We choose $\lambda = \sigma$ from the transmission condition, so that we finally have two equations:

$$a\left(u_F, \; v_F\right) + \langle \lambda, \; v_F^- \rangle = \int_{\Gamma_t} \bar{t} \; v_F \; ds_y \tag{54}$$

$$\langle \mu, \; u_F^- - u_B^+ \rangle = 0 \tag{55}$$

where $\langle \; ., . \; \rangle$ denotes the integral of the product of the arguments over $\Gamma_0 = \partial\Omega_B \cap \partial\Omega_F$ because:

$$T[u_B^+] = \frac{\partial u_B^+}{\partial n} = 0 \text{ on } \Gamma_f \tag{56}$$

In addition, we need the boundary integral equation for the region Ω_B:

$$\frac{1}{2} u_B^+(x) = - \int_{\Gamma_0 \cup \Gamma_f} T(y,x) \; u_B^+(y) \; ds_y + \int_{\Gamma_0} U(y,x) \; t(y) \; ds_y \tag{57}$$

In operator notation we have:

$$\left(\frac{1}{2} I + K\right) u_B^+ = V \; T[u_B^+], \text{ for } x \in \Gamma_0 \cup \Gamma_f \tag{58}$$

in the case of elasticity, and

$$\left(\tfrac{1}{2} I + K\right) u_B^+ = V \frac{\partial u_B^+}{\partial n}, \text{ for } x \in \Gamma_0 \cup \Gamma_f \qquad (59)$$

in the case of the Laplace problem.

We require that both u_B and v_B satisfy the governing equation in Ω_B, and, therefore, we have the symmetry property:

$$\left\langle \frac{\partial u_B^+}{\partial n}, v_B^+ \right\rangle = \left\langle \frac{\partial v_B^+}{\partial n}, u_B^+ \right\rangle \qquad (60)$$

or in the elasticity case:

$$\left\langle T[u_B^+], v_B^+ \right\rangle = \left\langle T[v_B^+], u_B^+ \right\rangle \qquad (61)$$

Finally, in the case of elasticity, we arrive at the following formulation:

$$a\left(u_F, v_F\right) + \left\langle T[u_B^+], v_F^- \right\rangle = \int_{\Gamma_t} \bar{t}\, v_F\, ds_y \qquad (62)$$

$$\left\langle T[v_B^+], u_F^- - u_B^+ \right\rangle = 0 \qquad (63)$$

The test and trial functions are approximated by splines. There are two parameter groups, corresponding to the finite element and boundary element regions, respectively. The BEM parameters are expressed in terms of the interface FEM nodal parameters using Equation (63). This relationship, in turn, is used to eliminate the BEM parameters from Equation (62), which results in a stiffness relationship of the form:

$$A\alpha = F \qquad (64)$$

where α is the vector of FEM nodal displacements. It should be noted that the BEM parameters are those corresponding to the tractions $T[u_B^+]$. The corresponding BEM displacement functions are computed using the integral equation:

$$\left(\tfrac{1}{2} I + K\right) u_B^+ = V\, T[u_B^+] \qquad (65)$$

The bilinear form:

$$\left\langle\, T[v_B^+],\; u_B^+ \,\right\rangle \tag{66}$$

is symmetric and positive definite on the abstract level. Equation (65) is solved using the collocation method which introduces a symmetry error which is reduced by a fine BEM mesh.

The technical details may be found in the papers [01-04], [08].

We next discuss the numerical technique for decomposition which is illustrated in Figure 6. The vector α may be partitioned into two vectors, α_1 and α_2, where α_1 is the vector of nodal displacements from the region Ω_F, excluding those on the interface Γ_0, whereas α_2 consists of the nodal displacements on the interface Γ_0.

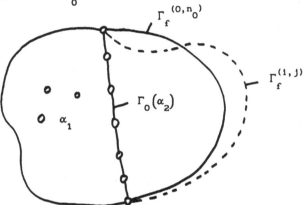

Figure 6. Numerical technique for decomposition

The system equations may be written in corresponding partitional form as follows:

$$A_{11}\alpha_1 + A_{12}\alpha_2 = F_1 \tag{67}$$

$$A_{21}\alpha_1 + A_{22}\alpha_2 = 0 \tag{68}$$

where F_1 is the vector of nodal forces associated with Γ_t.

Because of the symmetry it holds

$$A_{21} = A_{12}^T \tag{69}$$

It follows, from (68,69), that

$$A_{22}\alpha_2 = -A_{12}^T\alpha_1 \tag{70}$$

and, hence, from Equations (67) and (70), that

$$\left(A_{11} - A_{12}A_{22}^{-1}A_{12}^T\right)\alpha_1 = F_1 \tag{71}$$

We use St. Venant's principle (also demonstrated in the notch stress theory from Neuber [05]), i. e. that the perturbation in a stress field decays with distance from the local notch perturbation. This theorem leads to an efficient algorithm for the structural analysis following the incremental shape optimization process [09]. At each step of the incremental shape optimization process it is necessary to invert the decomposition equations (70) and (71). However, since the shape of Γ_f changes only slightly, we can expect, due to Neuber's principle [05], that the parameter group α_1 will change slowly over a number of steps. This leads us to a two-cycle algorithm for which α_1 is fixed in the inner cycle ($j = 1, n_1$) and is updated at the beginning of each outer cycle ($i = 1, n$):

$$\text{for } i = 1(1)n \tag{72}$$

$$\alpha_1^{(i)} = \left(A_{11} - A_{12}\left[A_{22}^{(i-1, n_{1-1})}\right]^{-1}A_{12}^T\right)^{-1}F_1 \tag{73}$$

$$\text{for } j = 1(1)n_1 \tag{74}$$

$$\alpha_2^{(i, j)} = -\left[A_{22}^{(i, j)}\right]^{-1}A_{12}^T\alpha_1^{(i-1)} \tag{75}$$

where $A_{22}^{(0, n_0)}$ is the submatrix of coefficients corresponding to the initial free boundary $\Gamma^{(0, n_0)}$, see Figure 6.

3 References

[01] Carmine, R.: Ein Kopplungsverfahren von FEM und BEM zur Berechnung ebener Spannungskonzentrationsprobleme. Doctoral Thesis, Karlsruhe University, 1989.

[02] Carmine, R., E. Schreck and E. Schnack: Experimental Results of Convergence Order for Two-Dimensional Coupling of FEM and BEM. ZAMM 70/6 (1990), T 699-T 701.

[03] Karaosmanoglu, N.: Kopplung von Randelement- und Finite-Element-Verfahren für dreidimensionale elastische Strukturen. Doctoral Thesis, Karlsruhe University, 1989.

[04] Karaosmanoglu, N. and E. Schnack: Three-Dimensional Structural Analysis with a Mixed FE-BE Method. ZAMM 70/6 (1990), T 705-T 707.

[05] Neuber, H.: Kerbspannungslehre. Grundlagen für genaue Festigkeitsberechnung, 3rd Edition. Springer-Verlag, Berlin, Heidelberg, 1985.

[06] Schnack, E.: Stress Analysis with a Combination of HSM and BEM. Mafelap 1984 Conference on The Mathematics of Finite Elements and Applications in Uxbridge/UK. In: The Mathematics of Finite Elements and Applications V. Ed.: J.R. Whiteman. Academic Press, Inc., London, Orlando, San Diego 1985, pp. 273-281.

[07] Schnack, E.: A Hybrid BEM-Model. Int. J. Numer. Meth. Eng. 24/5 (1987), 1015-1025.

[08] Schnack, E.: Macro-Elements Developed with BEM in Elasticity. MAFELAP 1990 Conference on The Mathematics of Finite Elements and Applications, 24 - 27 April 1990, Brunel University, Uxbridge/UK. Ed.: J.R. Whiteman. Academic Press, Inc. To appear.

[09] Schnack, E. and G. Iancu: Sensitivity Analysis with BEM. Lecture IV, CISM-Course on Shape and Layout Optimization of Structural Systems, 16 - 20 July 1990, Udine/Italy. To appear.

[10] Wendland, W.: On Asymptotic Error Estimates for Combined BEM and FEM. In: Finite Element and Boundary Element Techniques from Mathematical and Engineering Points of View. Eds.: E. Stein and W.L. Wendland. CISM Courses and Lectures 1988, No. 301, pp. 273-333.

Printed in the United States
By Bookmasters